T0334495

ECOSYSTEM SERVICE POTENTIALS AND THEIR INDICATORS IN POSTGLACIAL LANDSCAPES

ECOSYSTEM SERVICE POTENTIALS AND THEIR INDICATORS IN POSTGLACIAL LANDSCAPES

Assessment and Mapping

ANDRZEJ AFFEK

MAREK DEGÓRSKI

JACEK WOLSKI

JERZY SOLON

ANNA KOWALSKA

EWA ROO-ZIELIŃSKA

BOŻENNA GRABIŃSKA

BOGUSŁAWA KRUCZKOWSKA

ELSEVIER

Elsevier
Radarweg 29, PO Box 211, 1000 AE Amsterdam, Netherlands
The Boulevard, Langford Lane, Kidlington, Oxford OX5 1GB, United Kingdom
50 Hampshire Street, 5th Floor, Cambridge, MA 02139, United States

Copyright © 2020 Elsevier Inc. All rights reserved.

No part of this publication may be reproduced or transmitted in any form or by
any means, electronic or mechanical, including photocopying, recording, or any
information storage and retrieval system, without permission in writing from the
publisher. Details on how to seek permission, further information about the
Publisher's permissions policies and our arrangements with organizations such as
the Copyright Clearance Center and the Copyright Licensing Agency, can be found
at our website: www.elsevier.com/permissions.

This book and the individual contributions contained in it are protected under
copyright by the Publisher (other than as may be noted herein).

Notices
Knowledge and best practice in this field are constantly changing. As new research
and experience broaden our understanding, changes in research methods, professional
practices, or medical treatment may become necessary.

Practitioners and researchers must always rely on their own experience and knowledge
in evaluating and using any information, methods, compounds, or experiments
described herein. In using such information or methods they should be mindful of
their own safety and the safety of others, including parties for whom they have a
professional responsibility.

To the fullest extent of the law, neither the Publisher nor the authors, contributors, or
editors, assume any liability for any injury and/or damage to persons or property as a
matter of products liability, negligence or otherwise, or from any use or operation of
any methods, products, instructions, or ideas contained in the material herein.

Library of Congress Cataloging-in-Publication Data
A catalog record for this book is available from the Library of Congress

British Library Cataloguing-in-Publication Data
A catalogue record for this book is available from the British Library

ISBN: 978-0-12-816134-0

For information on all Elsevier publications visit our website at
https://www.elsevier.com/books-and-journals

Publisher: Candice Janco
Acquisition Editor: Candice Janco
Editorial Project Manager: Lena Sparks
Production Project Manager: Kamesh Ramajogi
Cover Designer: Matthew Limbert

Front cover photo credit: Andrzej Affek

Typeset by TNQ Technologies

Working together
to grow libraries in
developing countries

www.elsevier.com • www.bookaid.org

Contents

CHAPTER 1

Introduction

Contents

1.1 Theoretical assumptions, objectives, and scope of the study

The concept of ecosystem services (ES) describes the relations linking ecological systems with social systems by adopting the anthropocentric approach. In particular, it focuses on the benefits that man draws from nature. Although the concept originated in economic sciences, now it is definitely interdisciplinary, taking over the terminology and research methods from both the natural as well as the social and economic sciences. It was developed so that the inclusion of services provided by natural ecosystems into the global economic calculation was possible, in other words, to assess the value of nature in monetary terms.

After the first very general and widely criticized estimates provided by Costanza et al. (1997), it became clear that before final monetary valuation not only the concept itself needs to be clarified, but also the multifaceted, reliable, detailed, and region-specific recognition of service supply, use, and demand needs to be conducted. This study is a response to this expectation.

The theoretical and methodological basis for the research is the assumption that ecosystem services should be considered from both ecological and social perspective (Fig. 1.1). The ecological perspective

Ecosystem Service Potentials and Their Indicators in Postglacial Landscapes
ISBN 978-0-12-816134-0
https://doi.org/10.1016/B978-0-12-816134-0.00001-8
Copyright © 2020 Elsevier Inc.
All rights reserved.

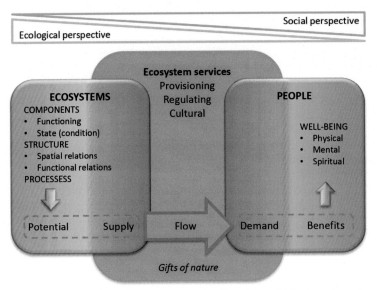

Figure 1.1 The relationships between nature and society within the ecosystem services concept.

focuses on the condition of ecosystems understood as specific dynamic structural and functional spatial systems composed of biocoenosis (the living world – plants, animals, and microorganisms) and biotope (a set of abiotic environmental conditions) (see Matuszkiewicz, 2001). Social perspective focuses on the benefits derived from ecosystem services and their impact on human well-being (physical, mental, and spiritual) (Fig. 1.1). Nowadays, the biggest challenge for the ES ecological-oriented research is to propose reliable tools to assess the potential of natural systems to provide services for humans. In turn, determining the demand and the actual use (flow) of services belongs to the domain of social and economic sciences.

In this book, particular emphasis was placed on investigating the potential of nature to provide ecosystem services, while the issues of demand and actual use were considered only in their impact on ES potential and its assessment. Hence, the key research objectives were as follows:

- review of current knowledge regarding ecosystem services;
- development of methodical solutions for estimating the potential of ecosystems to provide ecosystem services (definitions, typologies, indicators, methods of analysis and data synthesis);
- multifaceted assessment and mapping of nature potential to provide services in the test area; representative for the postglacial landscape;

- determination of ES bundles and similarities among ecosystems;
- determination of the impact of selected factors on the social assessment of ES potential.

It has been assumed that the monetary valuation of services and economic calculation will not be part of our research.

As part of the preparatory work, due to the remarkable dynamics of the concept development in recent years and the resulting conceptual and terminological chaos, we aimed to systematize the current knowledge on ecosystem services (Section 1.2). A review of scientific papers, books, and projects referring to ecosystem services was carried out. Key concepts have been clarified, often very inconsistently used in the literature. In the absence of satisfactory definitions, we introduced new ones.

Among various ES classification systems used in the literature (Section 1.2.2), we selected the most recent version of the *Common International Classification of Ecosystem Services* (CICES V5.1) (Haines-Young and Potschin, 2018) as the framework of our research. In contrast to many other classification systems, it includes only three main sections (provisioning, regulation and maintenance, and cultural) and relates only to final ecosystem services (details in Chapter 4).

Among numerous available definitions of the potential of ecosystems (Section 1.2.3), we followed the one proposed by Burkhard et al. (2012), referring strictly to the potential to provide ecosystem services. It states that the potential of ecosystems is the ability to provide services conditioned by natural factors (climate, terrain, habitat, and potential vegetation) and human activity (land use, pollution, etc.). However, for the purposes of assessment and mapping of services, it was often necessary to develop more detailed, operational definitions of the potential of ecosystems, for example, regarding services provided by bees - pollination and honey production (Affek, 2018).

To determine the potential of ecosystems to provide a whole range of ecosystem services, it was necessary to propose adequate and informative measures. To this end, a set of 35 indicators has been developed. All of them were characterized in the adopted theoretical assumptions, methods of construction, necessary source data, and measurement properties (including unit, scale, value range, and interpretation) (Chapter 6). As a rule, one indicator corresponds to one service. However, two exceptions were introduced: one in the case of provisioning service *edible biomass of wild animals* (separate indicators related to fish and game animals), and another in the case of regulating service *nursery habitat maintenance*, where due to inherent service complexity six different indicators were proposed.

Based on the developed indicators, a comprehensive assessment of the potential of ecosystems to provide services in the test area was carried out (Fig. 1.2). The selected area, representing the postglacial landscape, comprised three *gminas* (Nowinka, Giby and Suwałki) located in Poland in the Podlaskie Voivodeship, near the Wigry National Park (details in Section 2.2). The purposeful selection of the study area was made according to two basic criteria: (1) good representation of the rural postglacial landscape, and (2) high functional and structural diversity.

To determine the potential of nature for the provision of ecosystem services, two types of assessment were applied – expert and social. The former refers to direct measurements or estimations carried out by experts with the use of scientific knowledge (Sections 3.1.1 and 6.1), while the latter was based on the opinions expressed by service beneficiaries (residents and tourists) through questionnaire survey (Sections 3.1.2 and 6.2).

The basic spatial unit of assessment, both in our study and in the concept of ecosystem services, is the ecosystem. For the purposes of our study, we developed an extended typology of rural postglacial ecosystems, including 35 terrestrial and 7 lake ecosystem types. When distinguishing ecosystems account was taken not only of land cover, but also habitat conditions (fertility and humidity) and forest succession stage (Section 5.1). To

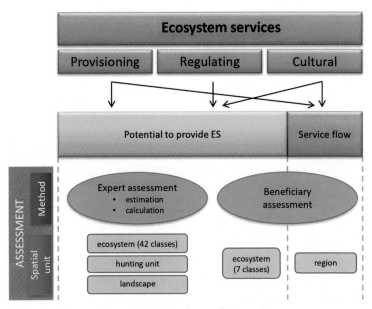

Figure 1.2 The scope of research.

investigate spatial interrelations between indicator values (ecosystem potentials for particular services), a map of ecosystems was developed for the study area. On this basis, all assessed service potentials were mapped and their spatial patterns were thoroughly analyzed.

As part of expert assessment, the values of indicators for individual types of ecosystem have been calculated, or in the absence of such possibility estimates were made based on the available literature sources. If the type of data did not give such a possibility, the potential of nature was estimated for spatial units other than ecosystems (i.e., hunting units, landscape units).

The already existing databases collected by various institutions in a standardized way were the main source of data. However, those databases were created for other needs than the analysis of ecosystem services, therefore required significant processing. Materials collected by the authors in the field (Section 3.1.1) served only as the complementary data source and were usually used for reference and calibration. The social assessment of the potential of nature carried out by direct ES beneficiaries covered seven basic types of ecosystem, which were determined through hierarchical aggregation of the detailed typology.

In the synthetic chapter, the results of expert and social assessment performed on the ecosystem level were summarized in the single assessment matrix and confronted with each other (Chapter 7). Then, spatial patterns of ES potentials were analyzed in line with the determination of ES bundles, synergies and trade-offs. We searched also for factors differentiating the level of potentials by applying the principal component analysis. By analyzing the distribution of values of the considered ES potentials, we aimed to show also the similarities among ecosystems and ecosystem clusters of comparable potentials. In the next step, an attempt was made to determine the overall aggregated potential of a given ecosystem. To estimate it, a dedicated method was applied adequate for the situation when not all possible single services had been assessed.

In the next chapter, we investigated the impact of various factors on the social assessment of ES potentials (Chapter 8). The considered factors include the actual use of ecosystem services, sociodemographic variables, and proximity to particular ecosystems. The data on the frequency of ES use, as well as sociodemographic characteristics of beneficiaries (e.g., age, gender, place of residence, and level of education) were obtained through questionnaire survey.

Based on the goals set and the literature review, as well as taking into account the scope of research outlined earlier, the authors formulated the following research hypotheses:

- ecosystems in postglacial landscape differ in their potentials to provide services;
- the results of the assessment depend on the method used (expert vs. beneficiaries);
- the more broadly the ecosystem service is defined, the less likely it is that the values obtained on the basis of various indicators will be consistent;
- distribution of ES potentials reflect the functional and structural diversity of ecosystems;
- social perception of ecosystem potential is determined by the way people interact with nature.

The authors hope that the methodological solutions presented in the book along with the verification of the formulated hypotheses will bring an original contribution to the development of the concept of ecosystem services.

1.2 Development of ES concept - state of the art

1.2.1 History of the concept and terminology of ecosystem services

Although the idea that human existence is highly dependent on Earth ecosystems dates back to the beginning of *Homo sapiens*, and the recognition of the way in which ecosystems provide complex services appeared already in the works of Plato about 400 BC (who knew, e.g., that deforestation can lead to soil erosion or drying up of water sources), the term "natural capital" was first used by Schumacher only in 1973.

The roots of the modern ES concept probably date back to the 19th century, when Marsh (1864) questioned the view that the Earth's natural resources are unlimited by pointing to changes in soil fertility in the Mediterranean. At the end of the 1940s, three authors: Leopold (1949), Osborn (1948) and Vogt (1948) began to promote the view of the close dependence of man on the environment. Several years later, Sears (1956) pointed out the key role of ecosystems in waste treatment and cycling of nutrients. In turn, Ehrlich and Ehrlich (1970) in their textbook drew attention to ecological systems and to a significant threat to human existence - the destruction of those systems by humans.

In the report *Analysis of key environmental problems* (Wilson and Matthews, 1970) the term "environmental services" was introduced for the first time. The authors listed services related to insect pollination, fisheries, climate regulation and flood protection. The term "ecosystem services", which eventually gained the widest acceptance of the scientific community, appeared for the first time in the context of extinction and replacement of populations and species resulting in the loss of benefits for humans (Ehrlich and Ehrlich, 1981; Ehrlich and Mooney, 1983).

Nevertheless, many authors of earlier works in the field of ecology and geoecology in the considerations regarding the basic components of the natural environment, especially soil and vegetation, as well as connections between these components and humans, included information about de facto ecosystem services, without using this term (e.g., Helliwell, 1969; Hueting, 1970; King, 1966; Odum, 1953).

The history of concept and terminology of ecosystem services until 1997 can be found in the excellent, multiauthor book edited by Daily (1997) *Nature's Services: Societal Dependence on Natural Ecosystems*. This is one of the first synthetic studies that reviewed many of the benefits and services offered to man by the natural environment and the close dependence of man on the natural systems. In particular, it presents a synthesis of the approach to ecosystem services and the preliminary valuation of their economic value consistent with the state of knowledge at the end of the 20th century. In subsequent chapters, services such as climate regulation, soil fertility, pollination, or pest control were considered, and the philosophical and economic valuation issues were addressed, referring to specific case studies of specific ecosystems and services they provide. The book also includes recommendations on actions that should be taken to solve the most urgent problems related to the natural environment and ES.

A very interesting description of the development of ES concept was published in 2010 (Gómez-Baggethun et al., 2010). This work focuses on the analysis of the development of economic theories, taking into account the growing role of the environment - from the classical economy (in which the land is only a factor of production) to the so-called ecological economics, in which natural capital supplements and complements human-generated capital. Against this background, the development of ES concept is presented and the most important, key publications and economic views are discussed.

The growing awareness of the benefits that ecosystems bring to society makes the ES concept worth spreading and developing (Foley et al., 2005). The major role in popularizing the idea of ecosystem services was played by Robert Costanza, an American scientist, specialist in the field of ecological economics and sustainable development (Costanza, 2008; Costanza et al., 1997). Also in European countries, including Poland, many works in this field have been created in recent years (e.g., Roo-Zielińska and Grabińska, 2012; Solon, 2008). The purpose of these publications is primarily to assess the impact of human activity on ES supply. However, they have different theoretical and practical value (Degórski, 2010). The biggest problem that arose because of the rapid development of the ES concept is the terminological chaos and freedom in defining the very concept (Degórski and Solon, 2014), as well as various ES classifications (Wallace, 2008). Despite the relatively long history of the concept, so far in the literature there were few attempts to clearly define ecosystem services (Barbier, 2007; Boyd, 2007).

For instance, Daily (1997) defined ecosystem services as "the conditions and processes through which natural ecosystems, and the species that make them up, sustain and fulfil human life". In this understanding, maintaining biodiversity and production of ecosystem goods, such as seafood, food, wood, biofuels, cellulose, substances of pharmaceutical value, or industrial raw materials, is an ecosystem service. Functions that support life processes, such as absorption and processing of waste, regeneration processes, and those which contribute to the provision of intangible (aesthetic and cultural) benefits to humans are also recognized as ecosystem services.

In recent years, further attempts have been made to standardize both terminology and methodology of ES research (EEA, 2013; Maes et al., 2013; TEEB, 2010; UK NEA, 2014), which resulted in better understanding and operationalization of the ES concept.

Ecosystem services in the general sense comprise all ecosystem outputs (e.g., wood, forest fruits, game animals) and functions (e.g., purification of water and air, oxygen production, providing space for recreation) used by the society (Costanza et al., 1997; Solon, 2008). In the widely cited reports of the *Millennium Ecosystem Assessment* (MEA), ES were simply defined as the benefits people obtain from ecosystems (MEA, 2005), which later led to difficulties in separating services from benefits. Ecosystem services are also defined as follows:

- components of nature, directly enjoyed, consumed, or used to yield human well-being[1] (Boyd and Banzhaf, 2007);
- the aspects of ecosystems utilized (actively or passively) to produce human well-being[2] (Fisher et al., 2008);
- direct and indirect contributions of ecological phenomena (functions) to human welfare[3] (TEEB, 2010).

According to the widely accepted cascade model (which is the conceptual framework for CICES approach), final ecosystem services are clearly distinguished from ecosystem goods and benefits (Haines-Young and Potschin, 2018, 2013). Final ecosystem services are the contributions that ecosystems (i.e., living systems) make to human well-being. They are the outputs of ecosystems (whether natural, seminatural, or highly modified) that most directly affect the well-being of people, but they still retain a connection to the underlying ecosystem functions, processes, and structures that generate them. In turn, ecosystem goods and benefits are things that people create or derive from final ecosystem services, and which ultimately have value for them. They are no longer functionally related to the systems from which they come.

Research on ES focuses mainly on services derived solely from narrowly defined ecosystems (Moser et al., 2011). However, more and more authors (e.g., Bastian et al., 2014; Solon, 2008) extend the ES concept to the entire landscape and understand it as a set of landscape outputs and functions that are useful for the society.

---

[1] In the understanding of Boyd and Banzhaf (2007), functions and ecosystem processes of an intermediary nature, as well as components of nature consumed indirectly are not recognized as ecosystem services.

[2] Here, the key points are that (1) services must be ecological phenomena and (2) that they do not have to be directly utilized (Fisher et al., 2008). Defined this way, ecosystem services include ecosystem organization or structure as well as processes and/or functions if they are consumed or utilized by humanity either directly or indirectly. The functions or processes become services if there are humans that benefit from them. Without human beneficiaries, they are not services. See Fisher et al. (2009) for the comparison of Fisher's and Boyd and Banzhaf's definitions.

[3] The *benefit* is the improvement of well-being resulting from satisfying needs (TEEB, 2010), and is not the synonym of service.

1.2.2 Classification of ecosystem services

Ecosystem services are grouped and ordered in various classification systems. *Millennium Ecosystem Assessment* and the TEEB initiative adopted and popularized the division into four major ES categories: (1) provisioning, describing the material outputs from ecosystems such as food, raw materials, fresh water, and medicinal resources, (2) regulating, resulting from the capacity of ecosystems to regulate climate, hydrological and biochemical cycles, earth surface processes, and a variety of biological processes, (3) supporting, representing the ecological processes that underlie the functioning of the ecosystem, such as the formation of soils and the circulation of elements, primary production, habitat function, and hydrological cycle, and (4) cultural, related to the nonmaterial benefits people obtain from contact with ecosystems through recreation, cognitive development, relaxation, and spiritual reflection (Maes et al., 2014, 2013; MEA, 2005, 2003; TEEB, 2010).

According to the classification proposed by Green et al. (1994), only two categories of services (provisioning and cultural) include outputs and structures directly useful for humans. The other two categories (regulating and supporting services) create a structural and functional framework affecting the overall integrity of the landscape system and the possibility of providing final services.

Costanza et al. (1997) proposed the classification of ecosystem services together with ecosystem functions underlying them. The scheme includes the following 17 types:

- Gas regulation (regulation of atmospheric chemical composition);
- Climate regulation (regulation of global temperature, precipitation, and other biologically mediated climatic processes at global or local levels);
- Disturbance regulation (capacitance, damping, and integrity of ecosystem response to environmental fluctuations);
- Water regulation (regulation of hydrological flows);
- Water supply (storage and retention of water);
- Erosion control and sediment retention (retention of soil within an ecosystem);
- Soil formation (soil formation processes);
- Nutrient cycling (storage, internal cycling, processing, and acquisition of nutrients);
- Waste treatment (recovery of mobile nutrients and removal or breakdown of excess or xenic nutrients and compounds);
- Pollination (movement of floral gametes);

- Biological control (trophic-dynamic regulations of populations);
- Refugia (habitat for resident and transient populations);
- Food production (that portion of gross primary production extractable as food);
- Raw materials (that portion of gross primary production extractable as raw materials);
- Genetic resources (sources of unique biological materials and products);
- Recreation (providing opportunities for recreational activities);
- Cultural (providing opportunities for noncommercial uses).

Although this classification is only of a historical relevance nowadays, it influenced for many years the way of thinking about services and their typologies.

Rudolf S. de Groot from the Wageningen University and Research Center, one of the leading theoreticians on the ES issues, presented a detailed ES classification and definitions dedicated for spatial planning and landscape management (de Groot, 1992; de Groot et al., 2010b, 2002). The classification covers 23 ecosystem services combined into four basic groups: (1) provisioning, (2) regulating, (3) habitat/supporting and (4) cultural. Ecological processes responsible for providing particular services were also identified, and indicators defining the efficiency of a given ES developed.

Other important and widely discussed classifications are as follows:
- MEA classification (MEA, 2005) - the most often cited, developed as part of the *Millennium Ecosystem Assessment*; aimed to help assess changes in human well-being (especially in the last 50 years) caused by the degradation of ecosystems; comprises four main categories: provisioning, regulating, supporting, and cultural.
- TEEB classification (TEEB, 2010) - developed within *The Economics of Ecosystems and Biodiversity* project and published in the report *TEEB Ecological and Economic Foundations*, replaces MEA supporting services with habitat services, emphasizes the role of ecosystems in providing nesting places for migratory species and protecting the gene pool.
- CICES (Haines-Young and Potschin, 2018) - recently most widely used *Common International Classification of Ecosystem Services* (CICES) developed by the European Environment Agency (EEA); first version presented in 2010 (Haines-Young and Potschin, 2010a) and published 3 years later (Haines-Young and Potschin, 2013). It is a strictly hierarchical system (four levels: section, division, group, and class), based on previous MEA and TEEB classifications, introduced for assessment

purposes, there are three categories on the highest section level: provisioning, regulation and maintenance (R&M), and cultural; TEEB habitat services were included into R&M section. The most recent version of CICES (V5.1) (Haines-Young and Potschin, 2018) serves as a framework for ES analysis in this study (see Chapter 4 for details).

All the mentioned systems, although directly or indirectly based on the cascade: ecosystem structure → ecosystem function → ecosystem service → human benefits → value, are focused on categorization of outputs obtained from ecosystems. Recipients of these outputs are not taken into account in classification schemes.

Different approach has been developed by the United States Environmental Protection Agency. The main idea of the two proposed classification systems is based on the combination of two independent dimensions: environmental types and beneficiary categories. In this approach, the final ecosystem good/service is identified through the intersection of the two dimensions. The first classification system identifies Final Ecosystem Goods and Services (FEGS) (Landers and Nahlik, 2013). The theoretical and methodical framework of this system is organized around the definition of FEGS[4] and two crucial questions: "Which ecosystems produce ecosystem services?" (environmental dimension) and "Who is the beneficiary and what are the FEGS?" (beneficiary dimension). For the purpose of the classification, beneficiaries are defined as follows: "the interests of an individual (i.e., a person, organization, household, or company) that drive active or passive consumption and/or ES appreciation resulting in an impact (positive or negative) on their welfare". The system identifies the following environmental classes and subclasses:

(1) AQUATIC - Rivers and Streams, Wetlands, Lakes and Ponds, Estuaries and Near Coastal and Marine, Open Oceans and Seas, Groundwater;

(2) TERRESTRIAL - Forests, Agroecosystems, Created Greenspace, Grasslands, Scrubland Barren/Rock and Sand, Tundra, Ice and Snow;

(3) ATMOSPHERIC - Atmosphere.

The second dimension (beneficiary categories) includes the following: (1) agricultural, (2) commercial/industrial, (3) government, municipal, and residential, (4) commercial/military transportation, (5) subsistence, (6)

---

[4] Final Ecosystem Goods and Services (FEGS) - components of nature, directly enjoyed, consumed or used to yield human well-being. The definition is based on the final ecosystem service and ecological endpoint concepts (Boyd and Banzhaf, 2007).

recreational, (7) inspirational, (8) learning, (9) nonuse, and (10) humanity. The 10 mentioned categories are further divided into 38 subcategories.

Through the combination and intersection process, 338 unique FEGS can be identified. The system is open in this sense that each environmental and/or beneficiary category might be further divided into categories on the lower level, giving as a result a greater number of FEGS understood in much narrower sense.

Similar, but not identical, and more sophisticated classification scheme is presented by the National Ecosystem Services Classification System (NESCS) (US EPA, 2015). It is based on the same theoretical background as the system described earlier, but the key concept of this approach is focused on identification and classification of Flows of Final Ecosystem Services (FFES).[5]

The NESCS structure consists of four groups divided into two classification systems:

- supply side (NESCS-S) covering (1) environmental classes that together cover the earth's surface, (2) classes of ecological end products, which are the biophysical components of nature directly used or appreciated by humans;
- demand side (NESCS-D) covering (3) classes of direct human use or nonuse appreciation of end products, and (4) classes of direct human users of end products.

According to the idea explicitly stated in the system description (US EPA, 2015), combinations across these four groups define FFES and can depict unique pathways that link changes in ecosystems with human welfare. Within each of these four groups, the system adopts a nested hierarchical structure so that each group can be represented at multiple levels of aggregation or detail. On the highest level of aggregation, main categories are generally very simple. Within NESCS-S, environmental classes and subclasses are taken from Landers and Nahlik (2013), while ecological end products cover water, flora, fauna, other biotic components, atmospheric components, soil, other abiotic components, composite end products, and other end products. Within NESCS-D, direct use/nonuse class distinguishes between extractive use, in situ use, and nonuse. Direct users are grouped into industries, households, and the government.

In the discussed two US EPA approaches services are not grouped into categories known from other classifications: provisioning, regulating

---

[5] Flows of Final Ecosystem Services - direct contributions made by nature to human production processes or to human well-being (US EPA, 2015).

(supporting), and cultural services. However, this division is partly reflected in detailed descriptions of each of FEGS or FFES.

1.2.3 Potential of the natural environment and ecosystem services

The term "potential" was introduced to landscape research already in 1949 by Bobek and Schmithüsen (1949), at first understood as "the spatial arrangement of natural development possibilities" (see Bastian, 2008). Neef (1966) defined a complex "gebietswirtschaftliches Potenzial" as an all-embracing economic capacity of the landscape or as the sum of all landscape features that create conditions for economic appreciation of the landscape with its materials and energy forming its structure. Haase (1978) defined "natural (natural-space) potential" as a "natural content of the landscape with its material properties, latent energies and processes, i.e. using its structure and dynamics the landscape has the ability to meet the needs of society" (see Kunáková, 2016). The proposed approach, although interesting from the theoretical point of view, was not fully applicable for practical solutions. Hence, for these purposes, Haase (1978, 1976) distinguished several specific "partial natural potentials" (germ. Partielle Naturraumpotenziale): biotic yield potential, water supply potential, waste disposal potential, biotic regulation potential, geoenergetic potential, and recreation potential. This concept has been further enhanced by different authors, so presently at least three additional partial potentials can be distinguished: raw material potential, construction potential, and climate regenerative potential (Kunáková, 2016).

For further refining practical application of the concept of partial potentials, Kolejka (2001) proposed to evaluate possibilities of actual utilization of the detected potential in a given region. He introduced the concept of the "free landscape potential" as a part of the detected potential that represents an actual territorial reserve in further development of the concerned activity, and "fixed landscape potential" understood as a proportion of the detected potential, which is already being utilized for the concerned activity. This distinction seemed to be very helpful in ecologically based spatial planning (Kunáková, 2016 and literature cited herein).

Similar to the idea of landscape potential was the concept of the so-called "benefits from nature", that is, all resources and forces of nature, which Bartkowski (1977) divided into two main groups: systems and subsystems of the epigeosphere mega-system and human-nature mega-system (including ecosystems). The author included favorable geographical location (e.g., the beauty of the landscape) to the group of substantial and energetic features of space, and natural conditions - to the group of relations and interactions occurring between them in the geographical space.

Przewoźniak (1991), referring to the concepts of Neef and Haase, distinguished three groups of potentials: self-regulatory resistant, resource utility, and perceptual behavioral. However, many authors pointed to large subjectivism of simple ranking of diagnostic features - especially in the case of a qualitative, intuitive - a priori approach (Pietrzak, 1998). Methodical attempt to objectify this problem was made by Kistowski (1996) who developed and modified the previously mentioned Przewoźniak's classification of potential. One step further in the perception of potentials was made by Solon (2004b). He suggested that the concept of landscape potential should not be limited to human—landscape relations, but should designate all resources a given population (including human) is willing to exploit. In his view, assuming any group of organisms, there are the following partial potentials of the landscape: self-regulating resistance, buffering, environment forming, and resource utilizing.

In the early 1990s, attention was drawn to the need to clarify the terminology: the landscape potential was to determine the possibilities, directions and conditions of its economic use (highly utilitarian aspect), and landscape function - to determine its ecological efficiency under conditions of specified use, whereby both terms were considered complementary. Bastian (1991) proposed to divide functions into productive (economic), regulatory (ecological), and for living space (social). A few years later, this division was used in the State Forest Policy (1997) of Poland, in which the productive (economic), ecological (protective), and social functions of the forest were mentioned. From the current perspective, it can be concluded that this classification was in a way an anticipation of the contemporary division of ecosystem services.

Solon (2008) noted that concepts of landscape potentials and ecosystem services are almost synonymous. However, some minor but significant differences refer to the way the natural environment resources are understood and how this understanding translates to practical applications (Table 1.1). Moreover, ecosystem services mainly relate to the use of resources produced or controlled by the biosphere, whereas landscape potentials - to the suitability of abiotic environmental components for various forms of land use. In recent years, however, the close-up of both approaches has been observed - on the one hand the potential to provide

Table 1.1 The main differences between concepts of landscape potential and ecosystem services (Solon, 2008).

Feature	Landscape potential	Ecosystem services
Use of resources	Possible	Actual
Types of resources	Mainly abiotic	Mainly biotic
Approach to resources	Wide	Narrow
Economic valuation	No	Yes
Taking into account the spatial context and scalability	No	Yes
Usefulness for spatial planning	High	Small
Usefulness for planning environmental compensation	Small	High
Cartographic presentation	Frequent	Rare

services is evaluated, and on the other hand there is a clearer differentiation between individual potentials, which makes it possible to find their equivalents in the ES typology.

The ES concept adopted in the EU distinguishes between ecosystem functions, basic ecological processes, and biophysical structures (Maes et al., 2013). The functions are created by various combinations of processes and structures and constitute the potential of ecosystems to provide services, regardless of whether they are currently used by people (TEEB, 2010). For example, basic production (process) is needed to keep the fish population alive (function), which can be further used to provide food (service). Possible misunderstandings may arise from the fact that many authors use the terms "function" and "process" interchangeably.

In contrast to ecosystem functions, ecosystem services assume the availability and presence of demand for meeting the diverse needs of people. Well-functioning intact natural ecosystems, which can be evaluated as having exceptional ecological value, can provide far less ecosystem services than transformed seminatural ecosystems located near large cities, just because there is huge difference in demand (e.g., the Carpathian Forest provides less recreational services per hectare than a city park). Nevertheless, natural ecosystems remain key components of the environment, providing a range of other important services (e.g., climate regulation, water retention, erosion prevention) and constituting a priceless natural heritage for many societies. That is why it is so important to consider a whole range of services and many dimensions of assessment when estimating the overall value of nature.

1.2.4 Biodiversity and ecosystem services

Links between biological diversity and individual ecosystem services or their bundles is a very large field of research, and the results so far are quite ambiguous. This is due to the lack of common definitions of particular services and biodiversity and the use of various, often incomparable methods of measurement (Liquete et al., 2016; Zhang et al., 2015).

According to the Convention on Biological Diversity - biodiversity is the diversity of all living organisms found on Earth in terrestrial, marine, and freshwater ecosystems and in ecological systems that they are part of. It relates to intraspecies diversity (genetic diversity), between species diversity and the diversity of ecosystems.

The same definition of biodiversity was adopted for Mapping and Assessment of Ecosystem Services (MAES) in Europe within the MAES project (Maes et al., 2014; Science for Environment Policy, 2015). However, in numerous detailed works carried out or inspired by the MAES project, biodiversity is understood more broadly and - according to the published conceptual scheme (Maes et al., 2013) - includes the following: (a) ecological processes, (b) functional traits, (c) biophysical structures, (d) genetic diversity, (e) species richness, and (f) biotic interactions. It is worth noting that such a division is neither fully disjunctive nor comprehensive.

Other authors understand the concept of biodiversity similarly broad. For example, Elmqvist et al. (2010) consider genetic variability, population sizes and biomass, species assemblages, communities and structures, interactions between organisms and their abiotic environment, interactions between and among individuals and species all as the components of biological diversity. In turn, according to Balvanera et al. (2016), biodiversity in a broad sense encompasses the number, abundance, functional variety, spatial distribution and interactions of genotypes, species, populations, communities, and ecosystems. Moreover, in many publications the term "biodiversity" is used instead of the term "species richness", which introduces a lot of confusion (see the discussion in Lugnot and Martin, 2013).

Although there is plenty of reports in the literature confirming the positive relationship between biodiversity, ecosystem functioning, and ecosystem capacity to provide particular services (Berbeć, 2014; Cardinale, 2011; Egoh et al., 2009; Harrison et al., 2014; Isbell et al., 2011; Kremen, 2005; Mace et al., 2012), it is not entirely clear what kind of links these are and what their dynamics is (Harrison et al., 2014; Liquete et al., 2016). At the same time, it has been demonstrated several times that biodiversity is a direct source of benefits derived from the biotic environment (Daily and Ehrlich, 1995).

Mace et al. (2012) also point out the unclear relationship between biodiversity and ecosystem services, stressing that such a situation makes the foundation of a coherent ecological policy difficult. Intensive development of interdisciplinary research on ecosystem management is therefore required, including, among others, ecologists, nature protection specialists, and economists. The previous literature reviews on the links between ecosystem services and biodiversity indicate that from year to year the number of publications devoted to these issues is growing (Balvanera et al., 2005; Solon, 2008). Over the past several years, the growing trend even accelerated (Balvanera et al., 2016).

Recent research in this field develops in several, overlapping directions, which can be described as follows: (a) attempts to identify precisely the providers of individual services, (b) theoretical considerations on the impact of particular aspects of biodiversity on the availability and delivery of services, (c) assessment of the role of landscape diversity in the provision of services, (d) development of theoretical and empirical mathematical models determining the relationship between biodiversity and service provision, and (e) fieldwork, including field experiments.

Work on the precise identification of individual service suppliers has led to the development of the concept of ecosystem service provider (ESP). According to this concept, the service availability depends only on specific organisms and their attributes (species diversity and richness, genetic diversity, functional groups, etc.). The provider of the service can be any set of living organisms, including, for example, a species, another systematic group, a specific functional group distinguished by any criteria, life form, trophic level or any other segment of biocoenosis (ecosystem) exclusively and directly responsible for the production of the service (Balvanera et al., 2016; Kremen, 2005). The concept of functional traits (of species or assemblages) is another one used to identify direct service providers and their complex relations with final services. This concept is related to two other concepts: (1) the specific effect function, which is the per unit capacity of a species to influence an ecosystem property or service, and (2) the specific response function, which is the ability of a species to maintain or enhance ecosystem services quantity in response to a specified change in the abiotic or biotic environment or to invade environment afresh (Díaz et al., 2013). Kremen (2005) demonstrated that the concept of ESP is related to the idea of the service providing unit introduced by Luck et al. (2003), which refers to the segment of a population or populations providing services in a given area. The latter concept allows measuring or estimating service supply.

Using the ESP concept, Kremen (2005) identified the providers of many essential, but relatively narrowly and precisely defined services, and proposed to use providers to indicate ES supply. For example, the providers of the carbon sequestration regulating service are tree species and biomass accumulation rate is the indicator. Providers of the pollination regulating service are assemblages of bees and bumblebees, and the indicator is the amount of pollen deposited daily. In the case of pest control regulating service, the providers are parasitoids of the pest insects, and the indicator - the proportion of infected insects. This approach - although methodically correct - is not always applicable due to often very difficult and laborious determination of indicator values. In such cases, Kremen (2005) suggested to make use of surrogate indicators. An example of such an approach can be indicating circulation of elements (from the group of supporting services) by the rate of organic matter decomposition (instead of counting the number of microorganisms).

Another approach to the identification of service providers was proposed by Harrison et al. (2014). Based on a systematic review of 530 peer-reviewed publications on the links between various attributes of broadly understood biological diversity and 11 ecosystem services they distinguished seven types of service provider: (1) single population, (2) two or more populations, (3) single functional group, (4) two or more functional groups, (5) dominant community, (6) single community/habitat, and (7) two or more communities/habitats. This classification allows better understanding the relationship between different aspects of biodiversity and the supply of services. Their analysis showed that timber production and inland fishery (angling) depends on two or more populations, water supply, and water treatment are regulated by single ecosystems or their groups, and different ecosystem types are responsible for regulating the atmosphere and carbon sequestration. In turn, pollination is mainly regulated by individual functional groups, while pest control is regulated by the entire set of biodiversity components, including functional groups, populations and ecosystems. In case of cultural services, analysis has indicated the main role of ecosystem diversity (aesthetic values) and populations of different species (recreation aimed at observing species, e.g., bird watching). In comparison to the ESP approach, the discussed proposal recognizes services a bit wider, including services provided also by the landscape system (which are in fact not covered by the ESP methodology). Thus, it has a more intuitive character and is easier for practical applications.

In turn, Ridder (2008) indicated that the resilience to environmental changes is also an important aspect of service supply, and proposed to divide ES into three groups:

- services provided by species functional groups, not dependent on particular species (e.g., carbon sequestration, water supply, and erosion control);
- services dependent on species resilient to environmental changes and often supported by humans (e.g., timber production, agricultural production);
- services dependent on species sensitive to environmental changes, often rare and endangered (mostly cultural services).

This division introduces a broader problem of maintaining the ability to provide services because of changes in species richness and the so-called redundancy of species, some of which may or may not have a significant role in service supply. Different authors indicate several mechanisms of such dependencies. The first one is the effect of complementarity, according to which the group of species uses resources more efficient than each species separately, which leads to a larger supply of the goods/services (this applies mainly to provisioning services). The second one is the selection effect. As a result, the dominance of a species with specific properties strongly determines the functioning of the entire ecosystem, and thus affects (positively or negatively) service supply (Elmqvist et al., 2010; Loreau and Hector, 2001). Elmqvist et al. (2010) emphasize the specific effect caused by keystone species whose spread or disappearance plays a disproportionately large role in service supply compared to the effect of randomly selected species (even from the same ESP group). Balvanera et al. (2016) mentioned also the effect of asynchronous reaction, according to which the nonsimultaneous and different reaction of different species to fluctuations and environmental stress results in a higher and more stable supply of services (see Elmqvist et al., 2010). A good example of the role of the effect of complementarity and asynchrony are the results of experimental studies, which show that the higher the number of algae species in streams, the higher the reduction in the content of nitrogen compounds in water. This is related to the filling of various ecological niches and higher resistance to external changes (Balvanera et al., 2014). In the same sense Kremen (2005) writes about the stabilizing and compensating effect, indicating at the same time the general model of "statistical weighing" because of random changes in the abundance of individual species.

The role of overall species richness and the benefits of species redundancy were investigated by Isbell et al. (2011). This team showed that 84% of the 147 species of plants tested in 17 experiments had a positive effect on the ability to provide ecosystem services at least once. Different species were responsible for this effect during different years, at different places, for different services, and under different environmental change scenarios. Furthermore, the species necessary to provide one service during multiple years were not the same as those necessary to provide multiple services within 1 year. The authors concluded that although species may appear functionally redundant when one function is considered under one set of environmental conditions, many species are needed to maintain multiple functions at multiple times and places in a changing world.

Lugnot and Martin (2013) also reached similar conclusions. They pointed out that individual species and functional groups are responsible for different services, what causes that diversity within the ecosystem and between ecosystems jointly positively impact the overall potential to deliver services. A similar picture emerges from other works. Brandt et al. (2014), based on independent mapping of nine service potentials representing all four groups (provisioning, regulating, supporting, and cultural) and the species richness of four taxonomic groups (mammals, birds, amphibians, and trees), analyzed the links between them. Significant correlations were found between overall service supply and species richness of all the analyzed groups. Such results strongly support the thesis that multifunctionality of the region strongly correlates with the general richness of the species.

Regardless of the works devoted to detailed identification of service providers and their determinants, an important theme of the research is to identify which aspects of the widely understood biodiversity best indicate the potential to provide services. It is widely accepted that for plant-dependent services, the first approximation is species richness (Balvanera et al., 2016). Similarly, Quijas et al. (2010) showed a positive relationship between plant species richness (called biodiversity) and services such as the provision of goods derived from plants, erosion control, and resistance to invasions and pathogens. Links with the control of soil fertility and pests were less important.

However, further analyses indicated a more complex relationship pattern. For the restored wetlands in China, Zhang et al. (2015) conducted an analysis of the relationships between various measures of biodiversity and 11 ecosystem services. As indicators of diversity, they used the measures of dominance, richness, Pielou's evenness, and Shannon's diversity as well as Simpson's diversity separately in relation to taxonomic diversity and the functional diversity of plant species. The results showed that the indicators of functional diversity better correlate with service supply than the taxonomic indices, showing strongest relationships with the dominance index, followed by the richness and diversity indicators. For crop pollination by native bees, Winfree et al. (2015) found that abundance fluctuations of dominant species drove ecosystem service delivery, whereas species richness changes were relatively unimportant because they primarily involved rare species that contributed little to that service. Other examples are given by Balvanera et al. (2016), indicating that the evenness within the species composition is related to resistance on invasions, and for animal-dependent services species number and composition in mammalian communities are associated with regulation of infectious disease.

The presented brief review shows that there is no universal measure linking species diversity with the capacity to provide services. This is highlighted, among others, by Braat (2014, 2013), who warns against considering single indicators as supply measures of entire packages of service, as this may, for instance, lead to overestimation of several species and ecosystems due to provision of food or wood, while neglecting the role of other aspects of biodiversity in the supply of different services at the ecosystem and landscape level.

Ecosystems providing services are part of a landscape, whose spatial structure may strongly affect the overall service supply and its assessment. This was clearly demonstrated by Anderson et al. (2009) who studied the match between places with the highest biodiversity (richness of protected species) and places with the highest potential to provide services. It turned out that the match depends on the size of mapping unit (square cell). Moreover, in different regions of England, the pattern of connections between services and species was different. The conclusion of this study is that the relationships between species, services, and space should be analyzed within small areas.

Similar, though nonidentical, conclusions arise from the work of Verhagen et al. (2016). They found (based on the analysis of literature data and cartographic models) that landscape heterogeneity plays an important role in determining the potential to provide certain ecosystem services. It is important to refer to the scale of analysis, because, for example, the effects visible in the analysis of artificial square cells or small catchments disappear when translated into large regional units. To explain the results obtained, the authors divided the heterogeneity of the landscape into two commonly extracted components: composition - defining the presence and extent of specific types of ecosystem (land cover), and configuration - defining the arrangement of individual patches in space. According to Verhagen et al. (2016), heterogeneity affects environmental benefits in the direct way (e.g., through the control of flows and retention of water and other substances), and indirectly - through the impact on species richness and composition. The authors distinguished four aspects of configuration affecting ES potential: (a) specific location of land cover patch (defined for example as the proximity to the closest patch of a given type), (b) the structure and distribution of multiple patches, determined, for example, by the distance to the nearest neighbor or connectivity index, (c) the structure of single patches, expressed, for example, by size, edge or shape metrics, (d) the presence of linear elements.

Their analysis showed that among all the services included in the CICES system, sediment retention accumulation of nutrients, pollination, and landscape aesthetics are the most dependent on landscape configuration. According to these authors, however, there is no clear evidence of the impact of configuration on agricultural production, flood control, and pest control. However, in the case of carbon sequestration, timber production, and livestock, this dependency most probably does not occur.

In addition, Zhang and Gao (2016) showed a clear relationship between patch size, fragmentation index, connectivity index, and the potential to provide many services. Similarly, Syrbe and Walz (2012) pointed out the role of landscape structure in shaping the potential of services. They indicated the necessity of applying various composition and configuration metrics, including connectivity and connectance metrics.

As emphasized by various authors, landscape fragmentation is an important aspect of spatial structure. The influence of fragmentation on services supply was investigated by Mitchell et al. (2015). In their model, the overall landscape potential depends on the potential of individual ecosystems and flows between them. The decrease in landscape potential can occur according to three schemes, that is, (a) linear - proportional to fragmentation, (b) with the shape of a decreasing exponential function, where initially the decline is very fast, and with the further increase in fragmentation the decrease of potential is much slower, (c) with the shape of a declining logistic curve where initially the decline is very slow, and with the further increase of fragmentation the decrease of potential accelerates significantly. On the other hand, the influence of fragmentation on the flows within the landscape can be neutral, negative, or positive. The combination of those patterns allowed the authors to identify three basic categories of relationships:

- fragmentation has negative impact on the potential of ecosystems and flow of services, which causes a sharp decline in the overall service supply (the potential in the landscape scale decreases); this case applies to water supply and regulation of water flows that are dependent on the size of the patches and their connectivity;
- fragmentation gives a compensative effect, that is, the impact of fragmentation on ecosystem potential is opposite to impact on flows, because of which the maximum landscape potential for a given service occurs at intermediate levels of fragmentation; such model applies to recreational, cultural, and aesthetic services of the landscape, genetic resources, pollination, and pest control;
- flows within the landscape are independent of the degree of fragmentation, service supply then depends only on the impact of fragmentation on the potential of individual patches (ecosystems); carbon sequestration is an example of such a service.

Several authors look for relationships and functional patterns to estimate the potential to provide services with the use of various biodiversity measures. Moreover, although in most cases simple linear relationships are assumed, some authors indicate that there are also dependencies described by nonlinear functions (e.g., bell-shaped or logarithmic - see Balvanera et al., 2016).

Braat and ten Brink (2008) proposed a general ideological scheme of dependence between the potential availability of different categories of services (provisioning; cultural-recreation; cultural-information; sum of all ES) and biodiversity measured by the MSA index (mean species abundance). The authors adopted land-use intensity (categories: natural, light, extensive, intensive, degraded, and urban) as the "proxy" indicator of biodiversity.

According to their scheme, the overall supply of provisioning services is maximal under intensive land use (with quite low species richness), the sum of cultural-recreation services is the highest under light land use (with high, but not the highest possible number of species), while cultural-information benefits gradually decrease as the intensity of land use increases. This pattern shows that the overall service supply is the highest at light and extensive use, although in such conditions none of the specific categories of service is maximized.

A more detailed analysis was conducted by Vos et al. (2014), who defined the functional models of the relationship between categories of service and species richness (Table 1.2). However, they did not take into account all species, but only richness within the group of service providers (ESP).

The associations presented in the table, as well as the proposals included in the other works, are essentially empirical and are rarely underpinned a well-reasoned cause and effect mechanism. Therefore, there is no certainty whether these are dependencies of a general nature or whether they only describe phenomena at selected local scales. As argued by Balvanera et al. (2016), assumption of the direct dependence of provisioning and regulating services on at least some aspects of biodiversity is quite weak because of the limited evidence. This lack of evidence is likely due to a lack of adequate experiments and testing, rather than a lack of dependence. On the other hand, Balvanera et al. (2016) emphasized that hundreds of field experiments and dozens of theoretical studies have demonstrated that decreasing plant diversity can alter ecosystem functioning in directions that would likely reduce ES supply.

1.2.5 Measures and indicators of ecosystem services

The interdisciplinary character of the ES concept causes that the proposed measures and indicators of ecosystem services represent different ways of describing reality. Indicators of ES potential and supply (the provider side, i.e., ecosystems and landscapes) belong to a wider group of indicators relating to the condition of the natural environment, while indicators showing demand and benefits (the consumer side, i.e., people) are part of widely understood social and economic measures. Due to the thematic scope of the book, only the first group of indicators will be the subject of further considerations.

In the literature dedicated to ecosystem services, many different definitions of indicators are formulated (see general discussion and review of definitions in Roo-Zielińska et al., 2007). In the more detailed considerations one can distinguish: (a) a measure - a direct result of the measurement of the state, quantity or process obtained, (b) an indicator - the transfer of measurement results in a recipient-oriented manner and adapted to the information purpose, (c) an index - a set of measures compiled (recalculated/converted) in such a way as to facilitate the transfer, make a synthetic approach, or present an issue that cannot be described by a single indicator (Balmford et al., 2008; Brown et al., 2014). Hence, ES indicator is a piece of information that presents in an efficient and understandable way the status and trends of service supply, flow, or demand.

Table 1.2 Selected types of relationship between species richness and potential as regards ecosystem services.

Service	Species group (ESP)	Ecosystem type	Landscape conditioning	Dependency type
Carbon sequestration	Long-lived species	Forest and permanent grasslands	Lack	Growing, with the level of saturation
Water purification	Nonleguminous species, algae	Wetlands, grasslands	Lack	Growing rectilinear
Soil fertility	Leguminous, soil biodiversity	Grasslands, crops	Lack	Growing rectilinear
Pest regulation	Birds, mammals, predating insects, parasitizing wasps	Flower rich and woody vegetation	Landscape mosaic area structure proximity (1–2 km)	Logistic
Pollination	(Wild) bees, hover flies, butterflies	Flower rich and woody vegetation	Landscape mosaic area structure proximity (1–2 km)	Logistic
Aesthetic appreciation	Appealing and charismatic species	Different	Landscape mosaic structure	Growing rectilinear

Based on Vos, C.C., Grashof-Bokdam, C.J., Opdam, P.F.M. 2014. Biodiversity and ecosystem services: does species diversity enhance effectiveness and reliability? A systematic literature review. In: Statutory Research Tasks Unit for Nature & the Environment (WOT Natuur & Milieu). WOt-Technical Report, vol. 25, Wageningen.

Indicators can be divided into direct and indirect. The former refer precisely to the tested object at each measurement level, while the latter are used instead of direct indicators due to greater simplicity of measurement or availability of data. They are often called surrogate indicators (Miguntanna et al., 2010). The basic condition for their applicability is that they must be measurable with adequate accuracy and must correlate with the direct indicator. In many papers also the term "proxy indicator" is used without being further defined, while the dictionary definition and context of applications show that in most cases it is an equivalent term with the concept of surrogate (indirect) indicator and these terms can be used interchangeably.

A completely different classification of indicators, based on the relation to the assessed object, was presented by Egoh et al. (2012). The authors promote the concepts of primary and secondary ES indicators. It should be emphasized that their terminology is not intuitively clear and may raise some doubts. In their view, primary indicators reflect measures used to quantify ES (e.g., tourist attractiveness), while secondary indicators provide data necessary to build a primary indicator (e.g., accessibility and naturalness for tourist attractiveness).

In addition to simple indicators (in terms of structure), directly corresponding to one type of measurement, complex indicators (sometimes called indices) are used, being a mathematical combination of simple indicators. Usually, the normalization of partial indicators is used. Normalization follows the formula:

$$X_{norm} = (X_{obs} - X_{min})/(X_{max} - X_{min}),$$

where X_{obs} is the observed value of the indicator; X_{min} means either the theoretically lowest indicator value or the lowest value in a specified dataset; X_{max} is either theoretically the maximum value, either the highest desirable value or the highest value in the specified dataset. The meaning of X_{max} and X_{min} depends on the theoretical data model and the purpose of the analysis. This approach is useful when different indicators are expressed in different scales and different units, require mutual comparison, or they are included in complex indicators.

Synthetic (integrated) ES indicators based on standardized values are used for various purposes, and they are called differently in the literature, for instance:

- Multiple Ecosystem Services Landscape Index (MESLI) - the sum of normalized partial indicators (Rodríguez-Loinaz et al., 2015);

- Total Ecosystem Service Indicator (TESI) - the mean value of normalized partial indicator values (Dick et al., 2014);
- Ecosystem Services Composite (ESC) indicator - intended to present in an integrated way a group of services, with the general formula: $ESC = (\sum(X_{norm}i \times w_i))/N$, where w_i is the weight of the service in the total bundle of N services (Alam et al., 2016).

The advantage of integrated indicators and weighing is the possibility to reduce a large number of partial, analytical indicators to a small number of synthetic indicators, better appealing to the recipient. On the other hand, their weakness is the lack of knowledge of the actual links between partial indicators and the subjectivism of weighting (not weighing is in fact also weighing with the weights equal to 1).

Regardless of the used definition and indicator construction, some scientific and practical criteria for indicator selection and evaluation always need to be adopted. A comprehensive review of this issue in relation to most environmental indicators is provided by Roo-Zielińska et al. (2007), while in the case of ES indicators, the following scientific and practical criteria are most commonly adopted (Kandziora et al., 2013; Wiggering and Müller, 2004):

- An indicator or set of indicators must ensure scientific correctness by (1) unequivocal representation of the indicated phenomenon, (2) a well-proven causal relationship between the indicator and the indicative phenomenon, (3) optimal sensitivity of representing the phenomenon, (4) information appropriate to the space-time scale, (5) the possibility of spatial and temporal aggregation, (6) a high degree of validity and representativeness of the data sources used, (7) a high degree of comparability with other indicators, (8) good compliance with the statistical requirements for verification, repeatability, representativeness, and validation of results.
- An indicator or set of indicators must provide high practical relevance through (1) high importance in making political/practical decisions, (2) a direct reference to practical activities, (3) the ability to determine normal states and formal standards, (4) high intelligibility and social transparency, (5) focus on environmental goals, (6) appropriately good measurability, (7) good availability of necessary data, (8) providing information on long-term trends, (9) usefulness for early warning purposes.

Sometimes formal requirements in relation to indicators are formulated in a more general way. In the important study by Maes et al. (2014), only two criteria were explicitly mentioned: data availability and suitability for providing relevant information to managers. Indirectly, however, the credibility, appropriateness, and validity of the assessment were also mentioned. In this context, Heink et al. (2016) indicate the failure to determine the validity of a significant part of indicators used in ES research. According to Heink et al. (2016), validity refers to the degree to which an indicator captures the meaning of an object that is to be measured or evaluated (e.g., regional climate regulation, bequest value). An indicator is considered valid if evidence and theory support the interpretation of the results achieved by means of this indicator.

Apart from the general criteria of scientific correctness and practical usefulness, additional postulates are often formulated from the point of view of the usefulness of information provided. For instance, Albert et al. (2016) indicate five specific requirements for sets of indicators to be useful in decision-making and land management. They should:

1. Show the ES supply in relation to areas of particular demand (e.g., higher importance of water supply and recreation near large cities). For some ecosystem services, no particular spatially explicit demand can be defined (e.g., global demand for carbon sequestration).
2. Explain if changes in the actual use of particular ecosystem services are caused by changes in ES potential and delivery, changes in ecosystem condition, changes in human inputs, or changes in demand.
3. Have the ability not only to determine current production and flow of ES, but also the ability of ecosystems to provide them in the future.
4. Have the ability to provide decision support in cases of insufficient information to precisely assess and evaluate ES supply and/or demand.
5. Distinguish precisely between the natural potential (the internal feature of the ecosystem) and the real ES supply that often originate from a conjunction of natural and anthropogenic contributions.

Brown et al. (2014) formulated more general criteria for a good indicator: (1) appropriate to the needs of the recipient, (2) understandable in construction, presentation, and interpretation, (3) useful for reporting, identification of changes, early warning, etc., (4) scientifically justified, that is, resulting from recognized theories and relationships between the indicator and the indicated object and based on reliable data, (5) adequately sensitive to changes in the measured services, (6) practical and cost-effective. These authors critically assessed the achievements in this field so far, stressing that the creation of a system of ES indicators is a big challenge because of the following reasons:

1. the ability of indicators to provide information on services is generally low, although different in relation to various service;

2. indicators available for individual services are not comprehensive and often are not suitable for characterizing the complexity of services and benefits;
3. adequate input data is missing for well-constructed indicators;
4. the indicators on regulating and cultural services proposed so far are on the significantly lower level than the indicators for provisioning services.

This critical evaluation is consistent with the results of the meta-analysis that Egoh et al. (2012) conducted on the basis of 67 studies that have mapped ES using various indicators. All indicators identified in these studies were grouped into primary indicators (reflecting the proxy used to measure ES) and secondary indicators (providing the necessary information used to compose the primary indicator), and ascribed to four classes of services (provisioning, regulating, habitat/supporting, cultural). According to their analysis, cultural services were evaluated more often by primary indicators than provisioning and supporting services, while regulating services - the most often Secondary indicators showed the same trend as the primary indicators. Regulating services had the greatest number (no less than 90 different types) of secondary indicators. This result could be explained by the fact that regulating services (such as carbon sequestration/storage or water flow regulation) are modeled using many different input data, which are just secondary indicators. Egoh et al. (2012) underlined that land cover proved to be an important secondary indicator for all four categories of services, comprising 16% of all secondary indicators used in different studies. Other common secondary indicators were nutrient fluxes and soil characteristics (6% of the total secondary indicators) and vegetation type (5%).

A more recent systematic analysis of 405 articles (Boerema et al., 2016) showed that each of the 21 ES analyzed had on average 24 different measures, which may indicate the complex reality of ES or suggest a lack of agreement on the content and scope of individual services. The authors also found that for regulating ES, ecosystem properties and functions (ecological aspects) are more commonly quantified (67% of measures), while for provisioning ES, benefits and values (socioeconomic aspects) are more commonly quantified (68%). Cultural ES are predominantly quantified using ranks (35%). The authors concluded that measurement and assessment of services are still poorly and unevenly developed, as often only one side is considered (either the ecological or socioeconomic side) and indicators are often oversimplified.

Here, similarly, it is worth emphasizing the uncertainty as to the value of using biodiversity indicators. Some of them explicitly reflect ecosystem services (e.g., abundance and diversity of cultivated plant species, or other harvestable species), others define phenomena that are in an ambiguous relation to services (e.g., cultural value of native plant species) (Boykin et al., 2013). The lack of well-defined relations between indicators and the actual level of ES supply, flow, or demand is also present in the European MAES project (Maes et al., 2014), in which indicators of different characters are proposed for the determination of different services. This inconsistency in the construction of indicators results from the assumption that only available data collected in individual countries in the same standardized way should be used to assess ecosystem services on the European scale.

The way to overcome these limitations can be a consistent linking of ES indicators to the DPSIR model, which is the framework recommended by the European Environment Agency to develop environmental indicators (EEA, 1999). This approach was proposed, among others, by Haines-Young and Potschin (2010b) and Kandziora et al. (2013) who presented a cascade including the following elements: biophysical structures and/or processes (measurable ecosystem properties) → ecosystem function (ecological integrity) → ecosystem services → human benefits, describing the impact on social, economic, and/or personal well-being → value (relative importance of individual components). Each segment of this cascade should have its own set of indicators, and the segment "ecosystem services" should only refer to the flow of services from "nature" to "society". The segment "ecosystem functioning (ecological integrity)" deserves special attention. The authors suggested indicating the entire set of features defining the level of ecological integrity, such as exergy accumulation, level of entropy, storage capacity, nutrient cycling, biotic water flows, metabolic efficiency, space heterogeneity, and biotic diversity. It was only in the next step that links of such synthetic indicators (though sometimes measured by simple proxy indicators) with specific services or their groups were determined.

The cascade model was used in a bit different way for the development of indicator system for Finland (Mononen et al., 2016). It was assumed that services do not form an independent block, but they are a process that covers all segments. In addition, the meaning of individual segments was slightly changed. In this way, the following cascade scheme was obtained: (a) structures and processes that are the basis for ecosystem functioning and reflecting the spatial perspective of the potential of ecosystems → (b) functions necessary for the provision of services, including temporal perspective → (c) benefits, that is, part of the ES potential (both tangible and intangible) that is used → (d) the value (social, health, economic, and other) of the benefits achieved. In that scheme, segments (a) and (b) relate to the ecosystem and biodiversity perspective, which form the basis for determining the potential of ecosystems, and segments (c) and (d) relate to the well-being of people and society. Based on this cascade and the CICES list of services modified for the national needs, a scheme was developed covering 28 services (10 provisioning, 12 regulating and 6 cultural). For each of the service, a set of four indicators (112 in total) corresponding to particular stages in the cascade was also proposed (Mononen et al., 2016). It is by now the only system of indicators that has consistently and unequivocally captured both the (potential) supply and actual flow of ecosystem services.

Despite evident progress in defining of services in recent years and increasing scientific rigor in the use of indicators, there is still considerable freedom in this field and there are no approaches generally accepted as a standard.

1.2.6 Spatial and temporal dimension of ecosystem services

The supply of a given service calculated in different spatial units (e.g., ecosystems, landscapes) is usually also different. Moreover, different services are delivered in different spatial scales. Having that in mind, Costanza (2008) proposed ES classification based on spatial characteristics of service providers and spatial relations to service consumers. He distinguished the following types of service:

- global nonproximal (does not depend on proximity), for example, climate regulation, carbon sequestration;
- local proximal (depends on proximity), for example, disturbance regulation, protection, waste treatment, pollination;
- directional, flow related, for example, flood protection, water supply, erosion control;
- in situ (point of use), for example, soil formation, food production;
- user movement related (flow of people to unique natural features), for example, genetic resources, recreation potential.

Scale dependence is especially important for regulating services. Hein et al. (2006) defined regulating service as an ecological process that has (actual or potential) economic value because it has an economic impact outside the studied ecosystem and/or it provides a direct benefit to people living in the area. Ecological processes involved take place at certain, ecological scales, most relevant of which for the regulating services are as follows: global (e.g., carbon sequestration); biome-landscape (e.g., regulation of the timing and volume of river and ground water flows); ecosystem (e.g., pollination, regulation of pests and pathogens); plot-plant (e.g., protection against noise and dust, run-off control, biological nitrogen fixation) (Hein et al., 2006).

The research on ecosystem services is conducted on many spatial scales: from the local (Allendorf and Yang, 2013; Lavorel et al., 2011) and regional (Casado-Arzuaga et al., 2013; Plieninger et al., 2013; Vihervaara et al., 2010), through the national (Mononen et al., 2016; Norton et al., 2012; Turner et al., 2014) and continental (Haines-Young et al., 2012; Okruszko et al., 2011; Paracchini et al., 2014), to worldwide (Luck et al., 2009; Naidoo et al., 2008). The scale depends on the nature of the services analyzed. Limburg et al. (2002) indicate that the ES assessment method is determined by the ES type and spatio-temporal scale of analysis. They emphasize that determining the scale in which the production and delivery of a given ecosystem service takes place is often very difficult. For example, the service associated with the mineralization of nutrients is mainly provided by microorganisms living in soil, water, or sediments. Plants and animals use these nutrients regardless of the time scale and geographical space. Nevertheless, the authors believe that including scale in the ES classification helps to better understand them and evaluate.

Some ecosystem services provide tangible or intangible benefits for all mankind, regardless of the location or sociocultural context. For instance, carbon sequestration by terrestrial and aquatic ecosystems affects air quality and climate of the whole Earth (Turner et al., 1998). Global scale investigation of this service enables showing flows between regions, countries, and continents (Luck et al., 2009; Maes et al., 2012b). On the other hand, water-related services require investigations at the catchment or river basin scales (Morri et al., 2014). In turn, ecosystem potentials related to pollination are usually analyzed on a local or regional scale (Affek, 2018; Albert et al., 2016; Naidoo et al., 2008; Vihervaara et al., 2010). Most of the provisioning and cultural ecosystem services tend to be strongly related to the region (Vihervaara et al., 2010). This relationship is particularly well visible in social studies on cultural ecosystem services (e.g., Gould et al., 2014; Kowalska et al., 2017). Supraregional research on cultural ecosystem services mainly concerns recreation and tourism (Paracchini et al., 2014). In turn, Swift et al. (2004) pay attention to the highly restricted temporal and spatial scales in which the research on agricultural systems has to be conducted.

The spatial scale in ES studies depends on several factors, the most important of which is the purpose of investigation. Typically, studies related to spatial planning and cost—benefit analysis are conducted at a regional or local level, while studies on general patterns related to the spatial distribution of services - at the above-national level (Maes et al., 2012a). The study objectives influence the selection of data sources and methods of obtaining and using them. The data should be uniform for the entire study area. Acquisition of accurate data is easier on a local or regional scale, while the larger the spatial extent of analyses, the more difficult this process becomes (Maes et al., 2012b). Naidoo et al. (2008) found that worldwide data are only available for four ecosystem services. At the European level, the CORINE Land Cover map is commonly used in ES research (Metzger et al., 2006). However, its quality varies between countries, and when applying to regional research it should be supplemented with relevant local data (Vihervaara et al., 2010).

In addition, obtaining information on ecosystem services from land cover/land use or habitat maps (see Burkhard et al., 2009; Kienast et al., 2009) is only appropriate if the service is directly related to land use (e.g., agricultural crops) and when the purpose of the research is to determine the occurrence of a given service, and not the size of ES supply (Okruszko et al., 2011). Statistical data, useful in quantitative assessment, are usually collected in administrative units and available primarily for provisioning ecosystem services. Various proxy measures are used to estimate regulating services (Feld et al., 2009). They can be obtained through modeling (species—surface relationships SAR, Nelson et al., 2009; functional features of plants, Lavorel et al., 2011) or extrapolating data collected in the field (biomass measurements, Roo-Zielińska et al., 2016). Remote sensing data and GIS techniques are also very useful for this purpose (Nemec and Raudsepp-Hearne, 2013). The demand side of ecosystem services is usually evaluated through social research (Scholte et al., 2015), which due to its nature is conducted on the limited spatial scale: mostly local (Badola, 1998; Lewan and Söderqvist, 2002), but also regional (van Berkel and Verburg, 2014; Castro et al., 2011), and sometimes national (Kikulski, 2009). Some researchers conducted observations covering the same type of ecosystem in several regions/countries - for example, forests (Sodhi et al., 2010) or grasslands (Lamarque et al., 2011).

1.2.7 Mapping ecosystem services

Mapping ecosystem services is a complex problem and includes several related issues, the most important of which are the following: (1) selection of appropriate indicators for individual services, (2) source data availability, and (3) selection of appropriate mapping unit. This last point is especially important because the character of the unit has a very strong impact on map reliability and the possibility of its use for scientific and practical purposes. Syrbe and Walz (2012), based on a broad literature review, listed the most commonly used reference units. According to them, various units used in practice can be grouped into the following categories:

- single patches, parcels, or landscape elements (most widely used);
- the smallest common geometric units (generated in GIS by overlaying thematic layers);
- administrative units (preferred in spatial planning);
- watersheds (dedicated to water-related landscape processes);
- so-called natural units (reflecting the diversity of geology, soil, topography, etc.);
- landscape units (delineated based on natural environment and land use);
- regular artificial geometric units (e.g., raster grid).

Administrative units, although often used to define the boundaries of the study area, are seldom really suitable for ES analysis. In turn, purely natural units should be used as long as environmental characteristics determine a service or if the available data relate to them, whereas landscape units seem to be useful for assessing most services. They were used, for instance, as the spatial basis in conflict analysis among landscape services (de Groot, 2006), biodiversity investigation (ISCU-UNESCO-UNO, 2008; Naveh, 2007) and studies including several water-related processes (Maass et al., 2005). In turn, regular grids are most often used for rescaling data with very different resolution (see Nemec and Raudsepp-Hearne, 2013).

Even in the above-outlined framework, a large variety of reference spatial units is possible, depending primarily on the purpose of the research, the scale of the study and the available data. In scientific papers at local or regional level, the identification of services is usually carried out at the ecosystem level (Kruczkowska et al., 2017). Scientists investigated services provided by wetlands (Okruszko et al., 2011), grasslands (Lamarque et al., 2011), forests (Decocq et al., 2016; Grilli et al., 2016), localized and distinguished with the help of various spatial databases: Natura 2000 sites, thematic and topographic maps.

If ecosystems are used as reference units in pan-European analyses, they are usually distinguished only in relation to land cover or land use (Burkhard et al., 2012; Kienast et al., 2009; Metzger et al., 2006). The source of spatial information for terrestrial ecosystems is primarily the CORINE Land Cover map (Maes et al., 2012a). Recently, the so-called MAES classification of ecosystems proposed by the European Commission (Maes et al., 2013) is also used to achieve the objectives of the EU Biodiversity Strategy for the period up to 2020 (Maes et al., 2016). The classification includes seven types of terrestrial ecosystem and five related to inland and marine waters.

Analyzes of ecosystem services on a European scale are also conducted in artificial regular units (e.g., in 1 km × 1 km grid; Haines-Young et al., 2012), and their results are generalized and presented in administrative regions (NUTs) for statistical purposes. Natural units are sometimes combined with artificial units for which statistical data on the use of services (mainly provisioning) is available (Crossman et al., 2013). For instance, in the study of Morri et al. (2014), watershed division was combined with the division into communes to show the relationship between supply and demand for selected services (water retention and supply, soil protection, and carbon sequestration). In turn, Vilhervaara et al. (2010) in their research in Lapland combined land-use patches with reindeer-herding districts.

For some cultural services such as spiritual experience, finding a reference spatial unit may be difficult. Usually, places where such services are provided are marked with points (Raymond et al., 2009), because they are related to single landscape elements. The preferences of users of cultural services in relation to various elements (such as hedgerows and tree lines) and landscape structures (e.g., mosaic of fields and forests) were also studied by van Berkel and Verburg (2014). In many studies that make use of social methods, services are evaluated for the entire study area (national park, region), without any further divisions (Allendorf and Yang, 2013; Casado-Arzuaga et al., 2013; Castro et al., 2011; Kulczyk et al., 2016).

In recent years, many papers that synthesize the results of ecosystem mapping have been published. Several works discuss spatial relationships between physical geographic units (e.g., landscapes) and the flow of ecosystem services (Brown et al., 2015a,b; Burkhard et al., 2009). Capacity of ecosystems to provide services and ES flow to the society was mapped in different spatial units (Braat et al., 2013; Brown et al., 2015a,b; Liquete et al., 2015). The methodology used takes account of ecosystem multifunctionality and potential benefits for the society (Burkhard et al., 2014, 2012). Apart from ecosystem/landscape quantitative assessment and mapping, the key habitats and ecological corridors have been also often identified. The research carried out across all 27 EU member states showed that key areas of the so-called green infrastructure network, having the highest capacity to provide ES, cover 23% of Europe (Liquete et al., 2015). The MAES project - previously described in the text - played a huge role in mapping ecosystem services and their indicators. One of its results will be the MAES digital atlas, which aims to provide a map of ecosystems and services in Europe. The information collected, successively developed, will be published in the European, national, and subnational level. The completion of the MAES atlas is planned for 2020.

Spatial aspect of ecosystem services has been also investigated through empirical studies involving public participation GIS (PPGIS) method (Brown and Fagerholm, 2015). Participatory mapping is a group of methods and techniques combining traditional expert mapping with input from nonexpert participants. Maps have been often developed together with stakeholders during workshops or simply include data derived from interviews and standardized surveys.

Attention should also be paid to the latest work edited by Burkhard and Maes (2017) *Mapping Ecosystem Services*, in which the authors emphasize the importance of mapping ecosystem services, explain how ecosystems contribute to human well-being and what the impact of mapping is on management of natural resources. In this comprehensive study, they propose indicators for mapping and assessment of the potential to provide services by major ecosystem types (agricultural, forest, freshwater, and marine) in different spatial scales. In conclusions, the authors note that mapping ecosystem services has developed over the past years into a separate scientific field. They underline that different levels of government started to use the concept of ES as a "bridge" between nature and society and ES maps are recognized as tools to help policy and decision-making. Understanding ecosystem conditions, processes and services, and their spatiotemporal dimension is essential for sustainable management of natural resources. Mapping ES offers a framework for combining spatial data and transdisciplinary knowledge of different sources. As Burkhard and Maes (2017) write, "the book is not only a synthesis of the state-of-the-art of ES mapping but it provides a comprehensive overview and guidance for those mapping ES themselves or for those using ES maps".

1.2.8 Social perception of ecosystem services

Public opinion is increasingly being used in the implementation of projects related to the management of ecosystem services (Scholte et al., 2015). In this way, social awareness and the number of satisfied ecosystem users increase (Felipe-Lucia et al., 2015). When aiming to objectively and reliably evaluate the value of ecosystem services, it is very important to take into account the opinions of all potential users, that is, people who have real influence on ecosystems or may be affected by decisions related to it (Freeman, 2010). Research should be conducted on a sufficiently large sample of people representing a given community, diversified in individual characteristics (age, gender, place of residence, education, profession, income, etc.), social position, and the way of using ecosystems (Chan et al., 2012). This last criterion is particularly important when using public opinion to make decisions regarding ES management. Many studies demonstrated that the assessment of services depends to a large extent on the use of ecosystems and the benefits users derive from them (Affek and Kowalska, 2017; Calvet-Mir et al., 2012; Carvalho-Ribeiro and Lovett, 2011; Casado-Arzuaga et al., 2013; Maass et al., 2005; Scholte et al., 2015).

Assessment of services is also influenced by the perception of ecosystems, which depends not only on ecosystem physical properties, but also on generally accepted landscape concepts related to cultural identity and tradition (Terkenli, 2001). Emotional bonds with a given area resulting from social relations, ethnic origin, or important experiences play a significant role as well (Soini et al., 2012). The factor limiting the possibility of assessment may be a lack of knowledge about different services. Without proper information, the service may not be noticed and consequently will not be considered important (Bingham et al., 1995).

The methods of collecting public opinion about ecosystem services are diverse and depend on the scope of research. Many studies focus on identifying the most valuable services for a given area, the aim of the others is to create a hierarchy of services or to understand how the preferences of users have changed over time (Felipe-Lucia et al., 2015). The methods used can be divided into two groups. In the first one, data are collected indirectly through the analysis of written materials or observation of behavior; whereas, in the second one, the information is obtained directly from people (Scholte et al., 2015).

Observations may require direct involvement of users, as it is in the case of cultural services (Sagie et al., 2013; Tzoulas and James, 2010). For instance, the number of visits to a national park can be used as a measure of recreational attractiveness of the area.

In the analysis of published materials (so-called content analysis), opinions of individual people, groups or entire communities are obtained through the review of various text documents, images and other forms of presentation (Chadwick et al., 1984; Mehtälä and Vuorisalo, 2010; Oswald et al., 2013; Piwowarczyk et al., 2013; Pramova et al., 2012; Seamans, 2013). This type of analysis has been used in social studies related to the natural environment since the 1970s, when Ernst and Ernst (1972—78) conducted research on valuation of environmental costs. The literature on this subject distinguishes two approaches to content analysis: mechanistic and interpretative (Beck et al., 2010). The mechanistic approach includes automatic search and counting of words, phrases, and sentences in the text documents (Maczka et al., 2016). It is based on the assumption that the more often the word/phrase/sentence appears in the text, the greater its meaning. This approach allows efficient quantitative analysis of large number of documents. The interpretative approach is more time-consuming. It assumes the thorough semantic analysis of sentences concerning the investigated subject.

Another method - expert assessment - can be counted as both indirect and direct technique, because experts can be asked to conduct research, express their own arbitrary opinion, or to provide information about the preferences of other people. Experts are by definition better acquainted with technical problems and specific terminology, but their assessment, if arbitrary, is based on subjective experience and knowledge that may not fully reflect the opinion of the wider community (Edwards et al., 2012).

The opinions of direct ecosystem users can be collected through interviews, surveys, or workshops with stakeholders. The way of interaction with the respondents and the method of analyzing their answers are important. Written questionnaire surveys are the method that ensures high standardization and limit the influence of the researcher. Large amounts of data can be quantitatively analyzed in this way. The disadvantage of this method is the relatively high cost and low flexibility, which can lead to loss of potentially important information. In-depth interviews allow a more detailed analysis of user preferences and are often regarded as a pilot study to questionnaire surveys (Calvet-Mir et al., 2012). In turn, workshop method is widely used in areas where local communities are actively involved in environmental management (Goma et al., 2001). During organized group discussions, participants share their knowledge and try to work out a common solution.

The spatial and temporal scope of the assessment should be clearly defined when using each of the methods discussed. Setting the time frame is important due to the constant changes in biophysical properties and land use, which affect ecosystem condition and human perception thereof (Hein et al., 2006; Turner et al., 2003). Equally important from the point of view of ecosystem management is referring to a specific area (Hauck et al., 2013). To date the spatial diversity in research on the perception of ecosystem services has been rarely considered. Values have been usually assigned to the entire study area, without any further spatial divisions (Allendorf and Yang,

2013; Casado-Arzuaga et al., 2013; Castro et al., 2011). Another option is to indicate specific locations where the services are provided. However, this approach may be of limited validity, because respondents may mark first of all those places that they know well (Scholte et al., 2015).

1.2.9 Synergies and trade-offs among ecosystem services

The concepts of "trade-off" and "synergy" belong to one of the most important terms in the analysis of ecosystem services. In general, a trade-off is a situation in which the use of one ecosystem service directly reduces the benefits obtained from another service (Turkelboom et al., 2016). The reverse situation, when the use of one service results in an increase of the benefit from another, is called synergy (Turkelboom et al., 2016). However, in many studies only the term "trade-off" is used to designate all types of associations between services, and even more broadly - associations between service providers and consumers. This created a terminological chaos, which was attempted to be overcome by introducing different schemes of narrowly defined relationships.

Mouchet et al. (2014) show two possible approaches to the typology of such associations. The first one, developed as part of the *Millennium Ecosystem Assessment*, assumes trade-off division into four categories: (1) spatial trade-off - the spatial lag between the place of service production and the place of its supply (e.g., forest in the upper part of the catchment and its impact on the supply of water in the lower sections of the catchment), (2) temporal trade-off - the delay in the delivery of services produced resulting from human decisions or from natural processes, (3) reversible trade-off - the ability to return to the initial supply after a disturbance in ES production, and (4) trade-off among services - the positive or negative impact of the supply of one service on the supply of other ES.

The TEEB assessment (2010) proposed another classification, with similar terminology, but a slightly different definition of categories: (1) spatial trade-off - the spatial lag between the benefit and the cost related to the targeted ES, (2) temporal trade-off - the time lag between the benefit of a service and the associated cost because the deterioration of this or other ES in the future, (3) trade-off between beneficiaries - where beneficiaries can be either "losers" or "winners" depending on who bears the cost of or the benefit of ES supply, and (4) trade-off among ES - addressing management of one ES at the expense of another. As emphasized by Mouchet et al. (2014), the MEA classification focuses on the consequences of ecological trade-offs for ES supply, while TEEB's is framed in economic benefits and costs for ES demand.

The idea of ES spatial associations has been developed further into the concept of areas supplying services and using services (Fisher et al., 2009; Syrbe and Walz, 2012). According to this concept, service providing areas (SPAs) are places (ecosystems, landscapes), in which real production of services takes place. Service benefit areas (SBAs) are those, where services are used. In the case of significant distance between SPA and SBA, it is possible to delineate service connecting areas (SCAs), enabling the transfer of matter, energy, and organisms between SPA and SBA. Based on this conceptual framework, Fisher et al. (2009) distinguished four main types of spatial associations, corresponding to the first MEA trade-off category (spatial trade-off): (1) SPA and SBA overlap, which means that the production and use is in exactly the same area, (2) SPA is part of a much larger SBA area, and the use of the service does not require any particular flow, (3) SPA and SBA are spatially separated, with a connecting area (SCA) between them. The condition for the existence of SCA and flow is the gradient (in the physical sense) between SPA and SBA. Most often this applies to gravitational processes (cold air, water, and mass movements), (4) SPA and SBA are spatially separated, with a connecting area (SCA) between them, while there is no gradient—potential relationship. In such a case, the quality of SCA (and therefore the possibilities of service flow) depends on other conditions, often nonspatial (e.g., legal conditions). Of course, such a division is correct and meaningful as long as all areas and associations are considered in the same spatial scale. Otherwise, there are so-called scale trade-offs, meaning, for instance that service supply is local and the benefits are supraregional or vice versa (which happens much less often) (Syrbe and Walz, 2012).

The role of spatial scale in all types of association between services was repeatedly pointed out. For instance, on the basis of research in the Basque Country Rodríguez-Loinaz et al. (2015) clearly stated that synergies and trade-offs between services are of different nature when identified in a detailed and in supraregional scales.

Landscape composition and configuration are also an important aspect of spatial scale considerations. Mitchell et al. (2015) suggest that synergies and trade-offs between services may be shaped in a completely different way depending on the degree of landscape fragmentation, which of course involves the presence of different spatial associations (MEA first type trade-offs) and the impact of fragmentation on service supply within ecosystems along with the change in flows between ecosystems.

The papers dedicated solely to the associations between services are not very numerous, and the available results come from general works devoted to identification, quantification, and mapping of services in specific areas. Therefore, there is a lack of reliable data of universal character in this research field. However, clear negative relationship between provisioning (mainly from agricultural areas) and regulating services has been demonstrated several times, which, apart from land cover, may also be affected by the size and type of anthropogenic energy subsidy, recognized as one of the steering variables of provisioning services supply (Balvanera et al., 2014; Maes et al., 2014).

When considering many different services simultaneously, the term "ES bundle" is often used. Berry et al. (2016) stated that, according to most of definitions, the term "ES bundle" focuses (explicitly or implicitly) on spatial coincidence of the delivery of a range of services. The authors also recall some more complex ES bundle definitions, for example, a set of ecosystem services that repeatedly appear together across space or time (Raudsepp-Hearne et al., 2010) and the idea to separately consider ES bundles on the supply and demand sides (García-Nieto et al., 2013). Berry et al. (2016) were those who proposed definitions for ES supply/demand bundles: (a) ES supply bundle – a set of associated ecosystem services that are linked to a given ecosystem and that usually appear together repeatedly in time and/or space, and (b) ES demand bundle – a set of associated ecosystem services that are demanded by humans from ecosystem(s). It is worth to underline that in the same landscape (with the same ES supply bundles) different ES demand bundles could be demanded by different stakeholder groups. The concept of ES bundle is of great importance when considering multifunctional areas (ecosystems, landscapes), those that have the capacity to perform many different functions at the same time, and thus can provide many different services forming a bundle or bundles (Turkelboom et al., 2016). Although from a formal point of view, there is a requirement that there should be interactions between services in the bundle, this term is increasingly being used (misused) to designate simple spatial coexistence of services.

1.2.10 Major initiatives related to ecosystem services

Since the publication of the *Millennium Ecosystem Assessment* reports (MEA, 2005, 2003) prepared under the auspices of the United Nations, the popularity of the concept of ecosystem services has grown exponentially, both regarding scientific articles related to this subject, as well as to practical actions and commitments undertaken at the international, national, and local levels. One of the MEA alarming conclusions that deserved immediate action was that the capacity of world ecosystems to provide 2/3 of services, including the very essential ones such as clean water provision, air purification, and climate regulation, has been recently significantly reduced.

Another important milestone of ES research was the publication of *The Economics of Ecosystems and Biodiversity* report (TEEB, 2010). The TEEB initiative was established in 2007 by the German government, the European Commission, and the United Nations Environment Program (UNEP). The TEEB main aim was to draw attention of decision-makers and politicians to the growing costs of environmental degradation and biodiversity loss worldwide and to include the value of nature into the global economic calculations.

One of the main players interested in developing the ES concept and its practical application is the European Union. The EU Biodiversity Strategy[6], adopted by the European Commission in May 2011, defined a framework for EU actions in the coming decade to achieve the flagship goals of: "halting the loss of biodiversity and degradation of ecosystem services by 2020; restoring them as much as possible, and increasing the EU's contribution to preventing loss of biodiversity in the world". This strategy is an integral part of the EU action 6 and 7 in the field of environment and biodiversity - Decision No. 1386/2013/EU ("Europe 2020" COM (2015) 478 final).

As part of the Strategy, EU obligations under the Convention on Biological Diversity are implemented. The Strategy includes six mutually supporting objectives that address the main factors influencing the loss of biodiversity. Their fulfilment is intended to reduce threats to nature and ecosystem services in the EU. One of the objectives is to maintain and restore ecosystems and their services - as part of its implementation, it is recommended that by 2020 ecosystems and their services should be maintained and strengthened by establishing green infrastructure and rebuilding at least 15% of degraded ecosystems. The planned action is to improve knowledge about ecosystems and their services in the EU. By 2014, Member States, in cooperation with the Commission, were obliged to identify and assess the condition of ecosystems and their services on their territory (Brussels, 2.10.2015, COM (2015) 478 final).

In the EU Biodiversity Strategy, the UE vision for 2050 is also included. By 2050, biodiversity and ecosystem services have to be protected, valued,

---

[6] http://ec.europa.eu/environment/nature/biodiversity/comm2006/2020.htm.

and properly restored, due to their fundamental contribution in ensuring human well-being and economic prosperity. First of all, dramatic changes in ecosystem services caused by the loss of biodiversity should be avoided. The Strategy also includes a specific commitment of individual EU countries to assess the status and preparation of ecosystem maps on their territory, as well as to value the services they provide and strive to include this value in national accounts.

The most ambitious implementation of these objectives was the National Ecosystem Assessment in the United Kingdom and Northern Ireland (UK NEA, 2011). UK NEA was the first analysis of the environment in UK in terms of the benefits it brings to society and to the economy. The program involved government, scientists, NGOs, and business. Although the main conclusions of this study relate to the United Kingdom, they can easily be translated to other countries (UK NEA, 2014). A similar initiative, although on the smaller scale, was implemented in the Czech Republic (Frélichová et al., 2014). Six types of ecosystems (cropland, forest, grassland, urban, water, and wetland) were included in the research carried out in this country. It was found that the average value of ecosystem services in the Czech Republic is one and a half times as much as the value of the Czech GDP. However, not only EU member states make use of the developed ES frameworks. For instance, a nationwide ES assessment was recently conducted in Russia under the so-called TEEB-Russia project, run in cooperation with German scientists (Bukvareva et al., 2019).

In 2011, in response to the recommendations included in the EU Biodiversity Strategy, another major initiative was established - MAES (Mapping and Assessment of Ecosystems and their Services). The main objective of the program was to identify and characterize services provided by ecosystems, as well as the valuation of selected services and verification of available source data. The program aimed to be implemented in each country in four stages:

• determination of ecosystem types according to EUNIS classification (European Nature Information System) with the use CORINE Land Cover data;
• delimitation of the Basic Assessment Unit (BAU): a basic typological unit representing a given type of ecosystem designated on the basis of EUNIS classification[7];

---

[7] http://eunis.eea.europa.eu.

- elaboration of the list of relevant ecosystem services defined on the basis of CICES classification (Haines-Young and Potschin, 2018, 2013), adapted to national conditions;
- construction and mapping of indicators showing ecosystem capacities to provide selected ecosystem services.

Five major reports were published in 2013—2018 resulting from the work carried out within the MAES initiative, supplemented by carto-graphic visualization presented in the MAES digital atlas[8]. The results of practical applications of the developed framework in individual European countries have been published, among others by Jacobs et al. (2016), Mononen et al. (2016) and Albert et al. (2016).

The institution supporting research on ecosystem services at the Eu-ropean level is the European Environment Agency (EEA)[9], which provides information on the environment, deals with the development, adoption, implementation, and evaluation of environmental policy, also in a social context. The EEA works in close cooperation with the European Infor-mation and Observation Network and its 33 Member States. The EEA collects data and creates assessments on a wide range of environmental is-sues, especially those related to biodiversity. The collected and processed data cover the diversity of genes, species, and ecosystems that create life on Earth. The result of its work on the environmental economy is, among others, the previously described CICES system.

The growing popularity of anthropocentric perspective in environ-mental research has resulted in the establishment of international bodies working on the interface of science and policy. One of them is the Ecosystem Services Partnership (ESP)[10], launched in 2008 by the Gund Institute for Ecological Economics (University of Vermont, USA). Currently, it is being coordinated by the Environmental Systems Analysis Group (Wageningen University, the Netherlands) and supported by the Foundation for Sustainable Development (Wageningen, The Netherlands). ESP is a worldwide network dedicated to enhance the science, policy, and practice of ecosystem services for conservation and sustainable develop-ment. ESP connects over 3000 ecosystem services scientists, policy makers, and practitioners who work together in more than 40 working groups and a

---

[8] biodiversity.europa.eu/maes/maes–digital–atlas.

[9] http://www.eea.europa.eu.

[10] https://www.es-partnership.org.

growing number of national networks on all continents. ESP regularly organizes world and regional conferences.

Another such organization is the Intergovernmental Science-Policy Platform on Biodiversity and Ecosystem Services (IPBES).[11] It is an independent intergovernmental body, established by member States in 2012. The objective of IPBES is to strengthen the science–policy interface for biodiversity and ecosystem services for the conservation and sustainable use of biodiversity, long-term human well-being, and sustainable development. It operates under the auspices of the United Nations and four of its entities: UNEP, UNESCO, FAO, and UNDP.[12] Recent scientific publications of IPBES and ESP representatives clearly indicate that there is evident struggle for primacy in ES research between the two major scientific organizations dealing with ecosystem services, or according to the terminology used by IBPES, nature's contributions to people (Braat, 2018; Díaz et al., 2018).

A great contribution to the development of ES research is being continuously provided by scientific projects funded by the EU under Research and Innovation programs (FP7 for 2007–13 and Horizon 2020 for 2014–20). The project dedicated solely to the development of ES approach was OpenNESS[13], carried out in 2012–17 under FP7. It aimed to translate the concepts of natural capital (NC) and ecosystem services into operational frameworks that provide tested, practical, and tailored solutions for integrating ES into land, water, and urban management and decision-making. It examined how the concepts link to, and support, wider EU economic, social, and environmental policy initiatives and scrutinized the potential and limitations of the concepts of ES and NC. The very similar goals were set in OPERAs FP7 project[14], as the overarching aim of OPERAs was to bridge the gap between science, policy, and practice to better integrate the use of ES and NC concepts into EU policy.

The problem of ecosystem services was one of the main research topics of the FunDivEUROPE project (The Functional importance of biodiversity in European forests), also carried out under FP7.[15] It involved 24 partners from 15 European countries and was coordinated by the Faculty of Biology and Geobotany, University of Freiburg (Germany). The purpose of

---

[11] https://www.ipbes.net.
[12] https://www.iucn.org/theme/global-policy/our-work/ipbes.
[13] http://www.openness-project.eu.
[14] http://operas-project.eu.
[15] www.fundiveurope.eu.

FunDivEUROPE was to describe the relationship between the biodiversity of the main types of European forests and the range of ecosystem services they offer.

A large pan-European project funded under Horizon 2020 and dedicated to ecosystem services was ESMERALDA (Enhancing ecoSysteM sERvices mApping for poLicy and Decision-mAking).[16] The objective of this project, carried out in 2014−18, was to deliver a flexible methodology to provide the building blocks for pan-European and regional ES assessments. The ES mapping approach aimed to integrate biophysical, social, and economic assessment techniques. The developed methodology was built on already existing ES projects and databases (e.g., MAES, OpenNESS, OPERAs, MEA, and TEEB).

The very interesting and most recently started project devoted to ES and funded under Horizon 2020 is EKLIPSE.[17] It aims to develop a sustainable mechanism for supporting evidence-based and evidence-informed policy on biodiversity and ecosystem services. A large part of the budget of the EKLIPSE project is made available to the wider community through open calls.

Another important platform funding international ES research projects is BiodivERsA.[18] It is a network of national and regional funding organizations promoting pan-European research on biodiversity and ecosystem services. It is also financially supported by the EU within the Horizon 2020 program. Particularly, the three calls launched in 2014−18 were dedicated to support research on ecosystem services (e.g., to promote ES synergies and reducing trade-offs, to improve ecosystem functioning and ES delivery, and to develop ES scenarios).

Due to the limited volume of the book, only the most recognized international initiatives are mentioned, and certainly not all of them. The focus was primarily on those originating or implementing in Europe. Even though, the demonstrated abundance of organizations and initiatives dealing with ecosystem services, which emerged in the last decade, is certainly high and inevitably testifies to how much needed and useful the ES framework is. Time will tell whether it is only a fad, or approach that will dominate research at the interface between man and nature for decades.

---

[16] http://www.esmeralda-project.eu.
[17] http://www.eklipse-mechanism.eu.
[18] https://www.biodiversa.org.

CHAPTER 2

Postglacial landscape

Contents

2.1 European overview

The postglacial landscape in the northern hemisphere is morphogenetically connected with the occurrence of the last glaciation, which took place in this area. In Europe, the postglacial relief is the result of the presence of last Scandinavian ice sheet of the Weichselian glaciation (in Poland, the Vistulian glaciation) whose southernmost greatest extension (Last Glacial Maximum) marks the boundary between landscapes formed by the last and earlier glaciations. Attempts to reconstruct its course for Eurasia were made by many researchers (Becker et al., 2015; Diercke International Atlas, 2010; Svendsen at al., 2015, 2004). It is estimated that the advance and retreat of the last ice sheet took place in Europe about 25,000–10,000 years ago (Kleiber et al., 2000; Landvik et al., 1998; Lubinski et al., 1996; Polyak et al., 1997; Svendsen et al., 2004). In Poland, the Last Glacial Period occurred about 20,000 years ago (Kozarski, 1986, 1981; Mojski, 2005).

The consequence of morphogenetic processes that shaped the Eurasian postglacial landscape is landforms that constitute present relief with its specific lithological features. They are mainly composed of sedimentary rocks of glacial and glaciofluvial origin (boulder clays, glaciofluvial sands, clays), which, from the lithological point of view, form a specific spatial mosaic (Fig. 2.1). It consists of moraine plateaus made of basal loam, terminal

Ecosystem Service Potentials and Their Indicators in Postglacial Landscapes
ISBN 978-0-12-816134-0
https://doi.org/10.1016/B978-0-12-816134-0.00002-X
Copyright © 2020 Elsevier Inc.
All rights reserved.

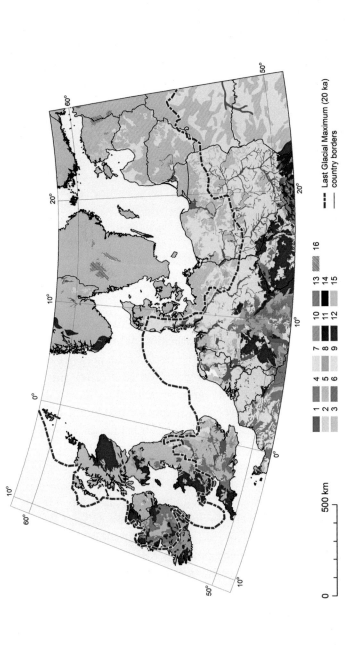

Figure 2.1 Lithology of the Eurasian postglacial landscape. 1 - river alluvium, 2 - marine alluvium, 3 - glaciofluvial deposits, 4 - calcareous rocks, 5 - soft clayey materials, 6 - hard clayey materials, 7 - sands, 8 - sandstone, 9 - soft loam, 10 - siltstone, 11 - detrital formations, 12 - crystalline rocks and migmatites, 13 - volcanic rocks, 14 - other rocks, 15 - organic materials, and 16 - unclassified (urban/water/ice). (Source data: (1) LANMAP2 (shp format) and its associated database received in 2009 from C.A. Mücher (Thanks!), rearranged and modified by the authors. Description of LANMAP2 see: Mücher, C.A., Bunce, R.G.H., Jongman, R.H.G., Klijn, J.A., Koomen, A.J.M., Metzger, M.J., Wascher, D.M., 2003. Identification and Characterisation of Environments and Landscapes in Europe. Alterra-rapport 832, Alterra, Wageningen; Mücher, C.A., Wascher, D.M., Klijn, J.A., Koomen, A.J.M., Jongman, R.H.G., 2006. A new European Landscape Map as an integrative framework for landscape character assessment. In: Bunce, R.G.H., Jongman, R.H.G. (Eds.), Landscape Ecology in the Mediterranean: inside and outside approaches. Proceedings of the European IALE Conference. IALE Publication Series 3, pp. 233–243. (2) Becker, D., Verheul, J., Zickel, M., Willmes, C., 2015. LGM paleoenvironment of Europe - Map. CRC806-Database.)

moraines of boulder clay, outwash plains made of glaciofluvial sands, eskers and kames made of glaciofluvial material, or kettle holes filled with clays.

Because of the short impact period of exogenic processes (weathering, erosion, and denudation), landforms are well preserved morphologically in the land spatial pattern, creating a varied mosaic of both erosive and accumulation forms. In terms of topography, the postglacial landscape is classified as lowland, although parts of it, especially terminal moraines of the last glaciation phases, exceed the height of 200—300 m above sea level. Greater hypsometric diversity, complex valley network, and the existence of a dense network of postglacial channels and kettle holes are among the features that clearly distinguish the postglacial landscape formed by the last glaciation from other landscapes with older glacial morphogenesis. It is also characterized by the presence of complex forms of glacial, slope, fluvial, and aeolian relief. The most characteristic are hills and terminal moraine belts with significant denivelations, vast undulated areas of ground moraine plateaus, outwash plains, kame hills, eskers, drumlins, and deeply cut subglacial channels, as well as water-filled potholes constituting dense network of lakes.

As a rule, the regions comprising many natural lakes are located north of the line of the Last Glacial Maximum (Fig. 2.2). For example, in Norway,

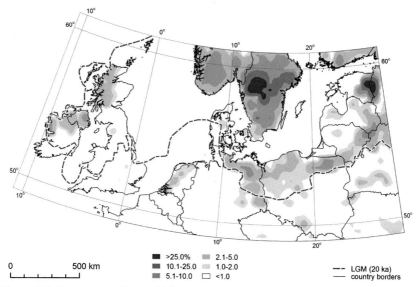

Figure 2.2 Percentage of natural lake area in relation to the Last Glacial Maximum. Lake area calculated in cells 50 × 50 km and interpolated by IDW method. *(Source data: (1) Corine Land Cover 2012 (EEA, 2012); OpenStreetMap for Russia, Belarus and Ukraine (http:// download.geofabrik.de/europe; accessed 4 June 2018); (2) Becker, D., Verheul, J., Zickel, M., Willmes, C., 2015. LGM paleoenvironment of Europe - Map. CRC806-Database.)*

Sweden, Finland, and the Karelo-Kola part of Russia, lakes cover approximately 5%—10% of the area of those countries, with concentration in some regions exceeding 25%. Large numbers of lake are also present in other countries around the Baltic Sea, as well as in Iceland, Ireland, and the northern and western parts of the United Kingdom. In the rest of Europe, south of the line of the Last Glacial Maximum, most natural lakes can be found in mountain regions, but they are generally smaller compared with those located in the north.

The river network of postglacial landscape is genetically linked to the morphological and hydrogeological conditions inherited from the Last Glacial Period. It is represented by river valley systems and river—lake systems. River valleys are mostly polygenetic - with sections of, inter alia, ice-marginal streamway, channel, and gorge character (Kostrzewski et al., 2008).

Postglacial morainic uplands and outwash plains are shaped by chemical denudation, defoliation, and aeolian accumulation and to a lesser extent by water erosion and suffosion. The value of chemical denudation is close to $62-90 \, \text{t km}^{-2} \, \text{year}^{-1}$, while the aeolian deposition exceeds even $600 \, \text{t km}^{-2} \, \text{year}^{-1}$ (Kostrzewski et al., 2008).

In the Holocene, morpholithological development of postglacial landscape had little impact on the relief as well as substrate properties. In this period, soil and vegetation cover started to develop, determined by geographical location and related climatic conditions and soil potential. For Central Europe, the most important pedogenic phase is the late Pleistocene and Holocene period (Bednarek, 1991; Degórski, 2002; Kowalkowski, 1986), in which the present soil cover started to develop. Lack of sediment accumulation in large parts of postglacial landscape caused the present soil cover to be developed using the resources of old soils (Manikowska, 1999). Pedogenic processes were highly influenced by the succession of plant communities and especially by the occurrence of forest vegetation (Catt, 1988). For example, pine expansion in Central Europe during the Younger Dryas (11,900 BP) and the Preboreal period (11,000 BP) favored podzolization processes (Friedrich et al., 1999), which intensified up to the Atlantic period (Degórski, 2002; Manikowska, 1999). During the late Pleistocene and Holocene, some soils were also formed from redeposited material transported by fluvial or aeolian processes (Degórski, 2007). In general, soil spatial pattern reflects substrate diversity of parent material. Brown soils (Cambisols) were developed on moraine plateaus, podzolic soils (Podzols) on outwash plains, alluvial soils (Fluvisols) in river valleys, and bog and peat soils (Histosols) on wetlands (see Section 2.2.4).

Present climate of Europe is determined by geographical location, landmass distribution, and topography. Latitude and topography affect the solar radiation balance. On this basis, the five main thermal zones can be distinguished from north to south: arctic, boreal, temperate, submeridional, and meridional. Climatic differentiation from west to east represents the oceanic–continental gradient (Fig. 2.3). Western Europe has an oceanic climate; in Eastern Europe, climate is of continental character, while the Central European plains have a transitional oceanic/continental climate. It

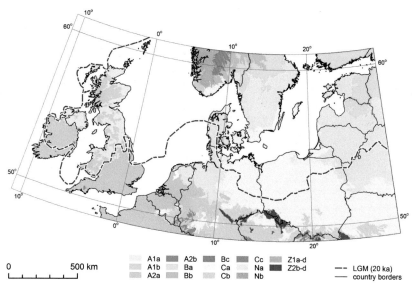

Figure 2.3 Climate types in particular altitudinal zones. A1 - Atlantic North, A2 - Atlantic Central, B - Boreal, C - Continental, N - Nemoral, Z1 - Alpine North, and Z2 - Alpine South; a - lowlands (0–300 m a.s.l.), b - hills (300–700 m a.s.l.), c - low mountains (700–1500 m a.s.l.), and d - mountains above the tree line (>1500 m a.s.l.). *(Source data: (1) LANMAP2 (shp format) and its associated database received in 2009 from C.A. Mücher, rearranged and modified by the authors. Description of LANMAP2 see: Mücher, C.A., Bunce, R.G.H., Jongman, R.H.G., Klijn, J.A., Koomen, A.J.M., Metzger, M.J., Wascher, D.M., 2003. Identification and Characterisation of Environments and Landscapes in Europe. Alterra-rapport 832, Alterra, Wageningen; Mücher, C.A., Wascher, D.M., Klijn, J.A., Koomen, A.J.M., Jongman, R.H.G., 2006. A new European Landscape Map as an integrative framework for landscape character assessment. In: Bunce, R.G.H., Jongman, R.H.G. (Eds), Landscape Ecology in the Mediterranean: inside and outside approaches. Proceedings of the European IALE Conference. IALE Publication Series 3, pp. 233–243. (2) Becker, D., Verheul, J., Zickel, M., Willmes, C., 2015. LGM paleoenvironment of Europe - Map. CRC806-Database.)*

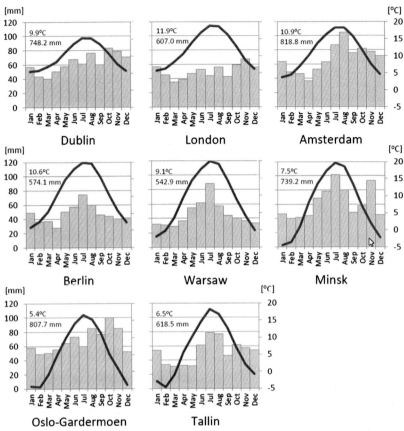

Figure 2.4 Climate charts for selected European cities for 1998–2017. *(Source data: https://www.weatheronline.pl.)*

is worth to underline that the climate of Western and partly Central Europe is milder compared with other areas of the same latitude due to the influence of the Gulf Stream (GS), which warms oceanic air masses, and affects Western and Northern Europe (despite GS gradual weakening) up to the Kola peninsula. As a result, the annual amplitude of mean monthly temperatures increases from west to east. Also, the distribution of precipitation changes from rather uniform in oceanic climate toward summer peak in continental climate (Fig. 2.4).

According to the phytogeographic division of Europe (Meusel and Jäger, 1992), most of the area of postglacial landscape belongs to temperate zone. Boreal zone is represented only by the northern part of Scotland and Scandinavia. According to a more general biogeographical division proposed by the European Environment Agency EEA (2002), with later modifications), boreal region is extended to the south (all Scandinavia, Estonia, Latvia, Lithuania, and northern part of Belarus), Atlantic region covers Atlantic Ireland and all Great Britain and sub-Atlantic provinces, while continental region corresponds to temperate zone but is extended to the south. As a separate unit, the Alpine region is distinguished, which is divided into several parts.

Temperate phytogeographic zone is further divided into provinces: Atlantic subzone - from the Atlantic coast to the line joining Paris (F), Essen, Hamburg (D), Viborg (DK), and Arendal (N); sub-Atlantic subzone - to the line Stuttgart, Magdeburg, Greifswald (D), Jonkoping (S), and to the north; Central European subzone - to the line Ivano-Frankivsk (UA), Brest (BY), Pskov, Saint Petersburg (RUS); Sarmatian subzone - east of that line. Boreal phytogeographic zone covers areas with climate types A1ab, Z1abc, Babc (climate codes as on Fig. 2.3), temperate phytogeographic zone (Atlantic and sub-Atlantic) is connected with A1, A2, and Ba climate types, Central European and Sarmatian zones cover areas with Ca and Na climates, while Submeridional zone is tied with Cb climate type.

Climate, soil, hydrology, and geographical ranges of species all together determine diversity of natural vegetation of Europe. According to a comprehensive synthesis of European vegetation elaborated by Bohn et al. (2000/2003), we may distinguish several main vegetation formations, most of which are of forest character (Fig. 2.5). However, only few of them can be found in postglacial landscape including hemiboreal spruce and fir-spruce forests with broad-leaved trees (Db code on Fig. 2.5), middle and southern boreal to hemiboreal pine forests (De), hemiboreal and nemoral pine forests, partly with broad-leaved trees (Df), species-poor acidophilous and oligo- to mesotrophic oak and mixed oak forests (Fa), mixed oak-ash forests (Fb), mixed oak-hornbeam forests (Fc), lime-pedunculate oak forests (Fd), species-poor acidic beech and mixed beech forests (Fe), and species-rich eutrophic and eu-mesotrophic beech and mixed beech forests (Ff). From the phytosociological point of view, Db and De forest types are represented by communities of the *Vaccinio-Piceetea* class; Fa comprises communities of *Quercetea robori-petraeae* class as well as *Vaccinio-Piceetea* class, while all the remaining vegetation types are represented by communities belonging to *Querco-Fagetea* class.

Db. HEMIBOREAL SPRUCE AND FIR-SPRUCE FORESTS WITH BROAD-LEAVED TREES (dominant tree species: coniferous - Norway spruce *Picea abies*, Siberian spruce *Picea obovata*, Siberian fir *Abies sibirica*; broad-leaved -

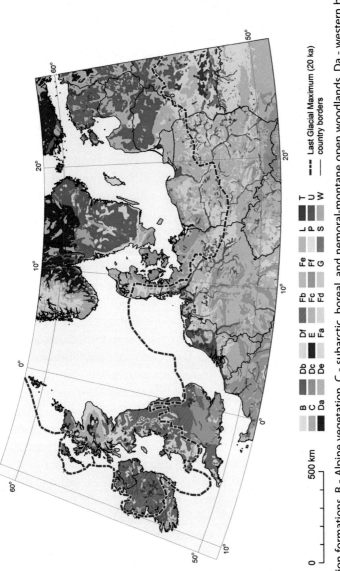

Figure 2.5 Main vegetation formations. B - Alpine vegetation, C - subarctic, boreal, and nemoral-montane open woodlands, Da - western boreal spruce forests with birch or alder, Db - hemiboreal spruce and fir-spruce forests with broad-leaved trees, Dc - montane and submontane fir and spruce forests in the nemoral zone, De - middle and southern boreal to hemiboreal pine forests, Df - hemiboreal and nemoral pine forests, partly with broad-leaved trees, E - Atlantic dwarf shrub heaths, Fa - species-poor acidophilous and oligo- to mesotrophic oak and mixed oak forests, Fb - mixed oak-ash forests, Fc - mixed oak-hornbeam forests, Fd - lime-pedunculate oak forests, Fe - species-poor acidic beech and mixed beech forests, Ff - species-rich eutrophic and eu-mesotrophic beech and mixed beech forests, G - thermophilous mixed deciduous broad-leaved forests, L - forest steppes, P - coastal vegetation and inland halophytic vegetation, S - mires, bogs, fens, and transitional mires, T - swamp and fen forests, U - vegetation of floodplains, estuaries, freshwater polders, and other moist or wet sites, W - lakes, reed beds, and sedge swamps. (Source data: (1) Bohn, U., Gollub, G., Hettwer, C., Neuhäuslová, Z., Raus, T., Schlüter, H., Weber, H., 2000/2003. Map of the Natural Vegetation of Europe. Scale 1:2 500 000. Federal Agency for Nature Conservation, Bonn, modified. (2) Becker, D., Verheul, J., Zickel, M., Willmes, C., 2015. LGM paleoenvironment of Europe - Map. CRC806-Database.)

English oak *Quercus robur*, small-leaved lime *Tilia cordata*, Scotch elm *Ulmus glabra*, Norway maple *Acer platanoides*).

This unit includes either forest communities with boreal coniferous and nemoral broad-leaved tree species mixed in the tree layer or vegetation topographic complexes of boreal coniferous forests on the lower slopes and valley floors and nemoral broad-leaved forests, which occupy flat ridges and upper slopes. This type of natural vegetation forms a continuous belt, from Southern Scandinavia and East Baltic countries to central part of Russian lowlands. Forest communities (and complexes) within this unit show marked floristic differentiation along north–south and west–east gradients. They occupy different habitats[1] in terms of moisture and fertility. It results in high diversity of herb layer physiognomy and composition, e.g., moss-, dwarf shrub-, or herb-rich forms (Bohn et al., 2000/2003).

DE. MIDDLE AND SOUTHERN BOREAL TO HEMIBOREAL PINE FORESTS (dominant tree species: Scots pine *Pinus sylvestris*).

This unit includes pine forests on poor, acid, and most often sandy soils, differing in moisture. All forests of this group are similar to each other regarding their species composition and represent one order of *Cladonio-Vaccinietalia* within *Vaccinio-Piceetea* class. Division into ecological (moisture) groups is clearly marked, but geographical differences are also visible. We can distinguish several physiognomic-floristic types of the herb layer, e.g., lichen-, cranberry-, blackberry-, greenmoss-, peat moss-, grass-rich forms, depending of the soil moisture. From phytosociological point of view within this unit, one may distinguish several associations. The driest places are domain of *Cladonia*-type pine forests. In Baltic countries, the most popular association is known as *Vaccinio vitis-idaeae-Pinetum*. Swamp pine forests are represented, among others by *Oxycocco-Pinetum* and *Vaccinio uliginosi-Pinetum* (Bohn et al., 2000/2003).

DF. HEMIBOREAL AND NEMORAL PINE FORESTS, PARTLY WITH BROAD-LEAVED TREES (dominant tree species: Scots pine *P. sylvestris*).

This unit represents transitional forms between the boreal and nemoral forests and is to some extent similar to the previous one (also from typological phytosociological point of view). The main difference in the herb layer is the occurrence of species of different character: boreal forest species,

---

[1] In this book, we use the word *habitat* in the narrow sense to denote a set of abiotic environmental conditions (biotope) suitable for particular plant communities and related animals, not to confuse with *Natura 2000 Habitats* denoting whole ecosystems (biotope + biocenosis).

nemoral, particularly thermophilous broad-leaved forest species, and xerophytic plants of open forests, dry grasslands, and meadow steppes. It is possible to distinguish several associations within this forest type. On the southern extremes of Scandinavia, in places, which are relatively warmer, these are *Saniculo-Pinetum*, *Convallario-Pinetum*, *Seslerio-Pinetum*, and *Melico-Pinetum*. The Central East Europe is the area of domination of two regional associations: *Leucobryo-Pinetum* and *Peucedano-Pinetum*. The first of them is observed in the western part of the area, encompassing Southern Poland and all of Germany and reaching the Netherlands and Austria. The second of the communities has a subcontinental character, and it occurs, in particular, in central and north-eastern parts of Poland, in Lithuania and Belarus. The associations mentioned do not cover all the diversity of the Central European pine forests. Other types have very limited geographical range, e.g., *Empetro nigri-Pinetum*, which occurs in the belt of coastal dunes in Poland and Germany. Swamp pine forests are represented, similarly as in the previous unit, by *Vaccinio uliginosi-Pinetum* and similar communities (Bohn et al., 2000/2003; Solon, 2003).

FA. THE SPECIES-POOR ACIDOPHILOUS AND OLIGO- TO MESOTROPHIC OAK AND MIXED OAK FORESTS (dominant tree species: English oak *Q. robur*, sessile oak *Quercus petraea*, Pyrenean oak *Quercus pyrenaica*, Scots pine *P. sylvestris*, European weeping birch *Betula pendula*, downy birch *Betula pubescens*, European chestnut *Castanea sativa*).

This unit is characteristic for Atlantic and sub-Atlantic regions, but with extensions into East and Southeast Europe. Acidophilous oak and mixed oak forests are characterized by the dominance of oaks, usually English oak *Q. robur* and sessile oak *Q. petraea*, in the tree layer. The shrub and herb layers are usually well developed and build by calcifuge plants of a middle European or temperate Eurasian geographic distribution type, but these are relatively species-poor compared with deciduous forests on more base-rich sites or in climatically more favorable areas. The herb layer is usually composed of hemicryptophytes, primarily grasses, woodrushes, and sedges. Geophytes are generally rare, but ferns play an important role, particularly in the Atlantic and sub-Atlantic regions, forming dense covers and displacing almost every other herb species. The soils are permeable or intermittently moist, oligotrophic, acidic, often sandy, and sometimes shallow or rocky. From the floristic point of view, there is a very clear division into communities of the sub-Atlantic and the Central European character. From phytosociological point of view, forests of this group belong to the seven alliances within the *Quercetea robori-petreae* class. Five of them are of very western or south-western character, and only two (*Agrostio capillaris-Quercion petraeae* and *Vaccinio myrtilli-Quercion petraeae*) extend to the east, vanishing gradually and being replaced by communities of pine—oak dominance, belonging to the *Dicrano-Pinion* alliance within the *Vaccinio-Piceetea* class (Bohn et al., 2000/2003).

FB. MIXED OAK-ASH FORESTS (dominant tree species: European ash *Fraxinus excelsior*, English oak *Q. robur*, Scotch elm *U. glabra*, sessile oak *Q. petraea*).

This group of forests is characteristic of the British Isles; additional stands occur also in Western Pyrenees, the Cantabrian Mountains, western coast of Norway, and in North-West France. Forests occupy base-rich, often calcareous, and moderately acidic to neutral brown soils. Herb layer is well developed and rich in species. This type of forests replaces oak-hornbeam forests and beech forests outside their natural range. From the phytoso-ciological point of view, this unit represents the order *Fagetalia sylvaticae* of the *Querco-Fagetea* class and most likely the following two alliances: *Carpinion betuli* on the most typical places in the south of Great Britain and *Alno-Ulmion* in damp locations (Bohn et al., 2000/2003).

FC. MIXED OAK–HORNBEAM FORESTS (dominant tree species: common hornbeam *Carpinus betulus*, English oak *Q. robur*, sessile oak *Q. petraea*, small-leaved lime *T. cordata*).

This type of forests is zonal vegetation in climatically moderate conti-nental areas of lowland, hilly, and lower montane landscapes in the temperate and submeridional zones of Central and Southeast Europe. They occupy meso- to eutrophic sites on moderately dry to moist soils. Forests are generally shady, due to dense canopy, and are characterized by rich species composition of the herb layer, with the presence of geophytes. From the phytosociological point of view, forests represent the *Carpinion betuli* alliance and are differentiated into associations, differing in species composition and geographic range: *Endymio-Carpinetum* - Atlantic, in Western Europe (Southern England, Northern France, Belgium, the Netherlands); *Stellario-Carpinetum* - sub-Atlantic, in rainy and cool summer areas of the north-west part of Central European lowlands; *Galio sylvatici-Carpinetum* - moderately continental, in warm summer and dry summer areas of the south-west and central parts of Central Europe; and *Tilio-Carpinetum* - subcontinental, in the eastern part of Central Europe and in Southeast Europe (Bohn et al., 2000/2003).

Fᴅ. Lɪᴍᴇ–ᴘᴇᴅᴜɴᴄᴜʟᴀᴛᴇ ᴏᴀᴋ ꜰᴏʀᴇꜱᴛꜱ (dominant tree species: English oak *Q. robur*, small-leaved lime *T. cordata*, sometimes with Norway maple *A. platanoides*, field maple *Acer campestre*, Scotch elm *U. glabra*).

Lime-pedunculate oak forests replace oak–hornbeam forests in the Eastern Europe, as a result of the absence of other forest-forming broad-leaved tree species (as common beech *Fagus sylvatica* or common hornbeam *C. betulus*). In comparison with oak–hornbeam forests, lime-oak forests are characterized by the simpler structure and lower species richness. Boreal species does not exceed 10%–15% in the herb layer, and species of more eastern or south-eastern types of geographic distribution are also present. From the phytosociological point of view, forests from this unit represent the *Querco-Fagetea* class and more specifically the *Fagetalia sylvaticae* order and the *Querco roboris-Tilion cordatae* alliance. At least two associations could be distinguished: *Mercurialo perennis-Quercetum roboris* (= *Querco roboris-Tilietum cordatae*), which is widely distributed and occurs over the entire hemiboreal region up to Ural Mts., and *Trollio europaei-Quercetum roboris*, which occurs only in the western part of the hemiboreal region (Bohn et al., 2000/2003; Smirnova et al., 2018).

Fᴇ. Sᴘᴇᴄɪᴇꜱ–ᴘᴏᴏʀ ᴀᴄɪᴅɪᴄ ʙᴇᴇᴄʜ ᴀɴᴅ ᴍɪxᴇᴅ ʙᴇᴇᴄʜ ꜰᴏʀᴇꜱᴛꜱ (dominant tree species: common beech *F. sylvatica*).

Over its entire range, beech forests occupy the most extensive habitats with mesic (to moderately dry), well-drained, and more or less deep soils and are characterized by the natural dominance of beech (common beech *F. sylvatica*) in the tree layer with only a few subdominant tree species in the first or second tree layer, as the beech is exceptionally competitive and intolerant toward other tree species. The degree of crown cover is at least 50%, but in most cases, it is over 90%. That is why forests are generally dark, and the cover of herb layer is low. Within the range of all beech forests, we may always distinguish two separate types: lowland forests and upland/montane ones, differing by the presence of defined species groups. Species-poor acidic beech and mixed beech forests are connected mainly with base-poor sands (on lowlands) or with acid rocks of different origin (upland/montane forms). Geographically, this unit is connected with Western and Central Europe, having the dominant position in Germany. From phytosociological point of view, forests represent no less than 15 associations from 3 suballiances (*Ilici-Fagenion* of Atlantic and sub-Atlantic character, *Luzulo-Fagenion* of Western and Central European character, and *Luzulo pedemontanae-Fagenion* of South-west Alpic and Apennine character) of *Deschampsio flexuosae-Fagion* (= *Luzulo-Fagion*) alliance (Bohn et al., 2000/2003).

FF. Species-rich eutrophic and eu-mesotrophic beech and mixed beech forests (dominant tree species: common beech *F. sylvatica*).

Species-rich eutrophic and eu-mesotrophic beech forests are extremely diversified floristically, what results in richness of phytosociological units. In these forests, we can distinguish no less than 42 associations from 10 alliances (some of them further divided into suballiances) of the *Fagetalia sylvaticae* order. Species-rich beech forests differ from species-poor beech forests not only by the total number of species and soils occupied but also by the more complex vertical and horizontal structure of forest patches. Geographically, these forests are dominant in Western Europe (France), and their role becomes smaller and smaller eastwards (Bohn et al., 2000/2003).

Because of its geographical (latitudinal) location and glaciation history, Europe is rather poor in animal species compared with other continents.[2] In Europe, for the most groups of terrestrial animals, there is a general gradient of species richness from the north (low values) to the south, but for different taxa, there are different areas of maximum richness.[3] For example, there are 219 terrestrial mammal species, of which 59 species (26.9%) are endemic. The smallest numbers of species (below 20) are observed in the boreal zone and the highest richness (over 65 species) in the Alps and the Carpathians. It is worth to underline that large herbivores (8 species) occur mostly in Central Europe (Mitchell-Jones et al., 1999).

For reptiles (151 species in Europe), there is a clear gradient of increasing species richness from the north (1 species in North Scandinavia) to the south, with the greatest richness being found in the Balkan Peninsula (up to 35 species). Among 84 amphibian species, whereof nearly 55% are endemic in Europe, only 1 species occurs in boreal Scandinavia, but almost 20 in the temperate zone from Western France to South-east Poland, and then to less than 13 in the Mediterranean area.

European ichtyofauna includes 546 native species of freshwater fish. The highest levels of species richness are found in the lower course of rivers flowing into the Black and Caspian Seas. However, a number of species with restricted ranges are also encountered in the Alps, in Great Britain and Ireland, and around the Mediterranean and Black Seas. Millennia lasting human activity almost all over the Europe transformed natural environment. Because of this, natural forests occupy only a small portion of the

---

[2] https://biodiversitymapping.org/wordpress/index.php/home.

[3] The Biodiversity Information System for Europe https://biodiversity.europa.eu/topics/species.

available habitats. As a result of all natural and anthropogenic processes, secondary forests and nonforest anthropogenic ecosystems were formed, constituting present complex landscape mosaic. Spatial pattern of nature potential to provide different ecosystem services reflects well the diversity of environmental components described above.

2.2 Case study area

2.2.1 General characteristics

The study area of 792 km^2 is located in North-east Poland in the postglacial Suwałki Lakeland and Augustów Plain. It ranges from 53 degrees 51'48"N to 54 degrees 11'02"N and from 22 degrees 46'15"E to 23 degrees 31'00"E. The study area lies in the Podlaskie Voivodeship (NUTS-2) and covers 3 rural-type *gminas* (NUTS-5; principal units of the administrative division of Poland): Suwałki, Giby, and Nowinka (Fig. 2.6). It is one of the least populated regions in Poland with an average population density of less than 30 people per km^2 with a predominant agricultural and tourist function (Table 2.1).

Forest is the dominant land cover type (61.77%), including coniferous and mixed coniferous forests - 42.85%, oak–hornbeam forests - 13.38%, swamp pine and similar forests on peat - 3.72%, alder carrs - 1.25%, and alder-ash riparian forests - 0.57%. Other land cover types are arable fields (13.57%), lakes (5.96%), built-up areas (2.36%), and wetlands (0.54%). Although settlements constitute a small percentage of the study area, their diversity and characteristic features are representative for the Podlaskie Voivodeship - a multicultural region rich in historic and modern wooden architecture (Fig. 2.7). Forests are in majority managed by the state. Four forest districts administered by State Forests National Forest Holding are at least in part located in the study area: Suwałki, Szczebra, Głeboki Bród, and Pomorze. The remaining state forests are protected within the Wigry National Park (WNP) (see Section 2.2.8).

2.2.2 Climate

A characteristic feature of present climate of the region in which the study area is located is its transitional character, resulting primarily from the fact that three air masses with different thermal and moisture properties meet there: wet from the North Atlantic (polar), cool from Greenland (Arctic), and relatively dry from Eastern Europe and Asia (polar continental). In

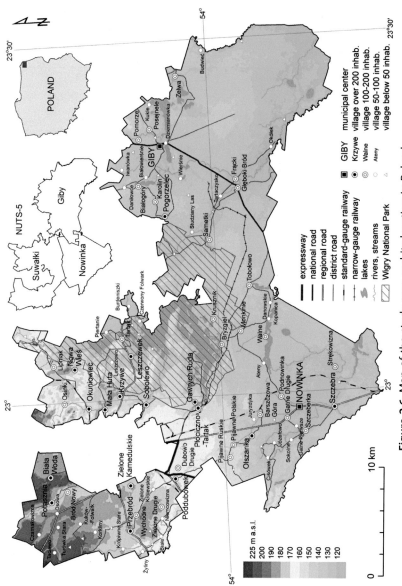

Figure 2.6 Map of the study area and its location in Poland.

Table 2.1 Population of the study area.

Gmina	Area (km^2)	Population	Population density (people km^{-2})
Giby	323.57	2952	9.12
Nowinka	203.84	2827	13.87
Suwałki	264.82	7346	27.43
TOTAL	792.23	13,125	16.57

Figure 2.7 Modern and traditional settlements in the Suwałki Lakeland. *(Photo: A. Affek, J. Solon.)*

addition, the geographical location on the western boundaries of the vast land area mainly determines its continental character (Górniak, 2000; Stopa-Boryczka et al., 2013). The main measure of thermal continentalism is the average annual air temperature amplitude, reaching 21.5°C in Suwałki (Woś, 2010), which is one of the highest in Poland. In Suwałki, the average temperatures of the coldest months (January $-4.4°C$, February $-4.3°C$), as well as the entire winter ($-3.8°C$), are the lowest in Poland (not including strictly mountain areas). Also, the average air temperatures in spring ($5.7°C$), autumn ($6.7°C$), and the whole year ($6.2°C$) are the lowest in the country (Woś, 2010).

The severity of the climate in the study area is also confirmed by the distribution and number of days with specific thermal characteristics, as certain threshold temperatures are very important for agriculture and natural vegetation. According to the thermal criterion ($T_{mean} < 0°C$), winter in Suwałki lasts the longest in Poland - on average 115 days (November 25—March 19). This region is also characterized by the shortest and the latest early spring ($0°C \leq T_{mean} < 5°C$) - 24 days (March 20—April 12) in Poland - and by the shortest and earliest prewinter ($5°C \geq T_{mean} > 0°C$) - 30 days (October 26—November 24) (Woś, 2010). These values are directly reflected in the length of the growing season, which in Suwałki is the shortest in Poland (except from mountain and submountain areas) and on average lasts only 196 days, compared with 230 days in the postglacial regions of Western Poland and 240 days in Brandenburg, i.e., around 1.5 months longer. The unfavorable thermal conditions during the year are also evidenced by the smallest number of warm days ($T_{min} > 0°C$) in Poland - 229, while the largest of cold days ($T_{max} < 0°C$) - 62.4 and very cold ($T_{max} \leq -10°C$) - over 4 (Woś, 2010).

Precipitation, in contrast to quite flat distribution of cloudiness and relative humidity, is characterized by high variability in time and space. The predominance of continental climate in this region is visible in the distribution of average monthly rainfall in the average annual sum, which in Suwałki is 587 mm, with higher summer over winter precipitation (ratio 2.2:1). Such an asymmetry of the annual rainfall is characteristic to pluvial continental climate (Kożuchowski and Wibig, 1988). The severity of climatic conditions causes that a significant part of precipitation reaches surface in the form of snow. The average annual number of days with snowfall is over 70, which corresponds to approximately 40% of days with precipitation ≥ 0.1 mm (Lorenc, 2005). In Suwałki, snow cover appears on average on November 18 and disappears on April 8 (at the earliest and latest in all of lowland areas of Poland). The average annual total number of days with snow cover is 96 (the longest in lowland Poland) and with the highest average thickness (over 9 cm) (Woś, 2010).

The high share of the Arctic and polar continental air masses causes that the climate of Suwałki—Augustów region is considered the most severe in Poland and comparable only with mountain areas.

2.2.3 Morphogenesis

Present topography of the study area and its postglacial character is a result of the sequence of geological and geomorphological processes that have been taking place since the genesis of the Precambrian East-European Platform. The platform paleomorphology, influenced by tectonic movements, both pre-Pleistocene and those associated with later glacial isostatic processes, is characterized by the occurrence of glacial depressions (e.g., melting basin of Lake Wigry) and specific arrangement of sediment layers of older glaciations probably referring to the formation of individual blocks of the crystalline basement subjected to various pressure (Ber, 2000). The course of geological boundaries of the crystalline basement with accompanying discontinuations and tectonic splits determined the pattern of subglacial glacier channels, and maximum and minimum ice sheet extent during the subsequent stadials and phases of the Last Glacial Period. The geological structure of the older bedrock also influenced the latitudinal course of the boundary between moraine uplands and outwash plain.

The genesis of landforms in the study area is directly related to geomorphological processes taking place during recent glaciations and the Holocene. Landforms created by older glaciations were strongly transformed as a result of glacial erosion during subsequent ice sheet retreats and led to their peneplanation. Older Pleistocene glaciations, up to the Saalian glaciation (Warta) according to Lindner et al. (2013), have not left visible traces on the land surface. Sediments from the Warta stadial (clay in the postglacial plateaus, sands and gravels with stone line on the flat erosive and accumulation area forming the lakeland paleosurface) are visible only in stratigraphic profiles (Ber, 2000).

The warmer period (Eemian interglacial) left relatively little traces in the landscape. It was not without significance that the Augustów Plain was a vast water reservoir, in which, as a result of glaciofluvial and shallow water sedimentation, river and organic deposits accumulated, and at the end of the interglacial period limnoglacial and fluvioglacial ones.

The last Vistulian glaciation, which finally shaped the present postglacial relief of the study area, began about 115,000 BP. The major landforms were already shaped by processes taking place during ice sheet advance and retreat in the main stadial of Vistulian glaciation. During the subsequent phases, the ice front split into two lobes (Masurian and Lithuanian), and these in turn into smaller lobes (Rospuda, Hańcza, Wigry, and Sejny) and small glacial tongues. A good benchmark of the activity and extent of individual lobes is the shape of the Wigry Lake shoreline and edge zones of glacitectonic disturbances (Ber, 2009). The glacial maximum of the Leszno stadial (Brandenburg) is most often delineated along the southernmost occurrence of ribbon lakes (Ber, 2000). In the proximal part of the ice sheet, no large outwash plains were created, which may mean that the ice sheet had a relatively small thickness and quickly began to melt, transforming into huge blocks of dead ice (Bogacki, 1985). The erosion caused by ablation waters resulted in the creation of not only the Augustów lakes but also some in meridionally oriented ribbon lakes. However, there are relatively few postglacial channels within the glacial extent of this stadial. The remains of deglaciation are sediments forming the lower level of boulder clay, deeper layers of fluvioglacial sands, and gravels deposited in the central and southern part of Augustów Outwash Plain, as well as uplifted terminal moraines on the southern boundaries of the Suwałki Lakeland.

The last advance of the Vistulian ice sheet in this area took place during the Pomeranian phase. Boulder clay identified with this phase builds postglacial plateaus and moraines. In turn, fluvioglacial sediments form top and youngest parts of the polygenetic Augustów Outwash Plain (especially in its northern part) and Sandur valleys in the plateaus, covering the older terminal moraines of Vistulian glaciation (Ber, 2000). Ice sheet fluctuated at that time - after a period of a relatively short-term retreat, there was another glacier advance (Wigry subphase). Its glacial maximum was almost parallel (1−2 km south) to the Pomeranian phase maximum (Ber, 1972; Lisicki, 1994).

The Last Glacial Period is reflected in the landscape in numerous glacial landforms. The group of ice-marginal features includes moraine hills (especially terminal moraines), which are commonly found in the Wigry Lake District and the Sejny Hills. The remains of dead-ice moraines are located, among others, by the Wigry Lake (Chmielewski, 1988). In the central and southern part of the study area, single hills made of erosion-resistant materials stick out from the glaciofluvial sediments of the Augustów Outwash Plain. There are also many isolated kame hills, formed as a result of disintegration of stagnating ice sheet into individual blocks of dead ice. In turn, eskers create distinct multi−kilometer-long sequences of embankment.

The outwash plains formed as a result glaciofluvial processes also contribute to morphological diversity of the postglacial landscape. Sediments transported by meltwater have led to the formation of both the extensive Augustów Outwash Plain and the narrow, elongated Sandur of the Czarna Hańcza Valley with a system of multilevel erosion terraces – intraglacial forms associated with individual retreat zones of the Vistulian glaciation. In the proximal part of outwash plain, where high-energy pulsating floods occurred, covering large areas, surface sands are poorly sorted, unwashed, and form complexes with gravels. In the distal part, where the water energy significantly weakened and the pulsating floods occurred only near the floodplain channels, well-washed and sorted fine sands predominate, poor in the colloidal fraction and nutrients (Biesiacki, 1982).

Also, valleys with characteristic NW−SE course, whose morphogenesis is related to the routes of former outflow of proglacial waters, are associated with outwash plains. Intensive transport and accumulation of sediments during warmer periods, combined with the permanent presence of long-term permafrost functioning as a permanent erosive base, effectively inhibited the progress of deep erosion. The melting water erosive activity was therefore directed only laterally. However, lateral erosion was quite specific because it has not only mechanical but also thermal character – waters with positive temperature affected the frozen ground and ground ice, in consequence degrading the long-term permafrost (Migoń, 2006). Braided rivers, forming a highly branched network of channels, flowed over the outwash plains, which led to the development of shallow valleys with swampy bottoms.

Subglacial channels of postglacial origin and kettle holes are further characteristic landforms of the study area. These channels often create multi-kilometer systems with radial pattern and course that indicate the direction of movement of the main lobe or smaller glacial tongues. The genesis of kettle holes is usually associated with dead ice blocks pressed into the ground by ice sheet during surface deglaciation.

A distinct example among all hollow forms of the study area is the Wigry Lake. The Wigry basin could originally comprised more than 20 lakes, which over time divided into separate water bodies as a result of filling with sediments and overgrowing. In terms of morphometry, four large melting depressions can be distinguished, which were created partly as a result of glacial exaration conditioned by tectonic structures of the ground and a distinct latitudinal subglacial channel. There are also several islands of various geneses within the Wigry Lake. These forms are built of terminal moraine and dead-ice moraine, glacial deposits accumulated in the land depressions, and remains of postglacial plateau (Chmielewski, 1988; Pie-czyński, 2012).

In the warmer periods of late Vistulian, an intensive process of dead-ice melting began, and thus the formation of a huge number of lakes - usually shallow, periodically freezing to the bottom and remaining outside initial system of river network. Sparse vegetation made possible the supply of clastic, mineral, multigrain-sized material to basins, while in the warmer periods, processes of biogenic and chemibiogenic sedimentation of gyttia and lacustrine chalk intensified. At the end of Alleröd and Bölling, the transition from mineral to organic sedimentation occurred. All small lakes disappeared at the beginning of the Subboreal period, and accumulation of peat took place on the surface of boggy depressions (Szwarczewski and Kupryjanowicz, 2008). Dead-ice melting in terminal moraine zones led to the interruption of the continuity of erosional dissections and extraction of subglacial channels, which were functioning in the Late Glacial over buried dead-ice blocks. In Bölling and Alleröd, the thickness of permafrost significantly decreased and thereby increased the depth of soil active layer. This led to intense deep erosion, which resulted in not only the creation of subsequent levels of fluvial terraces but also changes in the development of all river channels, from braided to meandering (Kondracki, 1998). In the Older and Younger Dryas phases in periglacial climate conditions, silty and sandy eluvial-type covers were formed in line with colluvial sediments filling depressions and covering slopes (Ber, 2000).

In the study area, wind activity also played a significant role in relief shaping processes, including sand accumulation in river valleys and outwash plains. In the next phases of Dryas, the aeolian activity was favored by cool and dry climate, presence of winds from relatively stable directions, and dominance of low vegetation: tundra (Oldest Dryas), park tundra with birch (Older Dryas), and park tundra with birch and steppe elements (Younger Dryas). Dune-forming processes were inhibited in Bölling and Alleröd along with warming and increased climate humidity and development of trees (Dylikowa, 1973).

With the beginning of the Holocene, most of the processes initiated in the late Vistulian had their continuation (Błaszkiewicz, 2010; Rotnicki and Starkel, 1999). The total disappearance of permafrost, rapid vegetation development, and limited accumulation of material in Sandur valleys have again led to the development of deep erosion and changes in the development of riverbeds. Further, ribbon lakes were interconnected by short gorges and became part of the river network, to which, with time, lakes of melting or ice-dammed origin were also included. Thus, characteristic polygenetic river—lake systems were created. Simultaneously, numerous small, shallow, and often endorheic water reservoirs began to overgrow, triggering eutrophication processes. This was caused by intensive increase of the role of biotic factor - an increase in resources and annual production of biogenic mass as well as the rate and capacity of biological circulation of matter (Ostaszewska, 2005). In addition, sandy and peaty muds and humiferous sands were accumulated in kettle holes and valleys of small watercourses. According to Kalinowska (1961), up to 70% of postglacial lakes could have disappeared at that time.

Soil cover, especially organic horizons, also underwent transformation. The process of humus formation significantly accelerated. Concurrently, the method of its accumulation in the soil has changed - cryogenic accumulation (in the entire permafrost layer) was replaced by biogenic one (in surface layer) (Ostaszewska, 2005). Intensive development of peat bogs initiated in the Preboreal period continued in the Atlantic period. This phenomenon was not only limited to land depressions but also occurred in valley bottoms and along elevated intervalley areas.[4]

In the Holocene, the dune-forming processes continued, but their course (secondary wind sweeping of aeolian sands in the late Vistulian) as well as the triggering factor (human activity, especially forest clearing) was significantly different. Nowadays, the share of both aeolian windswept sands and inland dunes is very small on the Augustów Plain.

Elevation differences between the highest moraine hills and the lowest part of the outwash plains (within the boundaries of the study area) reach up to 150 m - from 102 to 246 m above sea level (Fig. 2.6).

2.2.4 Soils

Present mineral soils of postglacial landscapes, developed usually on sediments with loamy sand, sandy and gravelly loam, or loamy texture, are characterized by deep dehydration of mineral and organic mineral substances (oxides, hydroxides, organomineral complexes) and stronger denaturation of humic substances supported by large amplitudes of air temperature (Wicik, 2005). A feature that differentiates the soil cover of moraine uplands and

---

[4] Currently existing peatlands are dominated by lowmoor peat with an average thickness of 1—4 m. Because of relatively small share of oligotrophic mires with typically ombrophilous water management, no layers are currently being exploited.

glaciofluvial areas (floodplains) is the content of calcium carbonate. In general, carbonate substrate dominates in ice-marginal zone. As it is not a uniform area, soil patches with high $CaCO_3$ content in profiles bottom parts intertwine with patches of decalcified soils. On outwash plains, carbonate soils occur mainly in the proximal part on poorly sorted sandy gravel deposits, while completely decarbonated or decalcified to a depth of 1.5 m in the distal part on well-sorted and rewashed fine sands.

The study area is dominated by zonal soils: Cambisols, Luvisols, and Podzols, which constitute the background for the mosaic of other soil taxonomic units (IUSS Working Group WRB, 2015). Cambisols, located mainly in hilly areas and hilly marginal zones, developed from materials with texture of loamy sands and sandy loams. Eutric Cambisols are common within carbonate-rich sediments, and Dystric Cambisols within decarbonated sediments. However, in Eutric Cambisols decalcification is often noticed (Haplic Cambisols), while in Dystric the features of podzolization processes.

Often Luvisols occur alongside Cambisols. These soils are mainly developed in sediments of sandy to heavy loam texture, usually with higher sand content (loamy sands). They are characterized by specific dichotomy of soil texture and the presence of eluvial (*luvic*) horizon (Et) developed as a result of clay mineral leaching (lessivage process). Nowadays, it is highly probable that on areas used for agricultural purposes, eroded Luvisols occur. Their resemblance to Cambisols is a consequence of plowing, resulting in Et horizon mixing with humus horizon (Świtoniak, 2014). In silty and dusty Luvisols, especially in land depressions with contested outflow, a frequent phenomenon is the appearance of pseudogley features caused by periodic stagnation of rainwater in surface horizons, i.e., in the contact zone above *argillic* horizon.

In turn, Brunic Arenosols predominate on forested areas of Augustów Outwash Plain. These soils were developed from fine and loamy sands, and their formation was favored by parent material poor in nutrients, percolative water regime, and impact of pine forests. In proximal part of Augustów Outwash Plain, Brunic Arenosol profiles show traces of brunification processes, while Brunic Arenosols in distal part have features characteristic for podzolization process. Local domination of specific pedogenic processes is determined by habitat type and fertility and the persistence of forest communities.

The last type of zone soils in the study area is poor in nutrients, Podzols, developed mainly from well-sorted fine sands. The secondary podzolization process is widespread in autogenic mineral soils of postglacial landscape. Eluvial (Es) and illuvial (Bs) horizons are often visible in morphology of Cambisols, Luvisols, and Brunic Arenosols; however, they do not meet all the criteria for subsurface diagnostic *albic* and *spodic* horizons.

On moraine hills formed from boulder deposits containing fragments of carbonate rock or distructed carbonates, intrazonal Eutric Regosols may occur, sometimes with features characteristic for secondary brunification process, being an evolutionary stage to subtype of Eutric Cambisols (Konecka-Betley et al., 1999).

Intrazonal soils also include Umbrisols and Phaeozems, developed from sediments of sandy loam to loamy sand texture, occasionally from silt or silt with admixture of clay fractions. Different subtypes of these soils occur on small areas, usually at the bottom of land depressions with high groundwater level and difficult natural drainage, at the foothills of moraine plateaus (colluvic material) in geochemically complexes coupled with deposits lying above and on river valley edges (Banaszuk, 1985). Intensive drainage of black earths leads to a gradual disappearance of hydrogenic features of the substrate, deterioration of humus quality, and *cambic* horizon formation - such soils evolve toward Cambisols (Prusinkiewicz and Bednarek, 1999).

Largest rivers valleys, which are typical geochemically dependent environments, are mostly filled with lowmoor peat and mineral sediments with loamy sand to sandy loam texture. Histosols and Gleysols developed from these materials in conditions of permanent moistening, while organic muck soils in periodically dry or dehydrated areas, where mineralization and humification processes occur, leading to disappearance of the original fibrous and sponge-like peat structure and formation of fine aggregate structure. The latter occur mainly on meadows (Bieniek, 2013).

Locally, Fluvisols developed as a result of fluvial accumulation processes. Nevertheless, the proportion of these soils in the study area is relatively small.

To sum up, the soil cover of the study area is characterized by evident zonation of autogenic soils - brown (Cambisols), lessive (Luvisols), rusty (Brunic Arenosols), and podzolic (Podzols), which remain in strong genetic relation with distribution and type of glaciofluvial and boulder Pleistocene sediments and relief diversity. The occurrence of semihydrogenic, hydrogenic, and inflow soils is determined mostly by water conditions.

2.2.5 Waters

According to the current hydrographic division of Poland[5], the study area is located within the Baltic Sea basin, which is a part of the Niemen river drainage basin (the Czarna Hańcza catchment) and the Vistula river drainage basin (the Narew catchment). The local river network is mostly determined by postglacial topography. Almost all deglaciation valleys (mostly NW−SE) are relatively shallow, wide, and characterized by locally swampy bottoms with peat and wet alluvial meadows. In turn, on moraine uplands in the northern part of the study area, there are narrower and deeper indented river valleys. Glacial channels were the natural zones of water outflow there. They were connected by short gorges and formed a common system of valleys, but it does not have a dendritic character typical of river networks because subglacial channels do not form a hierarchical system. The natural hydrographic network is complemented by over 2300 sections of drainage ditches. Most of these artificial hydrotechnical facilities are used for drainage of mid-forest and meadow eutrophic mires, and only a few are artificial links between lakes.

The study area comprises a total of 107 basic catchments (in whole or in part). River basins predominate (over 80%), but as many as 20 hydrographic units are direct and indirect catchment of lakes, among which the largest is the direct catchment of the Wigry Lake (over 78 km^2).

The Wigry Lake is also the largest (2118.3 ha) and the deepest (73 m) water reservoir in the study area, occupying the 10th and 5th place in Poland, respectively. It is also characterized by the highest shoreline development ratio (4.43). In total, there are 87 lakes with an area above 1 ha and 749 with an area below 1 ha (Jańczak, 1999).

The percentage of natural lake area (also including lakes below 1 ha) calculated on the basis of Map of Hydrological Division of Poland reaches 5.6%. This is a very high value, not only against the average for the entire country (0.9%) but also compared with the Masuria Lake District (3%−4.1%), which is the region with the largest lake area and volume in Poland (Choiński, 2007).

Several types of lake can be distinguished in terms of their origin and shape (Fig. 2.8):

---

[5] Map of Hydrological Division of Poland 1:10,000. National Water Management Authority, Warsaw, 2013.

Figure 2.8 Lakes in the Suwałki Lakeland with different origin and shape. *(Photo: A. Affek, B. Wolska, J. Wolski.)*

- Ribbon lake - long and narrow, deep, finger-shaped with steep shores, irregular bottom configuration, and usually poorly developed shoreline; created as a result of glacial erosion, i.e., ground exaration by a mass of mobile ice and the activity of subglacial waters under hydrostatic pressure; often occurs in characteristic sequences of evident directional location;
- Ground moraine lake - large with complex shoreline, numerous bays, peninsulas, and islands, low and gently sloping shores, diverse depth, and bottom shape; formed as a result of ice-block melting; ground moraine lakes are sometimes of polygenetic origin, like the Wigry Lake;
- Ice-marginal (proglacial) lake - usually elongated, parallel to terminal moraine embankments with relatively well-developed shoreline and asymmetrically elevated banks; formed as a result of dead-ice block

melting and accumulation of clastic material in the ice-marginal zone or by outflow obstruction of proglacial waters through terminal moraine embankments;

- Sandur lake - large and shallow, formed by ice melting; takes finger shape (like ribbon lakes) when located within channels eroded in glacio-fluvial sediments (polygenetic origin);
- Pothole lake - relatively small and deep with conical-shape bottom and round, simple shoreline; formed as a result of dead-ice block melting of large thickness or in-depth erosion due to vortices of proglacial water (evorsion).

The typology of lakes is supplemented by small and shallow water reservoirs of circular shape, usually peaty and covered with swamp vegetation. These lakes may have glacial origin, but they can also be found in blowouts or be the result of human activity. They are usually endorheic and ephemeral, in which water cycle takes place only through vertical exchange (precipitation, evaporation, percolation).

Lakes can be differentiated not only by origin and shape but also by the level of biological productivity (water fertility), the so-called trophic state, whose changes (harmonic succesion) are mostly conditioned by eutrophication. Quite barren α-mesotrophic waters (close to oligotrophic), characterized by low production of phytoplankton, high level of sediments mineralization, high transparency, and strong oxygenation, fill only few basins. The vast majority of water reservoirs in the study area, especially those with an area of >10 ha, are filled with relatively fertile waters with high abundance of nutrients (deeper β-mesotrophic and shallower eutrophic lakes).

In the study area, there are many lakes that are subject to disharmonic succession, in which a single factor gives the whole ecosystem a specific character. The first example is barren dystrophic lakes, characterized by negligible biological production, very low degree of sediments mineralization, low transparency, and often brown water color, associated with a large amount of humic substances causing strong acidification (pH < 6.5). These usually small, endorheic, mid-forest water reservoirs are surrounded by peat bogs. Water surface is mostly covered with floating vegetation, the so-called floating mat composed mainly of mosses, dwarf shrubs from the heath family, rannoch-rush *Scheuchzeria palustris* (L.), sundew *Drosera* (L.), and carex *Carex* (L.). There are also seedlings of pine, silver birch, and spruce, which grow very slowly due to the poverty of mineral nutrients. With time, as a result of natural succession, rapid deposition of sediments, and its minimal mineralization, these lakes undergo shallowing and disappearing and transform into oligotrophic mires (bogs) overgrowing with birch.

The alkalitrophic lakes - with rich in calcium alkaline water - are another example of aquatic ecosystems with disharmonic succession. The flow of water occurring there has decisive influence on water chemistry, causing significant loss of organic matter and a constant supply of calcium and magnesium compounds effluent from bottom spring fens. In summer, the phenomenon of biological decalcification of water (a process involving calcium carbonate precipitation during photosynthesis) occurs commonly and results in characteristic white precipitate on aquatic plants. In contrast to dystrophic lakes, alkalitrophic lakes transform into eutrophic mires (fens).

To summarize, lakes are distinctive features of postglacial landscape. Although almost all of them are in general of glacial origin, they are genetically related to various glacial activities and postglacial landforms and thus have different shapes. They also differ in terms of biological productivity, which is a consequence of both harmonic and disharmonic succession.

2.2.6 Vegetation

According to the geobotanical regionalization of Poland (Matuszkiewicz, 1993), the study area is located in the Central European Province, in the Northern Mazury-Belarusian Division and in the Augustów-Suwałki Land. The Northern Mazury-Belarusian Division is an area in which the ranges of the Central European hornbeam and boreal spruce overlap, and at the same time, there is no suboceanic beech. Sessile oak (*Q. petraea*), sycamore (*Acer pseudoplatanus*), fir (*Abies alba*), and larch (*Larix decidua*) are also absent in the natural tree stands. The specific composition of *dendroflora* and the biogeographic location cause a distinct specificity of a set of possible plant communities, especially forest ones. In comparison to other regions of Poland, the characteristic features of this area include the occurrence of lowland spruce forest *Sphagno girgensohnii-Piceetum* on peat substratum and moist mixed spruce-oak forest *Querco-Piceetum* from the *Eu-Vaccinio-Piceetenion* suballiance. An important feature of the region is also the presence of spruce in all types of forest communities and the occurrence of plant communities of boreal and continental character with almost complete lack of Atlantic and Mediterranean ones (Matuszkiewicz, 1993).

Diversity of abiotic conditions of the study area, related to the location on the contact of moraines and outwash plains with a number of the Holocene wetlands, is the reason for the development of a relatively rich set of 11 types of potential natural vegetation (PNV). In the northern part of the area, habitats suitable for oak-hornbeam forests dominate, while in the southern part the habitats suitable for coniferous and mixed forests. The northern and southern parts also differ in the share of other types of potential vegetation, which results in different types of potential vegetation landscape. For the majority of PNV types, at least part of the area is occupied by forest communities consistent with the potential vegetation. This mainly applies to poor and acidic habitats. Pine and mixed pine forests occupy over 50% of their potential coverage, and swamp spruce forests *Sphagno girgensohnii-Piceetum* and moist mixed forests almost 100%. Fertile habitats are mostly deforested because of their usefulness for agricultural purposes. Those suitable for oak-hornbeam forests are forested only in about 40%, and suitable for riparian in less than 10%. Habitat diversity as well as various forms of land use and different succession stages resulted in the development of highly diverse and typologically rich real vegetation in the study area. From the phytosociological point of view, the distinguished plant communities represent 16 classes, 27 orders, 39 alliances, and no less than 105 associations.

Because of the scope of this study, only the shortened characteristics of selected vegetation types are presented. The basic characteristics of plant communities are summarized in tables in relation to the types of ecosystem distinguished in the study area (see Fig. 5.1). The following sources were used to describe vegetation: (a) own information from field research and observations, (b) numerous published papers concerning mainly the area of the WNP and its surroundings (including Kostrowicki, 1988; Richling and Solon, 2001; Sokołowski, 1988), and (c) unpublished reports related to conservation plans (e.g., Sikorski et al., 2013a,b; Szneidrowski, 2014). The articles of Sokołowski (1980, 1968, 1966) were also used in relation to the forest communities of the Augustów Forest.

Forest communities occupy about 60% of the study area and are differentiated according to abiotic conditions (Fig. 2.9). They also differ in terms of species composition. They are alder forests, riparian forests, oak-hornbeam forests, coniferous and mixed coniferous forests, and swampy coniferous forests (Table 2.2). It should be emphasized that the majority of these forest communities are internally diverse, both in terms of the soil

Figure 2.9 Forests in the Suwałki Lakeland. *(Photo: J. Solon, B. Kruczkowska, B. Wolska.)*

conditions in which they occur and the age structure and floristic richness (Fig. 2.10).

The alder forests (*Ribeso nigri-Alnetum, Sphagno squarrosi-Alnetum, Dryopteridi thelypteridis-Betuletum pubescentis*) occur most often on peat soils within fens and transitional bogs, in depressions, river valleys, and shore zones of the lakes. They occupy small patches throughout all the study area. Currently, forests in the youngest tree stand age class dominate in the area (40%), while the oldest alder forests (over 120 years) occupy a negligible area (0.7%).

Riparian forests (*Fraxino-Alnetum*) are associated with hydrogenic habitats with slow water flow (surface or groundwater) and occupy 2.4% of the total area. These habitats occur in the periphery of lakes and in river valleys, often close to alder forests and in the vicinity of springs. Habitats suitable for riparian forests are mostly deforested and managed as grassland; however, some forests have been fragmentarily preserved. Tree stands 40–60 years old dominate in the study area (approx. 70%), the youngest stages (below 40 years) represent 17% of riparian forests, while the oldest ones (up to 120 years and above) occupy only small areas (6% and 4%, respectively).

Oak-hornbeam forests are represented by the *Tilio-Carpinetum* association and occupy slightly over 11% of the study area. Phytocoenoses of this association occupy diverse habitats in terms of fertility and moisture, hence

Table 2.2 Forest and shrub communities.

Ecosystem type (acronym)	Syntaxonomic unit	Habitat (biotope)	Main tree species	Main herb layer species
		Forest communities		
ALD 1–ALD 5	*Ribeso nigri-Alnetum*	Moderately fertile, slightly acid, neutral soils; distinct tuffy structure	Alder *Alnus glutinosa*	Nonclump ferns, sedges, and narrow-leaved grasses
	Sphagno squarrosi-Alnetum	Poorly expressed vertical water movement; poor, acid soils	Downy birch *Betula pubescens*, alder *A. glutinosa*	Abundant occurrence of nonclump sphagnum species *Sphagnum squarrosum*
	Dryopteridi thelypteridis-Betuletum pubescentis	Transitional bogs	Downy birch *B. pubescens*, Scots pine *Pinus sylvestris*, gray willow *Salix cinerea*	Typical species of transitional and raised bogs: marsh fern *Thelypteris palustris*, purple small-reed *Calamagrostis canescens*, crested buckler-fern *Dryopteris cristata*, gypsywort *Lycopus europaeus*
RIP1–RIP5	*Fraxino-Alnetum*	Slow movement of high-level groundwater; strongly hydrated, mainly peat soils (Histosols)	Alder *A. glutinosa*, downy birch *B. pubescens*, Norway spruce *Picea abies*, ash *Fraxinus excelsior*	Alder and partially rush species, as well as typical for moist meadows and swamps: elongated sedge *Carex elongata*, marsh bedstraw *Galium palustre*, yellow iris *Iris pseudoacorus*, gypsywort *L. europaeus*, bittersweet *Solanum dulcamara*

Continued

Table 2.2 Forest and shrub communities. (cont'd)

Ecosystem type (acronym)	Syntaxonomic unit	Habitat (biotope)	Main tree species	Main herb layer species
		Forest communities		
OAK1–OAK5	*Tilio-Carpinetum stachyetosum*	The most fertile, humid Histosols formed from peatland/fens	Pedunculate oak *Quercus robur*, ash *F. excelsior*, Norway maple *Acer platanoides*, small-leaved lime *Tilia cordata*, aspen *Populus tremula*, alder *A. glutinosa*	Touch-me-not balsam *Impatiens noli-tangere*, hedge woundwort *Stachys sylvatica*, alpine enchanter's-nightshade *Circaea alpina*, and species entering from riparian forests
	Tilio-Carpinetum typicum	Fertile, moderately moist habitats; mainly Cambisol	Small-leaved lime *T. cordata*, pedunculate oak *Q. robur*, hornbeam *Carpinus betulus*, Norway maple *A. platanoides*	Liverleaf *Hepatica nobilis*, ground elder *Aegopodium podagraria*, bugle *Ajuga reptans*, wood anemone *Anemone nemorosa*, asarabacca *Asarum europaeum*, and species entering from mixed coniferous forest
	Tilio-Carpinetum calamagrostietosum	Cambic Arenosols (less frequently Luvisols or Cambisols) made of Sandur sands with cover of fluvioglacial clay sands; more acid in relation to other oak-hornbeam forests/associations	Pedunculate oak *Q. robur*, small-leaved lime *T. cordata*, hornbeam *C. betulus*, Scots pine *P. sylvestris*, white spruce *Picea alba* (two last species as a result	Bilberry *Vaccinium myrtillus*, chickweed wintergreen *Trientalis europaea*, reed grass *Calamagrostis arundinacea*, bracken *Pteridium aquilinum*, and species entering from mixed coniferous forest

			of forest management)	
CON1–CON5	*Serratulo-Pinetum*	Cambic Arenosols (less frequently Cambic Arenosols with podzolization process) made of fluviglacial sands with a composition of fine calcic sands	Scots pine *P. sylvestris*, pedunculate oak *Q. robur*, Norway spruce *P. abies*	Coniferous and oak-hornbeam herb layer species
	Peucedano-Pinetum	Cambic Arenosols with podzolization process made of fine sands decalcified over the entire depth of the soil profile; low groundwater level	Scots pine *P. sylvestris*, pedunculate oak *Q. robur*, Norway spruce *P. abies*	Dwarf shrubs (mainly *Vaccinium* sp.)
	Querco-Piceetum	Small areas in local depressions mainly on Gleyic Podzols	Norway spruce *P. abies*, pedunculate oak *Q. robur*, aspen *P. tremula*, downy birch *B. pubescens*	In addition to pine/coniferous forest, species of deciduous forests are still present
SWP1–SWP5	*Vaccinio uliginosi-Pinetum*	Small patches in outflow depressions or on the periphery of dystrophic lakes, on peat soil of raised bogs (Histosols); in mature communities, water stagnates just below the peat surface	Scots pine *P. sylvestris*, downy birch *B. pubescens*	Coniferous and raised bog herb layer species: labrador-tea *Ledum palustre*, bog bilberry *Vaccinium uliginosum*; well-developed layer of bryophytes mainly composed of *Sphagnum* sp.

Continued

Table 2.2 Forest and shrub communities. (cont'd)

Ecosystem type (acronym)	Syntaxonomic unit	Habitat (biotope)	Main tree species	Main herb layer species
Forest communities				
	Sphagno girgensohnii-Piceetum	Most often at dystrophic lakes, less often in other depressions filled with acid mesotrophic peat of various thickness (Histosols); in mature communities, water stagnates just below the surface of the ground	Norway spruce *P. abies*, less frequently Scots pine *P. sylvestris*, downy birch *B. pubescens*, silver birch *Betula pendula*, alder *A. glutinosa*	Dwarf shrubs of *Vaccinium* sp. and *Lycopodium* sp. predominate. The strongly shaded bottom of the forest supports the abundant development of bryophytes, among which *Sphagnum* sp. dominates
Shrub communities				
Accompanying alder forests (ALD1–ALD5) and transitional bogs (the succession phase of alder forests)	*Salicetum pentandro-cinereae*	Like in alder forests, especially in small mid-field and mid-meadow depressions	Gray willow *S. cinerea*, bay willow *Salix pentandra*	Marsh fern *T. palustris*, lesser Pond sedge *Carex acutiformis*
	Betulo-Salicetum repentis	Like in alder forests, mainly in the transitional peat bog complex	Creeeping willow *Salix rosmarinifolia*, shrubby birch *Betula humilis*	Numerous species of *Scheuchzerio-Caricetea nigrae* class
Accompanying mixed coniferous forests (CON1–CON5)	*Rubo-Salicetum capreae*	Like in mixed coniferous forests, on older clearings and unmanaged young cultivations	Goat willow *Salix caprea*	Bramble (*Rubus* sp.) dominance

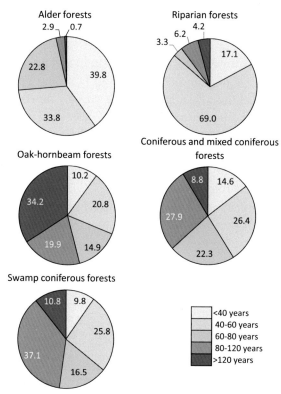

Figure 2.10 Percentage share of cover of tree stand age categories in forest communities.

the division of this association into several subassociations. Tree stand age is more evenly distributed compared with other forest types, with the domination (over 34%) of the oldest (over 120 years old) forests and the lowest share (slightly above 10%) of the youngest ones (up to 40 years old).

Three associations of coniferous and mixed coniferous forests occupy together the largest area among all forest communities (43%). Generally, these forests are at the intermediate stage of development, as the oldest coniferous tree stands (over 120 years) and the youngest one below (40 years) cover 9% and 15% of the forest area, respectively.

Swampy coniferous forests occupy about 4% of the study area. There are two associations in the study area: the continental swamp pine forests *Vaccinio uliginosi-Pinetum* and more boreal spruce swampy forest *Sphagno girgensohni-Piceetum*. The least frequent (about 10%) are the youngest stands as well as the oldest ones (about 11%). The largest areas among swampy forests occupy plots with tree stands of the age of 80–120 years (about 37%).

Shrub communities are strongly associated with forests, preceding them in succession or creating strongly degenerated forms of forests (Table 2.2).

An important group of plant communities constitute *associations with grass dominance* (meadows, pastures, dry grasslands) and floristically similar to them are *tall herb and fringe communities* (Fig. 2.11). This type of vegetation covers about 18% of the total area. They are strongly diversified in terms of both habitat and floristic composition (Table 2.3).

Another important group of plant communities consists of *wetland ecosystems* of peat bogs, fens, sedges, inland rushes, and reed beds. In total they occupy only slightly more than 0.5% of the study area but play an important ecological role. They represent six syntaxonomic units (in the rank of the phytosociological alliance or association) with different degree of maturity, habitat preference, and species composition (Table 2.4).

Figure 2.11 Agricultural landscape of Suwałki Lakeland. *(Photo: B. Wolska, J. Wolski.)*

Table 2.3 Meadows, pastures, and tall herb fringe communities.

Ecosystem type (acronym)	Syntaxonomic unit	Habitat (biotope)	Main herb layer species
		Meadow and pastures	
GRAS1	*Corynephoretalia canescentis*	Sandy location (mainly on outwash sands), on a substrate exposed as a result of degeneration or destruction of the coniferous forest	Brown bentgrass *Agrostis vinealis*, sheep's fescue *Festuca ovina*, sheep's bit scabious *Jasione montana*, cinquefoil *Potentilla collina*, gray hair grass *Corynephorus canescens*, sheperd's cress *Teesdalea nudicaulis*
	Arrhenatherion elatioris	Dry, moderately moist, fertile soils	False oat-grass *Arrhenatherum elatius*, marsh cranesbill *Geranium pratense*, wild parsnip *Pastinaca sativa*
GRAS2	*Calthion*	Mineral moist soils	Marsh marigold *Caltha palustris*, brook thistle *Cirsium rivulare*, marsh hawk's-beard *Crepis paludosa*
GRAS3	*Molinion*	Variable moist soils with calcium carbonate or more acid – refer to fens	Common betony *Betonica officinalis*, northern bedstraw *Galium boreale*, lilac pink *Dianthus superbus*
GRAS1–GRAS3	*Cynosurion*	Dry, moderately moist, fertile soils	

Continued

Table 2.3 Meadows, pastures, and tall herb fringe communities. (cont'd)

Ecosystem type (acronym)	Syntaxonomic unit	Habitat (biotope)	Main herb layer species
			Perennial rye-grass *Lolium perenne*, crested dog's-tail *Cynosurus cristatus*, daisy *Bellis perennis*
		Tall herb fringe communities	
Accompanying meadows (GRAS1–GRAS3)	*Filipendulo-Geranietum*	Wet, rich in humus, and fertile soils along watercourses; often as a secondary community on neglected and uncut moist meadows of the order of *Molinietalia*	Meadowsweet *Filipendula ulmaria*, marsh cranesbill *Geranium palustre*
	Lysimachio vulgaris-Filipenduletum	Wet, slightly acid, organic soil (Histosols)	Meadowsweet *F. ulmaria*, yellow loosestrife *Lysimachia vulgaris*
	Convolvulion sepium	Built by creepers/climbers at the edge of the riparian forest, alder trees, nitrophilous "veil" communities of fringes, wet soils on the banks of watercourses	Hedge bindweed *Calystegia sepium*, cleavers *Galium aparine*
	Urtico-Aegopodietum podagrariae	The periphery of the moderately moist habitat suitable for hornbeam forests	Ground elder *Aegopodium podagraria*, nettle-leaved bellflower *Campanula trachelium*, weasels-nout *Galeobdolon luteum*
	Trifolio-Agrimonietum	On the edge of drier oak–hornbeam forest; on fertile soils, in the spatial complex of drier forms of oak–hornbeam forests	Similar to dry meadows; agrimony *Agrimonia eupatoria*, zigzag clover *Trifolium medium*, bush vetch *Vicia sepium*

Table 2.4 Wetland communities.

Ecosystem type (acronym)	Syntaxonomic unit	Habitat (biotope)	Main herb layer species
BOG	*Ledo-Sphagnetum magellanici*	On the beds of peat with *Sphagnum* sp. and *Eriophorum* sp. in places fed by rainwater, most often in depressions completely overgrown or still existing dystrophic lakes	Labrador-tea *Ledum palustre*, small-fruited cranberry *Oxycoccus microcarpus*, sphagnum moss *Sphagnum* sp.
BOG	*Sphagnetum magellanici*	Nonforest peat bog	Sphagnum moss *Sphagnum* sp., small cranberry *Oxycoccus palustris*
BOG	*Rhynchosporion albae*	Unstable floating mat over dystrophic lakes; on the acid peat substrate (Histosols)	White beak-sedge *Rhynchospora alba*, oblong-leaved sundew *Drosera intermedia*
BOG	*Caricetum lasiocarpae*	On the surface of water reservoirs	Swards or sludge blanket; slender sedge *Carex lasiocarpa*, marsh scheuchzeria *Scheuchzeria palustris*
FEN	*Carici canescentis-Agrostietum caninae*	Fens - diverse and dynamically unstable group	Silvery sedge *Carex canescens*, velvet bent *Agrostis canina*
FEN	*Caricion davallianae*	Low sedge bog spring on limestone gyttja at the lakes	Davall's sedge *Carex davalliana*, globular-leaved valerian *Valeriana simplicifolia*

Arable lands, used mainly for grain and root crop production, cover approximately 14% of the total area (Fig. 2.11). They are associated with *weed communities*, differentiated according to habitat character and cultivated plant species (Table 2.5).

In addition to the above-listed groups of plant communities most important for the region (forests, shrubs, meadows, tall herb fringes,

Table 2.5 Arable field communities.

Ecosystem type (acronym)	Syntaxonomic unit	Habitat	Main herb layer species
ARB1—ARB3	*Vicietum tetraspermae*	Eutrophic Cambisols and Cambic Arenosols formed from clays and sands of different origins	Rye brome *Bromus secalinus*, smooth tare *Vicia tetrasperma*, pale persicaria *Polygonum tomentosum*
ARB1—ARB2	*Consolido-Brometum*	Fertile sites, most often on moraine forms	Rye brome *B. secalinus*, forking larkspur *Consolida regalis*, long-headed poppy *Papaver dubium*
ARB1	*Papaveretum argemones*	Permeable soils, quickly heating up and periodically too dry, with different carbonate content in the substrate	Thale cress *Arabidopsis thaliana*, prickly poppy *Papaver argemone*
ARB1	*Panico-Setarion*	Light, mesotrophic soils on clay sands	Cockspur *Echinochloa crus-galli*, wild radish *Raphanus raphanistrum*
ARB2—ARB3	*Veronico-Fumarietum officinalis*	Fertile, mainly eutrophic Cambisols and Cambic Arenosols	Common fumitory *Fumaria officinalis*, bugloss *Anchusa arvensis*
ARB2—ARB3	*Galinsogo-Setarietum*	Fertile and moist Histosols	Purple spurge *Euphorbia peplus*, quickweed cilié *Galinsoga ciliata*

wetlands, arable fields), there are also numerous ruderal communities associated mainly with the vicinity of built-up areas.

2.2.7 Fauna

According to the zoogeographical division proposed by Kostrowicki (1999), the analyzed area is located in the Boreal Province, which is part of Eastern Subregion within the Central European Region. Boreal Province is considered as the most western part of taiga biome. In relation to other regions of the zoogeographical division of Poland, it is distinguished by the presence of boreal species, often having the character of postglacial relicts, like mountain hare *Lepus timidus*, a very rare visitant Ural owl *Strix uralensis*, crustacean *Pallaseopsis quadrispinosa* reported in the Lake Wigry, and numerous insects, e.g., the taiga bumblebee *Bombus jonellus,* carabid beetle Menetries *Carabus menetriesi*, ant anemone cloak *Camponotus herculeanus*, and the butterfly of the arctic Jutta Arctic *Oeneis jutta*, of which about 500 individuals (that is, about 50% of the Polish population) occur in one bog.

Today's state of fauna and the structure of zoocenoses are the result of many factors, historical and present, among which the most important are zoogeographical location, habitat spatial diversity and their conservation status, and history of fish and wildlife management (e.g., hunting, legal protection). Data used in the study come mainly from studies and reports related to the WNP (Operat ochrony, 2014, 2013, 1999) and from the Standard Data Forms for Natura 2000 sites.

Based on the current knowledge, it can be assumed that in the study area there are over 3000 species of animals, with invertebrates dominating, and among them insects (Table 2.6). Among the many animal species, rare and protected species (by national or international law) related to particular groups of ecosystem deserve special attention.

Numerous invertebrates are associated with aquatic ecosystems, of which the part has been described for the first time (new species for science) from the Wigry Lake and adjacent lakes. This group includes 2 Acoela taxa (*Microdalyellia wiszniewskii* and *Microdalyellia lugubris wigrensis*), 20 species of Rotifers, and 1 species of Oligochaete *Paranais setosa*.

There are 31 species of fish, which is over 50% of freshwater native Polish ichthyofauna. Five of them, weatherfish (mud loach) *Misgurnus fossilis*, spined loach *Cobitis taenia*, stone loach *Barbatula barbatula*, bitterling *Rhodeus amarus*, and minnow *Phoxinus phoxinus*, are protected species. Spined loach, weatherfish (mud loach), and bitterling are on the list of Natura 2000 species. The largest fish richness is found in the lakes, which are characterized by the presence of vendace, whitefish, and lake trout. In bream-type lakes, bream, roach, rudd, tench, and bleak predominate. In shallow water reservoirs with well-developed underwater meadows, tench and pike have excellent life conditions. Extremely unfavorable conditions for fish prevail in crucian and dystrophic lakes.

There are also many bird species associated with water bodies and accompanying peat bog complexes, many of which are protected on the

Table 2.6 Species richness of selected faunal taxa.

Systematic group	Number of species	Species protected by national law	Species from Annexes II and IV of the Habitat Directive and from Annex I of the Birds Directive
Invertebrates total	>3000	72	
Rotifers, Rotifera	>300		
Platyhelminths	>40		
Annelida	>45		
Crustaceans	>120		
Molluscs	>55		3
Arachnids	>200		
Insects	>1100		
- Dipteran	260		
- Beetles	240		
- Hymenopteran	230		
- Butterflies	220		3
- Caddisflies	45		
- Dragonflies	39		2
Fish	31	5	3
Reptiles	5	5	
Amphibians	12	12	10
Birds	210 (160 nesting birds)	116	68
Mammals	53	29	20

basis of international agreements (species protected by the Natura 2000 system). Among them, there are whooper swan *Cygnus cygnus* (A)[6], reed warbler *Acrocephalus arundinaceus*, kingfisher *Alcedo atthis*, ferruginous goat *Aythya nyroca* (A, EN[7]), bittern *Botaurus stellaris* (A, LC), goldeneye *Bucephala clangula*, fusilier *Carpodacus erythrinus*, black tern *Chlidonias niger*, common tern *Sterna hirundo*, marsh harrier *Circus aeruginosus* (A), Montagu's harrier *Circus pygargus*, coot *Fulica atra*, snipe bush *Gallinago gallinago*, bluethroat *Luscinia svecica* (NT), little crake *Porzana parva* (NT), spotted

[6] The letter A means that during the nesting period, at least 1% of domestic population of that species inhabits a given area.
[7] Categories of threat used in the Polish Red Data Book of Animals (Głowaciński, 2001): *critically endangered* - CR; *endangered* - EN; *vulnerable* - VU; *near threatened* - NT. The remaining species is the LC (*least concern*) category, which means that they do not show a population regression and are not too rare.

crake *Porzana porzana*, little gull *Larus minutus* (LC), great crested grebe *Podiceps cristatus*, and horned grebe *Podiceps auritus*.

Some mammals are also in a special way associated with waters, and some of them are protected by international law: beaver *Castor fiber* and otter *Lutra lutra* and a pond bat *Myotis dasycneme* (EN), which mainly feeds on lakes.

The second, very important group of zoocenoses includes forest ecosystems and in particular extensive complexes of old-growth forests. A very large group of birds protected by international law is associated with the forest complexes of the Augustów Primeval Forest, partly located within the study area. These include red crossbill *Loxia curvirostra*, Tengmalm's owl *Aegolius funereus* (A, LC), lesser spotted eagle *Aquila pomarina* (A, LC), eagle owl *Bubo bubo* (A, NT), stock dove *Columba oenas*, roller *Coracias garrulus* (A, CR), white-backed woodpecker *Dendrocopos leucotos* (A, NT), black woodpecker *Dendrocopos martius*, middle spotted woodpecker *Dendrocopos medius*, red kite *Milvus milvus* (A, NT), capercaillie *Tetrao urogallus* (A, CR), hoopoo *Upupa epops*, hazel grouse *Bonasa bonasia*, red-breasted flycatcher *Ficedula parva*, pygmy owl *Glaucidium passerinum* (LC), honey buzzard *Pernis apivorus* (A), three-toed woodpecker *Picoides tridactylus* (A, VU), gray-headed woodpecker *Picus canus* (A), short-toed eagle *Circaetus gallicus* (A, CR), common crane *Grus grus* (A), black grouse *Tetrao tetrix tetrix* (A, EN), green sandpiper *Tringa ochropus*, and redwing *Turdus iliacus*.

Among the mammals, two predators, wolf *Canis lupus* (NT), and Eurasian lynx *Lynx lynx* (NT), due to their irreplaceable role in the food chain, deserve special attention. An important component of forest zoocenosis is also wild game mammals of great economic importance, including, above all, the so-called big game, i.e., red deer *Cervus elaphus*, roe deer *Capreolus capreolus*, elk *Alces alces*, wild boar *Sus scrofa*, and species of lesser economic importance.

Some of the species mentioned above are mammals that at least partly gain food outside forests (e.g., roe deer, hare, fox, and wild boar). Similar behavior is also characteristic of many bird species, e.g., white-tailed eagle *Haliaeetus albicilla* (A, LC), goosander *Mergus merganser*, black kite *Milvus migrans* (A, NT), spotted eagle *Aquila clanga* (CR), or black stork *Ciconia nigra* (A).

Among the bird species found in grassland ecosystems of the study area, two are very rare throughout Poland - the corncrake *Crex crex* and great snipe *Gallinago media* (VU).

Fauna of agricultural areas, including a mosaic of fields, meadows, coppice, thickets, and buildings, is much less specific and generally includes common species. Among them, however, there are species protected by national and international law, such as the relatively common white stork *Ciconia ciconia* and slightly rarer woodlark *Lullula arborea*, common nightjar *Caprimulgus europaeus*, red-backed shrike *Lanius collurio*, barred warbler *Sylvia nisoria*, and ortolan *Emberiza hortulana*.

The fauna of the study area is constantly changing as a result of not only anthropogenic or natural habitat change but also interspecies competition. Based on historical data, it can be assumed that in the 13th century, aurochs *Bos primigenius* became extinct, in the 17th/18th century tarpan *Equus gmelini* and European bison (wisent) *Bison bonasus*, in the 18th century brown bear *Ursus arctos*, and in the 20th century European mink *Mustela lutreola* and European pond turtle *Emys orbicularis*. Some species, once present in large numbers, have greatly reduced their abundance and are currently only occasionally found in the study area. These include eagle owl *B. bubo* and a few other birds of prey, capercaillie *T. urogallus*, as well as the noble crayfish *Astacus astacus*.

2.2.8 Nature conservation

In Poland, we distinguish the following categories of areas and objects legally protected under the Nature Conservation Act: national parks, nature reserves, landscape parks, areas of protected landscape, Natura 2000 sites, nature monuments, documentation sites, ecological lands, as well as nature and landscape complexes.[8] Zonal species protection of wild plants, animals, and fungi complements this list.

All conservation activities in the region are directly or indirectly related to the WNP established in 1989 (15,085.5 ha). The park comprises forest land (9464 ha), aquatic ecosystems (2908 ha), and other areas (2714 ha), mainly used for agricultural purposes (2229 ha). The strict protection areas cover 4% of the park (623 ha, including 283 ha of forests and 255 ha of aquatic ecosystems), partial protection (active) 75%, and landscape protection 21%. The park buffer zone covers 11,284 ha.

The Wigry Lake is the heart of the park. In 1975, it was entered by the International Union for Conservation of Nature (IUCN) on the list of the most valuable water bodies in the world (the Aqua Project). This largest

---

[8] In the study area, there are no landscape parks, documentation sites, ecological lands, and nature and landscape complexes.

water reservoir of the Suwałki Region is accompanied by 42 natural lakes, representing a wide range of limnological types differing in trophic state, thermal conditions, and concentration of humus compounds. The most peculiar are, described previously, dystrophic mid-forest lakes. The main river of the park and at the same time a popular canoeing trail is the Czarna Hańcza. Canoeing and other forms of water recreation are also extremely popular on most lakes of the study area (Fig. 2.12).

The floristic richness of the WNP consists of over 1000 taxa of vascular plants – native, alien but permanently domesticated and some cultivated plants (83 strictly protected and 15 partially protected), over 200 species of moss and liverworts, and almost 300 species of lichens. Lime-oak-hornbeam forests and subboreal mixed forests dominate among forest communities in the park. The nonforest communities that are unique in the

Figure 2.12 Water recreation in the Suwałki Lakeland. *(Photo: B. Wolska, J. Wolski, A. Kowalska, A. Affek.)*

country scale include *Sphagnum* bogs and turf and carbonate fens. The fauna of the WNP is represented by over 3000 species of animals, including 53 species of mammals, 210 species of birds, 12 species of amphibians, 5 species of reptiles, and 31 species of fish.

There are three nature reserves within the study area (Rąkowski, 2005):

- forest reserve "Pomorze" preserving over 200-year-old pine stands on the moraine hills and traces of the ancient settlement;
- aquatic reserve "Kalejty Lake" preserving natural values of the lake and peculiar features of the landscape, typical of a central part of the Augustów Forest with old-growth pine forests and dystrophic lakes among bogs;
- landscape reserve "Tobolinka" preserving the dystrophic lake, *Sphagnum* bogs, and swamp pine forests.

The study area comprises four areas of protected landscape: the Lake District of the North Suwałki Region, the Sejny Lake District, the Augustów Forest and Lakes, and the Rospuda River Valley. In spite of their lower rank protection, they play a significant role as ecological corridors, buffer zones of national and landscape parks, and Natura 2000 sites and act as an additional protection of forests. They are established primarily to protect (a) a seminatural landscape with a varied postglacial relief with numerous lakes, (b) Augustów Forest – one of the largest and most valuable forest complexes in Central and Eastern Europe, and (c) the biodiversity of natural habitats with a great number of rare species. Scenic value of landscapes is also important (Fig. 2.13).

A significant part of the study area is covered by the European Ecological Network Natura 2000 program, based on two EU directives dedicated to birds[9] and habitats[10], respectively. They consist of the lists of animal and plant species and of natural habitats (ecosystems), valuable and endangered on a European scale and characteristic of the main European biogeographical regions. Pursuant to the Birds' Directive, one Special Protection Area was designated in the study area – the Augustów Forest (PLB200002). Its reference area was an existing bird refuge of the

---

[9] Directive 2009/147/EC of the European Parliament and of the Council of November 30, 2009, on the conservation of wild birds (Official Journal of the European Union L 20, 26.01.2010, p. 7).

[10] Council Directive 92/43/EEC of May 21, 1992, on the conservation of natural habitats and of wild fauna and flora (Official Journal of the European Union L 206, 22.07. 1992, p. 7).

Figure 2.13 Scenic value of seminatural and cultural landscapes of the Suwałki Lakeland. *(Photo: J. Solon, J. Wolski.)*

international rank (Important Bird Area) - PL043 Augustów Forest constituting the key refuge in the country for capercaillie *T. urogallus* and hazel grouse *B. bonasia* and important breeding areas of lesser spotted eagle *A. pomarina*, marsh harrier *C. aeruginosus*, crane *G. grus*, corncrake *C. crex*, black- and green-backed woodpeckers *Dryocopos martius* and *Picus viridis*, hoopoe *U. epops*, and nutcracker *Nucifraga caryocatactes* (Wilk et al., 2010).

Four Habitat Areas (Sites of Community Importance) have been designated under the Habitats' Directive:

- Augustów Refugium (PLH200005) covering the area of almost the entire Polish part of the Augustów Forest with various valuable forest communities;

- Wigry Refugium (PLH200004) including the WNP with the Wigry Lake together with the whole complex of surrounding lakes, as well as the northern part of the Augustów Forest and a fragment of the Czarna Hańcza Valley;
- Sejny Lake District (PLH200007) characterized by a vivid postglacial relief and an exceptional accumulation of lakes in the Polish part of the Lithuanian Lake District;
- Jeleniewo (PLH200001) protecting the Poland's largest breeding colony of the bat nostril *M. dasycneme* and the area of its feeding sites in the Czarna Hańcza Valley.

Furthermore, there are a total of 42 nature monuments in the study area: 39 living nature (15 individual trees and 24 tree clusters and alleys) and 3 inanimate (granite erratic boulders). Moreover, there are also remarkable examples of zonal active protection of plants, animals, and fungi e.g., introduction of orchids in the WNP or 15 zones (796 ha) designated for the protection of breeding and the regular (or periodic) staying sites of the black stork *C. nigra*, white-tailed eagle *H. albicilla*, capercaillie *T. urogallus,* and carpentula *A. funereus.*

At the end, it is worth to highlight two initiatives that are the result of international agreements signed by Poland. One of them is entering of the WNP on the list of the Ramsar Convention, which currently comprises 2261 wetlands from 169 countries, including 13 from Poland. The second one concerns the project to create a UNESCO Geopark Augustów Canal - Augustów Outwash Plain, whose boundaries would cover almost the entire study area (Krzywicki and Pochocka-Szwarc, 2014). The concept of geoparks is associated with geotourism meaning tourism based on values and attractions related to the geological structure and relief. Glacial origin of the study area as well as numerous geological, geomorphological, and hydrogeological sites and objects observed there can certainly inspire to practice geotourism.

CHAPTER 3

Methods

Contents

3.1 Assessment of single ecosystem service potentials

To determine the potentials of postglacial ecosystems (landscapes) to provide services, two types of assessment were used: expert, based on scientific knowledge, and social, obtained from the opinions of direct ecosystem users (residents and tourists).

3.1.1 Expert assessment

The expert assessment of ecosystem (landscape) potential to provide services was primarily based on available environmental databases. These were cartographic and statistical materials collected by different agencies, in a standard way for the entire Poland, as well as data obtained directly from local offices in the study area. Detailed characteristics of materials used to develop individual indicators can be found in the dedicated sections of Section 6.1. In the absence of relevant data, instead of calculating the indicator value, the ES potential was estimated using the information from the literature.

To complement the existing databases, field measurements of selected ecosystem parameters were conducted in the study area. The field works were mainly focused on the recognition of vegetation and soil properties.

Ecosystem Service Potentials and Their Indicators in Postglacial Landscapes
ISBN 978-0-12-816134-0
https://doi.org/10.1016/B978-0-12-816134-0.00003-1
Copyright © 2020 Elsevier Inc.
All rights reserved.

The collected data were used primarily to verify the official databases and to determine the actual local reference values for literature data not specific to the study area.

The research on vegetation was used to identify ecosystem types (Section 5.1), to determine carbon storage in ecosystems, and to record invasive plant species and those producing edible wild berries (Section 6.1). The phytosociological relevés were collected in the 18 locations representing different types of plant community. All tree, shrub, herb, moss, and lichen species were recorded, depending on the plant community, within plots set at 400 m^2 in forests or 100 m^2 in grassland or wetlands. The horizontal structure of vegetation was described with the cover–abundance scale proposed by Braun–Blanquet (1964), which consists of a plus sign and a series of numbers from 1 to 5 denoting both the numbers of species and the proportion of the area covered by that species, ranging from + (sparse and covering a small area) to 5 (covering more than 75% of the area). The vertical structure was characterized through estimating the plant cover (in percents) of each forest layer.

To determine the organic carbon storage, samples of vegetation <0.5 m in height were collected from three 0.1 m^2 clip plots randomly placed within each of the relevés. The collected biomass samples were then separated into a living and dead biomass. The dead parts of plants were removed. Afterward, the samples were dried at 90°C for 24 h and weighted; total organic carbon (TOC) content was analyzed using Alten method and then converted to carbon storage estimates (Section 6.1.2.10).

The soil cover characteristics were also determined within the phytosociological relevé plots. Disturbed soil samples were collected from soil profiles 50 cm in depth. They represent the genetic soil horizons of the studied pedons: organic (O), humus (A), and enrichment (B) of autogenic soils as well as gley (G) and muck (M) horizons in semihydric soils. In total, soil material was collected from 49 genetic horizons. Comparative values as well as physical and chemical characteristics of soils for other not–studied ecosystem types were taken from the literature (Skorupski et al., 2011).

Disturbed mineral soil samples were processed for further analysis using standard procedures, including drying in 40°C and sieving through 2.0 mm sieve to remove skeleton fraction. Organic samples were dried to constant weight at 65°C and milled into powder. In the prepared samples, the following analyses were performed:

- loss on ignition was determined by igniting the samples at 550°C in muffle furnace;

- pH was measured potentiometrically in suspension with water and 1 mol dm^{-1} KCl in a soil: water/KCl ratio of 1:2.5 (pH–meter Elmetron CPC-401);
- TOC in mineral and organic mineral soil samples was determined by Tiurin method, while in organic samples, using the Alten method (Dziadowiec and Gonet, 1999);
- total nitrogen with the Kjeldahl method (van Reeuwijk, 2002);
- the content of Cu, Ni, Zn, and Mn was determined by microwave plasma atomic emission spectrometry (Agilent 4100 MP-AES) after digestion of ashed samples in aqua regia;
- exchangeable acidity was determined by Sokolov's method;
- hydrolytic acidity was measured using Kappen's method;
- the contents of exchangeable cations (Ca^{2+}, Mg^{2+}, K^+, Na^+) were analyzed by microwave plasma atomic emission spectrometry (Agilent 4100 MP-AES) after sample extraction in 1 mol dm^{-1}, pH = 7.0 solution of ammonium acetate.

The results of laboratory analyses were used in the calculations of several regulating ES indicators (Section 6.1.2).

3.1.2 Social assessment

The social assessment of ecosystem potentials to provide services was based on the anonymous questionnaire survey carried out over two seasons (summer 2014 and spring 2015) among residents living – and tourists staying – in the study area (for details, see the Assessment Method in Section 6.2). A time-consuming door-to-door method was applied, given that our earlier experience and observations with open participatory workshops, meetings, or remote surveys (delivered via the Internet or traditional post) failed to reach a considerable part of the ecosystem users, not least the elderly or tourists (Abildtrup et al., 2013; Scholte et al., 2015).

The survey was constructed around the key methodological assumptions that services taken account of derived solely from the local ecosystems and that a flow of services (actual use) takes place only when the said services are consumed directly (enjoyed) by the final recipient. This denoted that a job related to the exploitation of nature did not entail an ecosystem service, given that, for example, the harvesting of berries by a professional berry picker or wood by a woodcutter is of a different nature to the gathering or purchase of the same products by those who actually intend to make personal use of them. This assumption has the effect of avoiding double counting. To keep the ES list finite, we followed Haines-

Young and Potschin (2013) in excluding abiotic ecosystem outputs (metallic and nonmetallic mineral resources, fossil fuels, wind, and water energy) from provisioning services. Consequently (unlike in Common International Classification of Ecosystem Services [CICES]), we did not consider water itself as an ecosystem service, in contrast to water treatment (retention and purification) by biota, which represents an important regulating service.

The scientific term "ecosystem services" was absent from the questionnaire, with its place being taken by the more colloquial and intelligible old Polish phrase "dobrodziejstwa przyrody", hence the approximate translation here of "gifts of nature".

With such a framework kept in mind, we constructed a questionnaire survey divided into four major parts (description of part 1, 2, and 4 in Section 3.4). In the third part, oriented on the assessment of ES potentials, respondents were asked to indicate the capacity of ecosystems to provide 11 categories of the said "gifts of nature". The task was to indicate which of the services are provided by particular ecosystem types and to determine the importance of a given "gift of nature" in comparison with the other services included in the question. We applied Mapping and Assessment of Ecosystems and their Services (MAES) level 2 typology of ecosystems (Maes et al., 2013). In total, seven ecosystem types were distinguished (Section 5.2). Services from the second part of a questionnaire (Section 3.3) were grouped into 10 classes, of which 6 could be regarded as provisioning-related, and four cultural. We also added one regulating service (water purification and retention), which we assumed might be ranked variously as regards ecosystem potentials.

To analyze the collected data, a number of different statistical and graphic tools were used, adapted to the study purpose and the type of data. A detailed description of the data processing methods and comparative analysis is given in Section 6.2.

3.2 Analysis of assessment matrix

The obtained values of particular ES indicators developed for ecosystem units (from both the detailed and MAES-derived typology) were compiled into two corresponding databases - one with original and one with ranked values. Both databases were used for further synthetic work; rank values were used to calculate aggregated potentials, multiservice hotspots, and ES spatial patterns, while original indicator values served for the analysis of ES interactions and ecosystem similarities. Rank values (on a 1 - 5 scale) are presented in the form of a table (matrix of indicators and ecosystems) in Section 7.1. The adopted ranking scale and tabular setting is based on solutions proposed by Burkhard and others (2014, 2012, 2009). Thanks to this, the obtained results can be easily compared with the work of these and other authors, even if the indicator methodology and the study area are different.

3.2.1 Analysis of aggregated potential

To determine the overall potential of a given ecosystem, a dedicated method was applied adequate for the situation when not all possible single services had been assessed and when the number of assessed ES within each section is far from equal. As we assumed that each ES section (provisioning, regulation and maintenance, and cultural) has equal weight in the overall potential, we first needed to obtain the aggregate potential for each section. To do so, we took the average value out of services belonging to a given ES section. The way we calculated the aggregated potential can be described by the following formula:

$$P_{all} = \frac{1}{3}\left(\overline{P_p} + \overline{P_r} + \overline{P_c}\right),$$

where P_{all} is the overall aggregated potential, $\overline{P_p}$ is the provisioning potential, $\overline{P_r}$ is the regulating potential, and $\overline{P_c}$ is the cultural potential.

Additionally, we equaled the weights of the two ES subsections in the provisioning section: biomass for nutrition and biomass for materials and energy. In case there was more than one indicator describing a given service, the mean value was taken for further analysis to avoid overweighing of such a service (e.g., we took the mean value out of RICHNESS and NATURA indicators, as they both describe the same service: nursery habitat maintenance). We used the 1−5 rank values as input data. Zeros and Ns (not considered) were not included in calculating means. Because of the fact that very few ES indicators were estimated for settlement areas, this land use class was excluded from the analysis of aggregated potentials.

The obtained values were also used to map the aggregate potential (overall and for particular ES sections) in the study area. The scale range on all maps was unified by adjusting to the extreme values obtained, i.e., 1.82 and 4.00. Values were classified into seven equal intervals.

3.2.2 Multiservice hotspot analysis

An ES hotspot has been defined as an area (ecosystem) for which the potential to provide a given service was rated high or very high (4 or 5 rank). In turn, multiservice hotspot is an ecosystem for which the potentials to provide several services were rated high or very high.

In our research, we analyzed how many ES potentials were rated high and very high for a given ecosystem type. As we did not cover all possible services provided by the analyzed ecosystems and additionally have different number of analyzed services in each ES section, we also calculated the percent of services of each section and in total that received high or very high scores. In this way, we aimed to provide a more universal picture of whether a given ecosystem may be regarded as multiservice hotspot or not. We assumed 50% services ranked high or very high is the adequate threshold that separates true multiservice hotspots from other areas. To reliably calculate those percents, it was necessary to combine some provisioning services that were dedicated and closely linked to a given land use (i.e., forests – timber; arable fields – yields; grassland – livestock) into one service. Otherwise, hotspots for provisioning services expressed in percents would be falsely understated compared with other ES sections. As a result, 9 instead of 12 services constitute 100% of provisioning services and thus the maximum number of ES potentials that could be possibly rated high or very high. Also in this analysis, RICHNESS and NATURA indicators were averaged, and ecosystems that received the mean value equal or higher than 4 were considered as hotspots for maintaining nursery habitats. Apart from settlements, also lakes were excluded from hotspot analysis due to low number of evaluated services compared with terrestrial ecosystems. As a result, multiservice hotspot analysis was conducted for 34 ecosystem types and 25 services (9 provisioning, 12 regulating, and 4 cultural).

The obtained values (percents and proportions of highly rated ES potentials) were also used to map the distribution of multiservice hotspots (for all ES and for particular ES sections) in the study area.

3.2.3 Analysis of ecosystem service interactions

The aim of this analysis was to identify ES bundles that group services with similar levels of potential supply. To analyze associations between ES indicators, a relatively simple statistical tool was used – the visual analysis of correlation matrix. At the stage of preparatory work, methods using complex grouping and classification algorithms with a much greater degree of automation (e.g., K-means, fuzzy K-means, principal component analysis [PCA]) were also tested, but they did not contribute to better understanding of ES bundles than the manual analysis of correlation matrix.

We investigated ES synergies and trade-offs by means of pairwise correlations, as they are often used as proxy measures of positive and negative ES interactions (Turner et al., 2014; Yang et al., 2015). Although on their basis we could neither judge about any possible synergetic nor compete mechanisms and causal relationships, they give as reliable information on the coexistence of certain levels of ES potentials.

We conducted r-Pearson pairwise correlation analysis on all 29 ES indicators calculated in ecosystem units. We used original indicator values as input data to grasp the original differentiation of values. As original C/N indicator values were inversely proportional to the levels of potential, they were multiplied by -1. When a given ecosystem type was not considered in calculating a particular service (N label) or shown no relevant potential (0 value), then it was skipped when calculating a given pairwise correlation.

3.2.4 Analysis of ecosystem similarities

To determine ecosystem similarity, the hierarchical cluster analysis was applied together with its graphic representation in the form of a dendrogram. The Euclidean distance was used as the measure of distance/similarity and the Ward algorithm as the grouping method. This algorithm aims to minimize the sum of squared deviations of any two clusters formed at each grouping stage. To identify the most important dimensions ordering ecosystems in terms of ES potentials, we used PCA, one of the methods of factor analysis. Having in mind the Kaiser–Guttman criterion (eigenvalue > 1) (Legendre and Legendre, 1998) and observing the scree plot, we concluded that the first four PCA axes were sufficient to characterize the nonrandom structure of the data.

We conducted the analysis on 34 out of 42 ecosystem types included in the detailed typology. All types of lake and settlement areas were excluded, as only a small number of indicators were calculated for them. We used standardized (([value]−[mean])/[standard deviation]) original indicator values as input data to grasp the original differentiation of values and also to meet the requirements of multivariate analyses. As original C/N indicator values were inversely proportional to the levels of potential, they were multiplied by -1 before standardization. Also, the cells in the assessment matrix with N label (not considered) were recoded before standardization into 0 value for computational purposes. Statistical analysis and plots were generated in STATISTICA software.

3.3 Spatial analysis of ecosystem service potentials

Spatial analysis was conducted only for those ES, whose potentials were assessed in ecosystem units, and formed spatial patterns complex enough to make the analysis informative. Two independent approaches were used to analyze and evaluate spatial pattern of ES potentials in the study area. The first approach employs landscape metrics, while the second one is based on autocorrelation analysis.

Landscape metrics were used to describe spatial patterns of single ES potentials (see sections on indicator values in Chapter 6) and to conduct a joint spatial analysis of ES potentials (see Section 7.5). Autocorrelation analysis was used only for joint analysis.

Fifteen landscape metrics, covering different aspects of spatial pattern of patches, were selected (Table 3.1) and calculated for 16 indicators of ES potentials (corresponding to 15 ES). Metrics were computed with the help of Patch Analyst and Patch Grid toolbox for ArcGIS (Rempel et al., 2012) and OpenJUMP GIS[1]. All but two metrics (IJI and MPI) were calculated on vector maps. IJI and MPI were calculated on raster maps with 5 × 5 m grid.

Hierarchical clustering of spatial patterns of different ES potentials based on landscape metric values, Bray—Curtis similarity index, and paired group algorithm (UPGMA) was calculated in PAST (Hammer et al., 2001). The principal component analysis biplot of ES potential spatial patterns and landscape metrics was drawn in MULTBIPLOT software (Villardón, 2015).

To describe autocorrelation, Moran's I correlograms were calculated in GeoDa (Anselin et al., 2006). Correlograms show autocorrelation values for certain distance classes or the so-called lags. A lag distance of 200 m was selected for generating distance classes.

To investigate the relationship between spatial metrics and autocorrelation and to group autocorrelation curves, the method of decision tree was used according to the supervised learning algorithms ID3 and C4.5 in Tanagra software (Rakotomalala, 2005).

---

[1] http://www.openjump.org.

Table 3.1 Landscape metrics used to characterize spatial pattern of ecosystem service potentials.

Abbreviation	Name	Formula	Description
NumP	Number of patches	$\text{Num } P = n$	Total number of patches in the landscape
MPS	Mean patch size	$MPS = \dfrac{\sum_{i=1}^{n} a_i}{n} = \dfrac{A}{n}$	Average patch size. Mean patch size is a function of number of patches and total landscape area
PSCoV	Patch size coefficient of variance	$PSCoV = \sqrt{\dfrac{\sum_{i=1}^{n}\left[a_i - \left(\dfrac{\sum_{i=1}^{n} a_i}{n}\right)\right]}{n}} * \left(\dfrac{100}{MPS}\right)$	PSCoV measures relative variability about the mean (i.e., variability as a percentage of the mean)
LPI	Largest patch index	$LPI = \dfrac{\max(a_i)_{i=1}^{n}}{A}(100)$	The LPI is equal to the percent of the total landscape that is made up by the largest patch
i60	Index60	No formula	Minimal number of patches covering together 60% of the total area
HOLE	Hole	No formula	Number of holes (perforations) in the landscape
SEI	Shannon's evenness index	$SEI = \dfrac{-\sum_{i=1}^{k}(C_i * \ln C_i)}{\ln k}$	SEI measures the distribution of area among patch types and is expressed as the observed level of diversity divided by the maximum possible diversity for a given number of patch types

Continued

Table 3.1 Landscape metrics used to characterize spatial pattern of ecosystem service potentials. (cont'd)

Abbreviation	Name	Formula	Description
MSI	Mean shape index	$$MSI = \frac{\sum_{i=1}^{n}\left(\dfrac{p_i}{2\sqrt{\pi * a_i}}\right)}{n}$$	Shape complexity MSI is equal to 1 when all patches are circular (for polygons), and it increases with increasing patch shape irregularity
AWMSI	Area weighted mean shape index	$$AWMSI = \sum_{i=1}^{n}\left(\frac{p_i}{2\sqrt{\pi * a_i}}\right) * \left(\frac{a_i}{A}\right)$$	Shape complexity AWMSI differs from the MSI in that it's weighted by patch area, so larger patches will weigh more than smaller ones
FDIM	Mean patch fractal dimension	$$FDIM = \frac{\sum_{i=1}^{n}\left(\dfrac{2\ln p_i}{\ln a_i}\right)}{n}$$	A fractal dimension greater than 1 for a two-dimensional landscape mosaic indicates a departure from a Euclidean geometry (i.e., an increase in patch shape complexity). FDIM approaches 1 for shapes with very simple perimeters such as circles or squares and approaches 2 for shapes with highly convoluted, plane-filling perimeters
SSPI	Shumm's shape index	$$SSPI = \frac{\sum_{i=1}^{n}\left(\dfrac{\sqrt{a_i/\pi}}{0.5 * d_i}\right)}{n}$$	SSPI measures elongation of a patch as a ratio of diameter of a circle having the same area as a given patch to the longest axis of a patch
CNCV	Concavity	$$CNCV = \frac{\sum_{i=1}^{n}\left(\dfrac{a_i}{z_i}\right)}{n}$$	CNCV measures the shape irregularity as a mean (ratio of the patch area to the area of its convex hull)

	Formula	Description
IJI Interspersion and juxtaposition index	$$IJI = \dfrac{-\sum\limits_{i=1}^{m}\sum\limits_{k=1}^{m'}\left[\left(\dfrac{e_{ik}}{E}\right)*\ln\left(\dfrac{e_{ik}}{E}\right)\right]}{\ln(0.5[m(m-1)])}(100)$$	Measure of patch adjacency approaches zero when the distribution of unique patch adjacencies becomes uneven and 100 when all patch types are equally adjacent
MPI Mean proximity index	$$MPI = \sum_{i=1}^{n}\dfrac{a_i}{h_{ij}^{2}}$$	Measure of the degree of isolation and fragmentation MPI increases as the neighborhood (defined by the specified search radius) is increasingly occupied by patches of the same type and as those patches become closer and more contiguous and less fragmented in distribution
MNN Mean nearest neighbor	$$MNN = \dfrac{\sum\limits_{i=1}^{n}h_{ij}}{n}$$	Measure of patch isolation – the nearest neighbor distance of an individual patch is the shortest distance to a similar patch (edge to edge). The mean nearest neighbor distance at the landscape level is the mean of the class nearest neighbor distances

Symbols in formulas: A, total area of the landscape; a_i, patch area; C_i, spatial share of i-class (patch type) in the landscape; d_i, the longest axis of the patch; E, sum of all edges in the landscape; e_{iks} length of the common edge between patch i and k; h_{ij}, distance from the patch i to the patch j; k, number of patch classes (types); n, total number of patches; p_i, patch perimeter; z_i, area of the convex hull.

Metrics names, formulas and description partly based on McGarigal, K., 2015. FRAGSTATS Help https://www.umass.edu/landeco/research/fragstats/documents/fragstats.help.4.2-pdf; McGarigal, K., Marks, B.J., 1995. FRAGSTATS: spatial pattern analysis program for quantifying landscape structure. General Technical Reports PNW-GTR-351, USDA Forest Service, Pacific Northwest Research Station, Portland; Rempel, R.S., Kaukinen, D., Carr, A.P., 2012. Patch Analyst and Patch Grid. Ontario Ministry of Natural Resources. Centre for Northern Forest Ecosystem Research, Thunder Bay, Ontario; Steiniger, S., Lange, T., Burghardt, D., Weibel, R., 2008. An Approach for the Classification of Urban Building Structures Based on Discriminant Analysis Techniques. Transactions in GIS 12 (1), 31−59.

3.4 Assessment of ecosystem service usage

The questionnaire survey used to collect opinions on ES potentials (see Section 3.1.2) also aimed to quantify the actual use of local ecosystem services and to examine respondents' knowledge about benefits deriving from them. Information on the declared use of services was obtained from both the first and the second part of the questionnaire. The first part comprised open-ended questions concerning the respondents' use of local ES (one general question and four detailed questions corresponding to the main ES categories: provisioning, regulation and maintenance, and cultural). We intended to investigate what ordinary people living in and visiting the study area recognize as "gifts of nature". In the second part, 45 final ecosystem services were listed (Table 3.2), adapted to local conditions and practices, with a possibility to indicate frequency of use within the last 3 years (never, once, several times, regularly, I don't know/hard to say). We aimed to encompass all possible services from the provisioning and

Table 3.2 List of 45 provisioning and cultural ecosystem services ("gifts of nature") included in the second part of the questionnaire.

Ecosystem good/service	Division/Group in CICES V4.3
Provisioning	
1. Fish from nearby rivers and lakes	
2. Mushrooms from nearby forests	
3. Fruits from nearby forests or their preserves	
4. Venison	
5. Fruits from local gardens or their preserves	
6. Vegetables from local cultivations or their preserves	
7. Baked foods of local flour	Nutrition
8. Honey from local apiaries	
9. Meat from local farms	
10. Eggs from local farms	
11. Vegetable oil from local plantations	
12. Milk and its preserves from local farms	
13. Shellfish and molluscs (e.g., crayfish, snails)	
14. Timber from local forests (plywood, veneer, raw material)	
15. Wooden objects made of local wood	
16. Wicker of local willows	
17. Reed	

Table 3.2 List of 45 provisioning and cultural ecosystem services ("gifts of nature") included in the second part of the questionnaire. (cont'd)

	Ecosystem good/service	Division/Group in CICES V4.3
18.	Fodder (hay, etc.) from local meadows and cultivations	Materials
19.	Herbs/medicines from local nature	
20.	Ornaments made of natural elements (antlers, wreaths, Easter palms, etc.)	
21.	Compost, manure	
22.	Wool and animal hides	
23.	Wax from local apiaries	
24.	Seeds/seedlings/Christmas trees from local plantations	
25.	Fuel wood from local forests	
26.	Peat from nearby peat bogs and its derivatives	Energy
27.	Energy crops (e.g., willow), straw	

Cultural

28.	Nature observation	
29.	Hiking along nature, educational paths	
30.	Watching films/albums about the local nature	Education
31.	Observation of rare plants and animals	Inspiration
32.	Creative work (writing, painting, etc.) inspired by local nature	Spiritual life
33.	Visiting places of worship in nature (Calvary paths, springs, places of power, etc.)	
34.	Prayers/meditation close to nature	
35.	Fishing	
36.	Hunting	
37.	Walking/jogging/hiking	
38.	Canoeing/motor boating or rowing	
39.	Sailing	
40.	Photographing nature	Sport and recreation
41.	Sightseeing tours to enjoy nature	
42.	Swimming in the lake or river/sunbathing	
43.	Diving	
44.	Rest, relaxation (bonfire, grill, camping, etc.)	
45.	Mushroom picking	

CICES, Common International Classification of Ecosystem Services.

cultural sections that are known to be in use in the study area. The list was elaborated by reference to general ES classifications (MEA, TEEB, and CICES) as well as local authority data on various uses of nature. The final version was established during a brainstorm workshop at which both field experts and representatives of Wigry National Park's Scientific Council were present. We excluded regulating services from this list, as they relate mostly to the broader, rather than the local scale, and it is difficult to extract the actual use of those services provided specifically by local nature.

In the last part of the questionnaire, we gathered sociodemographic data regarding age, gender, education, source of income, place of residence, etc., with a view of the representativeness of the sample being verified and between-group comparisons made.

Responses to open-ended questions were standardized in terms of meaning, grouped and transformed into binary variables. In Section 8.1, bar charts show the frequency of listing ES used, in total by all respondents and broken down by permanent residents and tourists. Pearson's chi-squared test (χ^2) was used to evaluate significance of differences between respondents' subgroups in the frequency of ES listing. To analyze the relationship between the frequency of answers and respondents' age, the variable "age" was recoded to three classes: <20, $30-60$, and > 60 years.

Bar charts were used to visualize the results of the second part of the questionnaire as well. In the declared closed formula of the frequency of ES use, the nonparametric U Mann–Whitney rank test was used to determine the significance of differences between respondents' subgroups. The application of parametric tests was not possible by the used ordinal scale (not once, once, several times, regularly) and significant deviations of the results from normal distributions. The strength of the associations between the frequency of ES use and education level (three classes) or age was examined using the nonparametric Spearman's *rho* correlation coefficient.

To compare the ES usage declared in the open and closed formula, the answers from the second part of the questionnaire were transformed into binary variables (used - not used). The results, broken down into residents and tourists, were presented on the bar chart. The significance of differences between respondents' subgroups for newly reclassified variables was calculated using the Pearson χ^2 test.

3.5 Analysis of links between ES usage, user characteristics, and perceived ES potentials

Analysis of links between ES usage, user characteristics, and perceived ES potentials is presented in Section 8.2. We assumed that the intervals between numbers (from 1 to 11) assigned to services by the respondents in the third part of the questionnaire are equal (interval scale), with this permitting the use of parametric analysis (see Section 6.2). The Student t-test was used to compare answers between subgroups of respondents (men vs. women, secondary vs. higher education, residents vs. tourists), while values for r-Pearson correlation coefficients measured the strengths of relationships between perceived ecosystem potentials and the ages of respondents.

In turn, because of the ordinal scale of actual frequency of ES use (the second part of the questionnaire) and considerable deviations from normal distributions, the *rho* Spearman correlation coefficient was utilized to test the links between actual usage intensity and the estimated potentials to deliver services. Altogether 3465 pairwise correlations (7 ecosystem types × 11 services evaluated × 45 services actually used) were calculated and analyzed. To capture more general patterns, a cumulative correlation matrix was computed, showing the percentages of significant positive and negative correlations in each ES section (in case of potential) and each ES group (in case of usage).

CHAPTER 4

CICES V5.1 classification

The *Common International Classification of Ecosystem Services* (CICES) has been designed to help measure, account for, and assess ecosystem services. It has been used widely in ES research for designing indicators, mapping, and for valuation (Haines-Young and Potschin, 2018). CICES was intended as a reference classification that would allow translation between different ES classification systems, such as those used by the *Millennium Ecosystem Assessment* (MEA) and *The Economics of Ecosystems and Biodiversity* (TEEB), and in many aspects, it follows the concepts of these initiatives (see Section 1.2.2).

CICES aims to classify final ecosystem services, which are defined as the contributions that ecosystems make to human well-being. These contributions are framed in "what ecosystems do most directly" for people. Final services are distinct from the goods and benefits that people subsequently derive from them and from functions or characteristics of ecosystems that come together to make something a service. In principle, CICES covers contributions that arise from living processes. However, the latest version (V5.1) includes also the nonliving parts of ecosystems, for example, water, mineral substances, wind, and solar energy.

CICES V5.1, released in 2018, has been developed on the basis of the review of relevant scientific literature and feedback from CICES user community (e.g., expressed in the survey conducted by the European Environment Agency and during workshops held as part of the EU funded ESMERALDA and OpenNESS Projects). It retains four level hierarchical structure of the previous 4.3 version (Haines-Young and Potschin, 2013). At the highest classification level, services are grouped according into three sections: (1) provisioning, (2) regulation and maintenance, and (3) cultural (Table 4.1).

Provisioning section covers all nutritional, nonnutritional material and energetic outputs from living systems as well as abiotic outputs (including water). The division level makes a distinction between biomass-based (biotic) provisioning services and the aqueous and nonaqueous abiotic ecosystem outputs.

Ecosystem Service Potentials and Their Indicators in Postglacial Landscapes
ISBN 978-0-12-816134-0
https://doi.org/10.1016/B978-0-12-816134-0.00004-3
Copyright © 2020 Elsevier Inc.
All rights reserved.

Table 4.1 Common International Classification of Ecosystem Services ver. 5.1 (CICES V5.1) to the group level (only biotic ES).

Section	Division	Group	Group Code
Provisioning	Biomass	Cultivated terrestrial plants for nutrition, materials, or energy	1.1.1
		Cultivated aquatic plants for nutrition, materials, or energy	1.1.2
		Reared terrestrial animals for nutrition, materials, or energy	1.1.3
		Reared aquatic animals for nutrition, materials, or energy	1.1.4
		Wild plants (terrestrial and aquatic) for nutrition, materials, or energy	1.1.5
		Wild animals (terrestrial and aquatic) for nutrition, materials, or energy	1.1.6
	Genetic material from all biota (including seed, spore, or gamete production)	Genetic material from plants, algae, or fungi	1.2.1
		Genetic material from animals	1.2.2
Regulation and maintenance	Transformation of biochemical or physical inputs to ecosystems	Mediation of wastes or toxic substances of anthropogenic origin by living processes	2.1.1
		Mediation of nuisances of anthropogenic origin	2.1.2
	Regulation of physical, chemical, biological conditions	Regulation of baseline flows and extreme events	2.2.1
		Lifecycle maintenance, habitat, and gene pool protection	2.2.2
		Pest and disease control	2.2.3
		Regulation of soil quality	2.2.4
		Water conditions	2.2.5
		Atmospheric composition and conditions	2.2.6
Cultural	Direct, in situ and outdoor interactions with living systems that depend on presence in the environmental setting	Physical and experiential interactions with natural environment	3.1.1
		Intellectual and representative interactions with natural environment	3.1.2
	Indirect, remote, often indoor interactions with living systems that do not require presence in the environmental setting	Spiritual, symbolic, and other interactions with natural environment	3.2.1
		Other biotic characteristics that have a nonuse value	3.2.2

Regulation and maintenance section covers all the ways in which living organisms can mediate or moderate the ambient environment that affects human health, safety or comfort, together with abiotic equivalents. The division level covers the following: (1) the "transformation of biochemical or physical inputs to ecosystems" in the form of wastes, toxic substances, and other nuisances, and (2) the "regulation of physical, chemical, biological conditions, which categorizes the various ways in which living systems can mediate the physico-chemical and biological environment of people in a beneficial way".

Cultural section covers all the nonmaterial, and normally nonrival and nonconsumptive, outputs of ecosystems (biotic and abiotic) that affect physical and mental states of people. Cultural services are primarily regarded as the environmental settings, locations, or situations that give rise to changes in the physical or mental states of people, where the character of those settings is fundamentally dependent on living processes; they can involve individual species, habitats, and whole ecosystems. The classification of cultural services has been redesigned complied with V4.3, and the division level now separates in situ and remote opportunities.

Moving down from section to division, group, and class, the "services" are increasingly more specific but remain nested within the broader categories that sit above them. In comparison to V4.3, the nomenclature has been modified to ensure that it is more clearly seen as a "functional" classification. The group level descriptors are now framed in a way ecosystems are useful to people (e.g., nutrition), whereas the division level captures functional attributes, or the ecosystem properties under consideration that facilitate human use (directly or indirectly). Both these dimensions are now reflected in the class definition. Example services belonging to each class have been provided in the full CICES table[1], alongside the related benefits. These examples are intended to help users understand what the class entails, and to clarify the distinction between services and associated benefits.

However, when framing our research scope, we decided to focus only on biotic ecosystem outputs, without such abiotic outputs such as metallic and nonmetallic mineral resources, wind, and water energy. Consequently, we did not consider water itself an ecosystem service (as it is abiotic), in contrast to water treatment (retention and purification) by biota, which represents an important regulating service. To exclude also such biotic-based goods such as fossil fuels (oil, coal), we restricted further ecosystem services to those that are provided by animals, plants, and fungi that were formed in the Holocene. In Chapter 6, a list of the ecosystem services analyzed in this book is included, with reference to CICES V5.1 (Table 6.1).

---

[1] https://cices.eu.

CHAPTER 5

Spatial reference units

Contents

Natural space, being the overall ecosystem services (ES) provider, is a complex system of high spatial and temporal diversity. In assessing its diversity in ES potential, that is, the capacity to provide ES, most authors focus on spatial dimension (although temporal aspect has not been completely ignored-see, e.g., Mononen et al., 2016).

Two closely related functions of space division can be identified. The first relates to the very nature of the service. It is possible, therefore, to find links between provision of particular services and specific spatial units. For instance, plant biomass or carbon sequestration can be clearly associated with homogeneous ecological systems (ecosystems in a narrow sense, or the so-called ecosystems, distinguished for instance in the Mapping and Assessment of Ecosystems and their Services [MAES] classification). On the other hand, biomass from wild animals, nursery habitat maintenance and numerous services from the cultural ES section are provided by heterogeneous spatial units, covering a number of different ecosystems and forming complex landscapes, which can be identified and delimited in various ways. Moreover, the provision of at least some of the services (e.g., erosion control) depends simultaneously on spatial diversity in several spatial scales, e.g., in the local geotope scale (the area that is part of the ecosystem, homogeneous in soil, lithology, hydrology, topography, plant cover, and land use), or in the broader scales (landform, catchment, and landscape).

The second function of space division is linked with ES assessment. In general, the selection of reference unit depends on several factors, of which-apart from the ES provisioning scale-the most important are the extent of the study area and the desired spatial resolution, assessment goals, recipient of the results, and finally data availability.

Ecosystem Service Potentials and Their Indicators in Postglacial Landscapes
ISBN 978-0-12-816134-0
https://doi.org/10.1016/B978-0-12-816134-0.00005-5

Copyright © 2020 Elsevier Inc.
All rights reserved.

At the European level, the most common source of spatial information used for assessing ES from terrestrial ecosystems is CORINE Land Cover (CLC1990 and its updates: CLC2000, CLC2006, CLC2012) (Kruczkowska et al., 2017; Metzger et al., 2006). The quality of its data varies between countries and should be supplemented by local data at local or regional level (Vihervaara et al., 2010). However, the coarse spatial resolution with the minimum mapping unit of 25 ha and thematic generalizations of the most common CLC source data strongly limit the possibilities of working on a local or regional level (Burkhard et al., 2012, 2009; Vihervaara et al., 2010). The ES provision can be seen on four general levels of ecological systems: (1) global ($>1,000,000$ km^2), (2) biome ($10,000-1,000,000$ km^2), (3) ecosystem ($1-10,000$ km^2), and (4) plot-plant (<1 km^2). For each of these levels, various basic spatial reference units can be proposed that are conveniently grouped into two series, that is, a set of homogenous units (e.g., ecosystems) and of heterogeneous ones (e.g., landscapes) (Hein et al., 2006; Kruczkowska et al., 2017). In our study, four different types of reference unit were applied: ecosystem in the narrow sense, MAES–derived ecosystem, hunting unit, and landscape (Table 5.1).

Table 5.1 Spatial reference units used in ES assessment.

Spatial unit	Description	No. of types	No. of patches in the study area
Ecosystem in the narrow sense	The original author classification of ecosystems homogenous in plant community, tree stand age, and soil properties	42	3146
MAES–derived ecosystem	Typology of ecosystems based on MAES level 2 classification used for mapping and assessment of ecosystems on a European scale (Maes et al., 2013)	7	1153
Hunting unit	Official game management units delineated by Forestry Offices within each Forest District	—	15
Landscape	The original author division of space into heterogeneous units based on the configuration of ecosystems (in the narrow sense), and taking account of the diversity of potential natural vegetation and abiotic conditions	12	91

5.1 Ecosystems in the narrow sense

Ecosystems differ in their ability to provide services to potential recipients. For this reason, it has become a priority to create a classification that includes ecosystem types that best reflect the diversity of nature potential of postglacial landscape to provide services. The extended ecosystem typology applied was derived from ecosystem typology elaborated for the purposes of MAES at the European level (Maes et al., 2013). As the most detailed MAES level 2 typology distinguishes only seven types of terrestrial ecosystem (urban, cropland, grassland, woodland and forest, heathland and shrub, sparsely vegetated land, wetlands) and one type of freshwater ecosystem (rivers and lakes), ecosystem spatial pattern in local and regional assessments would be too homogeneous, and would be insufficient to fully grasp the actual ES diversity. In the proposed author classification, alongside land cover variety, also forest types, their successional stages and soil properties were taken into account, given the way these contribute much to the final diversity where the supply of most ecosystem services is concerned. By doing so we aimed to provide greater accuracy, reliability, and completeness of the map content compared to the more general pan-European classifications. The obtained classification was used, first of all, in expert assessment of ecosystem potentials to provide ES.

As a result, 42 ecosystem types were identified (Table 5.2). The classification of 25 types of forest ecosystems was based on forest habitat types, comprising alder forests, riparian forests, oak-hornbeam forests, pine (-spruce) and mixed pine forests, swamp coniferous forests, and the following five categories of tree stand age: 0−40, 40−60, 60−80, 80−120, and above 120 years. To specify the characteristics of grassland (three types), the following soil categories were distinguished: (1) dry mineral, (2) moist mineral, (3) peat and mud, and of arable fields (three types): (1) poor in nutrients and dry (usually sandy), (2) fertile mesic, and (3) fertile moist. Wetlands were divided into three categories based on peat accumulation and major hydrological characteristics (after Okruszko et al., 2011): (1) reed beds and sedge swamps accumulating peat or alluvium and permanently inundated by shallow water bodies, (2) fens formed in land depressions or river valley bottoms and fed mainly by groundwater, and (3) bogs formed in landscape locations separated from groundwater supply, where rainfall has no drainage network and locally inundates the surface for a long period. The classification of lakes (seven types) was based on the following criteria: area (ha), maximum and mean depth (m), conductivity, Ca content, fishing

Table 5.2 Types of ecosystem (in the narrow sense) used in ES expert assessment and their coverage in the study area.

No.	Ecosystem type	Ecosystem acronym	km²	%
1.	Settlement area	SETTLE	18.72	2.36
2.	Arable field on poor sandy soil	ARB1	77.65	9.80
3.	Arable field on fertile mesic soil	ARB2	27.81	3.51
4.	Arable field on fertile moist soil	ARB3	2.03	0.26
5.	Dry grassland	GRAS1	43.84	5.53
6.	Wet grassland on mineral soil	GRAS2	58.08	7.33
7.	Peat grassland	GRAS3	39.09	4.93
8.	Alder forest <40 years	ALD1	3.95	0.50
9.	Alder forest 40−60 years	ALD2	3.36	0.42
10.	Alder forest 60−80 years	ALD3	2.26	0.29
11.	Alder forest 80−120 years	ALD4	0.29	0.04
12.	Alder forest >120 years	ALD5	0.07	0.01
13.	Riparian forest <40 years	RIP1	0.77	0.10
14.	Riparian forest 40−60 years	RIP2	3.10	0.39
15.	Riparian forest 60−80 years	RIP3	0.15	0.02
16.	Riparian forest 80−120 years	RIP4	0.28	0.04
17.	Riparian forest >120 years	RIP5	0.19	0.02
18.	Oak-hornbeam forest <40 years	OAK1	9.23	1.17
19.	Oak-hornbeam forest 40−60 years	OAK2	18.73	2.36
20.	Oak-hornbeam forest 60−80 years	OAK3	13.43	1.70
21.	Oak-hornbeam forest 80−120 years	OAK4	17.95	2.27
22.	Oak-hornbeam forest >120 years	OAK5	30.83	3.89
23.	Coniferous and mixed forest <40 years	CON1	49.54	6.25
24.	Coniferous and mixed forest 40−60 years	CON2	89.78	11.33
25.	Coniferous and mixed forest 60−80 years	CON3	75.60	9.54
26.	Coniferous and mixed forest 80−120 years	CON4	94.75	11.96
27.	Coniferous and mixed forest >120 years	CON5	29.80	3.76
28.	Swamp coniferous forest <40 years	SWP1	2.88	0.36
29.	Swamp coniferous forest 40−60 years	SWP2	7.62	0.96
30.	Swamp coniferous forest 60−80 years	SWP3	4.87	0.62
31.	Swamp coniferous forest 80−120 years	SWP4	10.93	1.38
32.	Swamp coniferous forest >120 years	SWP5	3.18	0.40
33.	Wetlands-reed beds and sedge swamps	REED1	3.16	0.40
34.	Wetlands-fens	FEN	0.99	0.12
35.	Wetlands-bogs	BOG	0.12	0.01
36.	Large, deep mesotrophic lakes	LAKE1	26.44	3.34
37.	Large, deep eutrophic lakes	LAKE2	6.71	0.85

Table 5.2 Types of ecosystem (in the narrow sense) used in ES expert assessment and their coverage in the study area. (cont'd)

No.	Ecosystem type	Ecosystem acronym	km²	%
38.	Big eutrophic lakes of medium depth	LAKE3	7.80	0.99
39.	Medium-sized eutrophic lakes of medium depth	LAKE4	3.91	0.49
40.	Small, shallow eutrophic lakes	LAKE5	0.80	0.10
41.	Small dystrophic lakes	LAKE6	0.77	0.10
42.	Artificial lakes	LAKE7	0.77	0.10
TOTAL			792.23	100

type, fertility, and stratification (see Table 5.3). Our classification also includes settlement area category, comprising rural built-up areas, home gardens, and orchards. As the research refers to rural/silvo-agricultural landscape, urban ecosystems were not included in our classification (in MAES level 2 typology, they encompass urban, industrial, commercial and transport areas, urban green areas, mines, dumping and construction sites).

Ecosystem mapping was carried out by reference to several cartographic source materials (e.g., forest, soil, and geological maps), verified during field studies (2014—15). The following data sources were used:

- orthophotomaps 1:5000 (Head Office of Geodesy and Cartography);
- Database of Topographic Objects 1:10,000 (Head Office of Geodesy and Cartography);
- Map of Hydrological Division of Poland 1:10,000 (National Water Management Authority);
- Agricultural Map of Soils 1:25,000 (Institute of Soil Science and Plant Cultivation-State Research Institute);
- VMap Level 2 1:50,000 (Head Office of Geodesy and Cartography);
- Detailed Geological Map of Poland 1:50,000 (Polish Geological Institute-National Research Institute);
- Wetlands Database 1:100,000 (Institute of Technology and Life Sciences);
- Digital Forest Map (Forest Data Bank);
- DTED Level 2.

Spatial source data were processed using ArcGIS 10.2.2 software. The map was designed to occupy an intermediate position (in thematic detail

Table 5.3 Classification of lakes in postglacial landscape.

Acronym	LAKE1	LAKE2	LAKE3	LAKE4	LAKE5	LAKE6	LAKE7
Name	Large, deep mesotrophic lakes	Large, deep eutrophic lakes	Big eutrophic lakes of medium depth	Medium-sized eutrophic lakes of medium depth	Small, shallow eutrophic lakes	Small dystrophic lakes	Artificial lakes
Area (ha)	>50	>50	>40	10–45	<10	0.1–12	<50
Maximum depth (m)	>35	>20	10–20	4–24	<5	<10	No data
Mean depth (m)	>10	>7	<7	<7	<4	<4	No data
Conductivity (μS cm^{-1})	>180	>300	>180	>180	>180	<60	No data
pH	>8.2	>7.9	>7.9	>7.9	>7.9	<7	No data
Ca content (mg L^{-1})	>30	>30	>30	>30	>30	<8	No data
Fishery type	Coregonid type	Bream type	Bream type	Bream type and tench-pike type	Tench-pike type and Cruciancarp type	Not classified	Bream type
Trophy type	Mesotrophic	Eutrophic	Eutrophic	Eutrophic	Eutrophic	Dystrophic	No data
Stratification	Stratified	Stratified	Meromictic	Meromictic	Polymictic	Polymictic	No data

and spatial resolution) between maps showing the diversity of ecosystems in plant communities (which requires detailed field data) and maps developed very locally based directly on level 3 or 4 CLC units. The value of 5 ha was assumed to be the smallest mapping unit. This was a trade-off between available source data resolution and the need to depict complex spatial pattern of postglacial landscape. The technical challenge was to determine the common course of ecosystem boundaries, as thematic source maps were not harmonized. In addition, as the maps were produced in different spatial scales (1:10,000–1:100,000), different level of spatial generalization was applied resulting in different spatial precision. The complementary fieldwork carried out revealed that in many cases source maps did not match the actual conditions, thus demanded updates to prepare the final ecosystem map. Topological mismatches were sometimes substantial and concerned radically different ecosystem categories, for example, polygon classified during cartographic works as dry grassland appeared to be sedge swamp in reality. Nonetheless, in our opinion, the final map reflects well the ecosystem diversity in the study area (Fig. 5.1) and served as a basis for obtaining spatial pattern of ecosystem potential to provide services. The detailed ecosystem classification and spatial resolution of the map may be applied to any other site (at least in the European Union), as most of the EU countries have similar set of forest and soil-agricultural data. More detailed information about the performed ecosystem mapping can be found in Kruczkowska et al. (2017).

5.2 MAES-derived ecosystems

In the questionnaire survey aimed at assessing ecosystem ES potentials from the beneficiary perspective, due to the predicted respondent difficulties in distinguishing several dozen classes of ecosystems, a simplified division was introduced, more closely matching the MAES level 2 typology of ecosystems (Maes et al., 2013). Only forests (in line with the high proportion of the study area accounted for) were further divided into the three subtypes: deciduous forests, coniferous forests (both on a mineral soil), and swamp forests (mainly pine or spruce dominated). Urban ecosystems (settlement area) were not included in the survey due to their low representation in the study area and very high internal variability, in line with other two MAES level 2 terrestrial ecosystem classes (sparsely

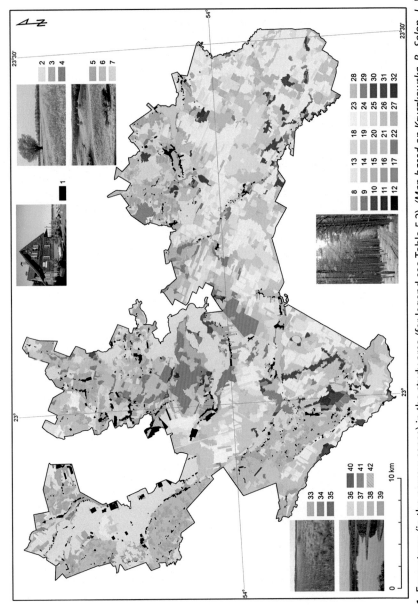

Figure 5.1 Ecosystems (in the narrow sense) in the study area (for legend see Table 5.2). *(Map based on Kruczkowska, B., Solon, J., Wolski, J., 2017. Mapping ecosystem services — a new regional scale approach. Geographia Polonica 90 (4), 503–520.)*

vegetated land and heathland and shrub), which does not occur at all in the analyzed area. In total, seven ecosystem types were distinguished (Table 5.4). Ecosystem spatial pattern presented in Fig. 5.2 was obtained because of the generalization of the detailed map of ecosystems (see Fig. 5.1).

Table 5.4 Types of MAES-derived ecosystem evaluated by direct beneficiaries, and their coverage in the study area.

No.	Ecosystem type	km^2	%
1.	Cropland	107.49	13.57
2.	Grassland	141.01	17.79
3.	Deciduous forest	104.59	13.22
4.	Coniferous forest	339.47	42.84
5.	Swamp forest	29.48	3.72
6.	Wetlands	4.27	0.53
7.	Rivers and lakes	47.2	5.97
TOTAL		773.51	97.64

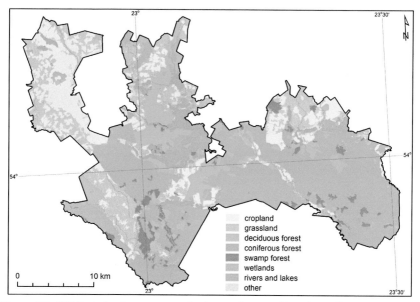

Figure 5.2 MAES-derived ecosystems in the study area. *(Based on Affek, A., Kowalska, A., 2017. Ecosystem potentials to provide services in the view of direct users. Ecosystem Services 26, 183—196.)*

5.3 Hunting units

According to Polish Hunting Law of 1995 (2015), hunting unit is an administrative unit with an area not less than 3000 ha, created for game management. Hunting units are established to meet optimally the needs for protection, preservation and development of preferred game species. Their boundaries should be easily detectable in the area, and if possible delimited without splitting water bodies. Delimitation and later modification of hunting unit borders are the responsibility of Voivodeship authorities (Okarma and Tomek, 2008).

Hunting units were used in expert assessment of ES potentials. Their spatial distribution in the study area is shown in Fig. 5.3. Some hunting units only slightly overlap with the study area. Therefore, the actual boundaries of neighboring hunting units have been modified to enhance

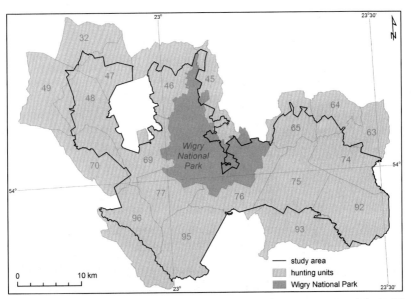

Figure 5.3 Hunting units covering the study area, and the Wigry National Park. *(Map based on Solon, J., Roo-Zielińska, E., Affek, A., Kowalska, A., Kruczkowska, B., Wolski, J., Degórski, M., Grabińska, B., Kołaczkowska, E., Regulska, E., Zawiska, I., 2017. Świadczenia ekosystemowe w krajobrazie młodoglacjalnym. Ocena potencjału i wykorzystania. Instytut Geografii i Przestrzennego Zagospodarowania PAN, Wydawnictwo Akademickie SEDNO, Warszawa (in Polish).)*

the cartographic presentation of the results. This can be justified because the included fragments do not differ significantly in land cover and game resources. Finally, 14 hunting units were surveyed within the study area (Table 5.5). The research also includes (as an additional unit) the Wigry National Park (WNP), as hunting units are not created within national parks, but the inventory of game species is carried out there with the same rules.

Hunting units can be divided into forest hunting units (in which forests constitute at least 40% of the total area) and open-land hunting units (with less than 40% forest cover). According to such division, eight forest and six open-land hunting units are in the study area (Table 5.5).

However, ecological diversity of these units shows that the division into three groups appears to be more appropriate. The first group, where over 80% of the area is occupied by arable fields and grassland, comprises five hunting units (47, 48, 64, 65 and 70). Units 46, 69, 96, and WNP have a more balanced structure, as forests cover there between 30% and 62%. The third group (units 74, 75, 76, 77, 92, and 95) is formed by highly forested hunting units (over 70% forest cover).

5.4 Landscapes

According to the European Landscape Convention (2000), the term "landscape" means "an area, as perceived by people, whose character is the result of the action and interaction of natural and/or human factors". A detailed analysis of this definition indicates that-despite its apparent simplicity-it integrates different, objectively existing and subjectively interpreted aspects of the landscape, including biotic and abiotic environmental components and human made objects, along with information on historical, utilitarian, cultural, and aesthetic values (resources).

In the most comprehensive approach, landscape can be described, classified, and evaluated as follows: (a) a set of physical objects, (b) a system of related processes, (c) a set of stimuli affecting all the senses, (d) a set of values, and (e) a system providing real and potential services (Richling and Solon, 2011).

In our research, landscape units were used in expert assessment of ES potentials. The methodology used to distinguish landscapes was initially

Table 5.5 Land cover in hunting units.

Official hunting unit number	Forest (F)/ Open-land (O) hunting unit	Land cover (%)								
		Alder forest	Riparian forest	Oak-hornbeam forest	Coniferous and mixed forest	Swamp coniferous forest	Grassland	Cropland	Wetlands	Lakes
46	O	3.6	1.8	9.1	14.9	0.6	40.5	22.7	1.5	5.3
47	O	0.8	3.6	0.6	5.3	0.0	34.6	50.9	0.0	4.2
48	O	0.4	0.7	2.3	3.0	0.0	25.3	66.3	0.2	1.7
64	O	0.0	1.7	12.7	1.0	0.0	25.8	58.4	0.4	0.0
65	O	1.4	0.0	2.3	1.2	10.6	50.2	30.6	0.6	3.1
69	F	0.3	0.1	6.7	51.8	0.0	18.8	19.9	0.2	2.2
70	O	0.1	0.3	3.5	0.6	0.0	54.4	40.7	0.4	0.0
74	F	0.4	0.6	10.1	65.4	2.6	6.4	3.9	2.4	8.1
75	F	0.7	0.6	8.4	69.1	2.5	7.2	8.6	0.6	2.2
76	F	0.1	0.7	12.1	60.6	0.8	8.9	15.0	0.1	1.6
77	F	0.0	0.1	4.9	57.1	9.1	19.5	8.8	0.5	0.0
92	F	1.1	0.3	4.2	83.2	7.0	1.1	1.2	0.7	1.2
95	F	2.5	0.0	6.5	58.5	15.6	8.2	1.3	0.9	6.1
96	F	3.5	0.1	11.3	33.0	2.1	33.5	15.7	0.2	0.6
Wigry NP	–	3.6	1.0	35.4	19.9	2.7	10.3	5.6	3.6	17.8

applied to delineate vegetation microlandscapes (Solon, 1993, 1983). Spatial units determined this way encompass a piece of land differing from the surrounding area in the following:

(a) physiognomy-resulting from the type of vegetation;

(b) qualitative and quantitative relations among existing plant communities;

(c) level of fragmentation;

(d) spatial pattern of plant communities.

In practice, the delimitation of landscapes was based on the following criteria[1]:

(a) the existence of spatial relationships between ecosystem types, detected on the basis of neighborhood analysis;

(b) shape diversity of ecosystem patches (e.g., linear, point, large area);

(c) affiliation of ecosystems to specific dynamic circles of vegetation;

(d) the existence of vicarious plant communities occurring in similar ecosystem types within the same land use.

It was done using the detailed map of ecosystems intersected with the map of potential natural vegetation.[2] In total, 91 individual landscape units were delineated in the analyzed area (Fig. 5.4). They were clustered in the share of particular ecosystem categories and grouped into 12 types (Fig. 5.5, Table 5.6). Four types (A1—A4) cover agricultural landscapes, differing in the role of settlement areas and the proportions between arable fields and grassland. The next five types (B1-B5) are highly forested landscapes, differing in the dominance of specific forest communities, while the last three types (C1—C3) are associated with lakes and adjacent wetlands. However, the post hoc landscape typology was not used in assessing landscape potentials to provide ES. Each individual patch was evaluated separately.

The distinguished landscape units are clearly different in their size (Table 5.7). Most of the small or very small landscapes of various types are grouped in the northern part within the WNP. In contrast, two largest

---

[1] The similar procedure for delimiting landscapes had already been carried out in Wigry National Park and its buffer zone (see Richling et al., 2001; Sikorski et al., 2013a,b).

[2] The map of the potential natural vegetation was compiled specifically for this project. It is based on the digital version of 1:300,000 general map (Matuszkiewicz, 2008), and was downscaled and updated using data from the Wigry NP Directorate and Forest District Offices.

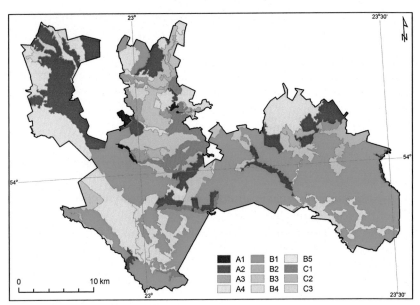

Figure 5.4 Landscape units in the study area (for legend see Table 5.6). *(Map based on Solon, J., Roo-Zielińska, E., Affek, A., Kowalska, A., Kruczkowska, B., Wolski, J., Degórski, M., Grabińska, B., Kołaczkowska, E., Regulska, E., Zawiska, I., 2017. Świadczenia ekosystemowe w krajobrazie młodoglacjalnym. Ocena potencjału i wykorzystania. Instytut Geografii i Przestrzennego Zagospodarowania PAN, Wydawnictwo Akademickie SEDNO, Warszawa (in Polish).)*

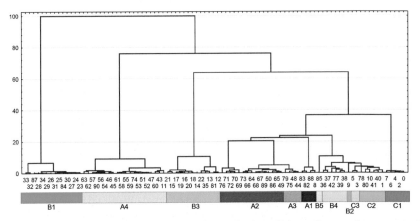

Figure 5.5 The cluster dendrogram (Ward's method, Euclidean distance-in percent of max distance) for landscape units in relation to the share of different ecosystem categories and distinguished landscape types.

Table 5.6 Codes and short description of landscape types.

Code	Short description
A1	Settlement
A2	Arable field-grassland-settlement
A3	Grassland-settlement (grassland less than 50% and settlement more than 30%)
A4	Grassland-settlement (grassland more than 50% and settlement less than 10%)
B1	Coniferous and mixed forests
B2	Alder forests
B3	Domination of oak-hornbeam forests
B4	Swamp coniferous forests
B5	Riparian forests
C1	With a very large share of lakes
C2	Lakes and swamp coniferous forest
C3	Lakes and grassland

Table 5.7 Area variability of landscape units.

Area (ha)	No. of landscape units
<100	14
100−200	18
200−400	18
400−1000	21
1000−5000	18
>5000	2

landscape units, covering respectively 69 and 108 km^2, represent type B1 (landscapes with the dominance of coniferous and mixed forests) and occur in the southeastern part of the analyzed area.

CHAPTER 6

Potentials to provide ecosystem services - analytical approach

Contents

This chapter is the main analytical part of the book and forms the basis for the following synthetic chapters. Here, we present the entire path of indicator development for the evaluation of ecosystem service (ES) potentials in the postglacial landscape, starting from the presentation of theoretical foundations, through the applied methodological solutions and ending with the final assessment of ES potentials. Altogether, 29 ES were assessed by means of 35 indicators (Table 6.1). Our original intention was to develop a

Ecosystem Service Potentials and Their Indicators in Postglacial Landscapes
ISBN 978-0-12-816134-0
https://doi.org/10.1016/B978-0-12-816134-0.00006-7
Copyright © 2020 Elsevier Inc.
All rights reserved.

list of analyzed ES using Common International Classification of Ecosystem Services (CICES) V4.3 classification and the theoretical framework presented in the associated technical guidance (Haines-Young and Potschin, 2013), but as the updated CICES V5.1 has been released in the meantime, we decided to review our services and reclassify them according to the new proposal (Haines-Young and Potschin, 2018). We adopted the approach, in which the main emphasis is on final ES to be clearly separated from the resulting benefits for humans. Less clear distinction has been thus achieved between ecosystem function and ES, particularly in the case of regulating ES. In some cases, the function is basically the same as the service.

In this chapter, services are arranged in the first place according to the type of assessment (expert/social). Further ordering of services follows CICES V5.1 classification and coding (Table 6.1). In expert assessment conducted by a team of scientists specialized in various elements of the environment, only provisioning and regulating services were taken into account. Cultural services were evaluated only through social assessment due to their high subjectivity.

A separate section is devoted to each service that is subject to expert assessment. In turn, services evaluated through social assessment are described together because all of them were assessed by means of the same questionnaire survey and respondent group.

In principle, potential to provide each service is indicated by one indicator. Because of particular complexity and multidimensional character, only nursery habitat maintenance ES is indicated by six various indicators. In the case of edible animal biomass, two complementary indicators were used – one for terrestrial and one for aquatic ecosystems. Some services and indicators, due to their nature, were assessed only in selected ecosystem types (e.g., edible biomass of cultivated plants only for agroecosystems, tree biomass for nonnutritional purposes only for forest ecosystems, FISH indicator only for aquatic ecosystems). In case of some other services, single ecosystem types were left not assessed, mainly due to lack of data.

All sections devoted to particular services are structured in the same way and comprise the following subsections: ES description, theoretical framework, assessment method, and indicator values. Each first subsection ends with a table presenting ES location in CICES V5.1, indicator metadata, and briefly mentioned data sources. The listed indicator metadata include indicandum (what is indicated), indicator name, acronym, short standardized description, type of logical link with the indicandum (direct/indirect), general characteristics of indicator construction (simple/compound/complex

Table 6.1 Ecosystem services (ES) and indicators of nature potential to provide them.

No.	ECOSYSTEM SERVICE	CICES 5.1 CLASS CODE	INDICATOR	INDICATOR ACRONYM	SPATIAL ASSESSMENT UNIT
EXPERT ASSESSMENT					
Provisioning ES					
1.	Edible biomass of cultivated plants	1.1.1.1	Annual cereal yield equivalent	YIELD	Ecosystem
2.	Tree biomass for non-nutritional purposes	1.1.1.2; 1.1.1.3; 1.1.5.2; 1.1.5.3	Volume of timber resources	TIMBER	
3.	Honey	1.1.3.1	Abundance of raw material for the production of honey	HONEY	
4.	Livestock and their outputs	1.1.3.1; 1.1.3.2; 1.1.3.3	Maximum effective density of livestock units (LSU)	MAXLSU	
5.	Edible wild berries	1.1.5.1	Cover of berry-producing plants	BERRY	
6.	Edible biomass of wild animals	1.1.6.1	Total biomass of large game animals	GAME	Hunting unit
			Annual net productivity of fish of commercial meaning	FISH	Ecosystem
Regulating ES					
7.	Erosion control	2.2.1.1	Vegetation capacity to prevent erosion	EROSION	Ecosystem
8.	Water retention	2.2.1.3	Soil water-holding capacity (field capacity – FC)	SOILH2O	
9.	Pollination	2.2.2.1	Abundance of nesting wild bees	POLLIN	
10.	Nursery habitat maintenance	2.2.2.3	Occurrence of rare and endangered habitats which are considered to be of European interest	NATURA	
			Species richness of herb-layer vascular plants	RICHNESS	
			Species diversity of game mammals	MAMMAL	Hunting unit
			Hazel grouse population density	GROUSE	
			Ecosystem diversity	DIVERSITY	
			Abundance of small-scale habitats	HABITAT	Landscape
11.	Invasive plant species control	2.2.3.1	Biotic resistance to invasive plant species	INVAS	Ecosystem
12.	Pest rodent control	2.2.3.1	Number of small rodents eaten by predators	RODENT	Hunting unit
13.	Soil formation	2.2.4.1	Degree of base saturation	SATURATION	Ecosystem
14.	Organic matter decomposition	2.2.4.2	Ratio of organic carbon to total nitrogen in soil	C/N	
15.	Oxygen emission	2.2.6.1	Annual net production of oxygen	OXYGEN	
16.	Carbon sequestration	2.2.6.1	Organic carbon stock in soil, herb layer and trees	CARBON	
17.	Plant aerosol emission	2.2.6.1	Efficiency of plant aerosol emission	AEROS	
18.	Atmospheric heavy metal accumulation	2.2.6.1	Concentration of heavy metals in soil	METAL	
SOCIAL ASSESSMENT					
Provisioning ES					
19.	Biomass for nutritional purposes	1.1.1.2; 1.1.2.2; 1.1.3.2; 1.1.4.2; 1.1.5.2; 1.1.6.2	Opinion of direct users on ecosystem capacity to provide...	NUTRITION	Ecosystem (7 classes)
20.	Biomass for medicinal purposes	1.1.1.2; 1.1.2.2; 1.1.3.2; 1.1.4.2; 1.1.5.2; 1.1.6.2		MEDICINE	

Table 6.1 Ecosystem services (ES) and indicators of nature potential to provide them.—(cont'd)

21.	Biomass for construction purposes	1.1.1.2; 1.1.2.2; 1.1.3.2; 1.1.4.2; 1.1.5.2; 1.1.6.2		CONSTRUCT	
22.	Biomass for the production of fertilizer and fodder	1.1.1.2; 1.1.2.2; 1.1.3.2; 1.1.4.2; 1.1.5.2; 1.1.6.2		FODDER	
23.	Biomass for ornamental purposes	1.1.1.2; 1.1.2.2; 1.1.3.2; 1.1.4.2; 1.1.5.2; 1.1.6.2		ORNAMENT	
24.	Biomass for energy production	1.1.1.3; 1.1.2.3; 1.1.3.3; 1.1.4.3; 1.1.5.3; 1.1.6.3		ENERGY	
Regulating ES					
25.	Water regulation (retention, purification/detoxification)	2.1.1.2; 2.2.1.3	Opinion of direct users on ecosystem capacity to regulate water	H2OREGUL	Ecosystem (7 classes)
Cultural ES					
26.	...sport and recreation	3.1.1.1; 3.1.1.2		RECREATION	
27.	...scientific investigation and education	3.1.2.1; 3.1.2.2	Opinion of direct users on ecosystem usefulness for...	EDUCATION	Ecosystem (7 classes)
28.	...creative work	3.1.2.4		CREATION	
29.	...spiritual experience	3.2.1.2		SPIRIT	

(Rows 26–29 grouped under: Characteristics of living systems that enable/support...)

and estimated/calculated), scale of measurement (ratio/rank), unit of measurement, spatial assessment unit (ecosystem in the narrow sense/MAES-derived ecosystem/hunting unit/landscape), the acquired value range in the study area, and value interpretation in terms of ES potential. ES potential was assessed on the 1—5 rank scale for all indicators and services, where 1 means very low potential, 2 - low potential, 3 - medium potential, 4 - high potential, and 5 - very high potential. In some cases, also 0 value was assigned, meaning no relevant potential. The 1—5 scale of potential was adjusted (stretched) to value ranges obtained in the study area.

As shown above, all of the proposed indicators were characterized on several independent dimensions. Three of the used dimensions need further clarification, as they are the author original proposals. The first one describes the logical link (function) between indicandum (ES potential) and indicator. If there is known causal relationship between the phenomenon measured by the indicator and a given ES potential, and other possible variables have negligible effect on the indicated potential, then we assume the indicator is DIRECT (in that case, the indicator usually has the same unit of measurement as the indicandum). In turn, we assume the indicator is INDIRECT when it is either partial indicator (there is causal indicator—indicandum relationship but also other variables strongly influence the indicandum) or proxy (surrogate) indicator - indicator and indicandum are correlated, but no cause-and-effect relationship is known. The second and the third differentiating dimensions refer to indicator construction. The second one shows the level of indicator complexity in terms of the variables and parameters considered. If the indicator is constructed using only one variable, it is SIMPLE; if constructed using a set of additive variables from the same domain (characterized by the same unit of measurement), it is COMPOUND; and if constructed using a set of nonadditive variables from different domains, it is COMPLEX. The third dimension shows whether the final indicator values are built on general impression based on individual/group experience (ESTIMATED) or direct measurements and/or data/theory-driven models (CALCULATED).

Each section ends with a map showing ES potential in the study area. For all services evaluated in the ecosystem scale (both in the narrow sense and MAES-derived), indicator values for each ecosystem type are listed in assessment tables. Additionally, all services assessed in the most detailed spatial scale (ecosystems in the narrow sense) have their spatial pattern described in the parametrized way by means of landscape metrics (for details, see Chapter 3).

6.1 Potentials to provide ecosystem service - expert assessment

6.1.1 Provisioning ecosystem services

6.1.1.1 Edible biomass of cultivated plants
Ecosystem service description

This service belongs to *Provisioning* section and *Biomass* division in CICES V5.1 (Table 6.2). It can be further described as the ecological contribution to the growth of cultivated, land-based crops that can be harvested and used as raw material for the production of food. The equivalent of annual cereal yield has been proposed to indicate the potential of agroecosystems to provide it.

Theoretical framework

Agricultural ecosystems both provide and receive many ES (Swinton et al., 2007; Zhang et al., 2007), but they are primarily managed to cultivate

Table 6.2 Ecosystem service in Common International Classification of Ecosystem Services (CICES) V5.1 and indicator metadata.

Ecosystem service		Edible biomass of cultivated plants
CICES V5.1	Section	Provisioning
	Division	Biomass
	Group(s)	Cultivated terrestrial plants for nutrition, materials, or energy
	Class(es)	Cultivated terrestrial plants grown for nutritional purposes
	Class code(s)	1.1.1.1
Indicandum		**Potential of arable fields to produce edible plant biomass**
Indicator		**Annual cereal yield equivalent**
Acronym		**YIELD**
Short description		Calculation of annual cereal yield equivalent using regression model linking the indicator of agricultural production spatial valorization (IAPSV) with cereal yields; IAPSV takes account of soil quality, climate, topography, and water conditions; applicable only to agroecosystems
Direct/indirect		Direct
Simple/compound/complex		Complex
Estimated/calculated		Calculated
Scale		Ratio
Value range		13.5–45.7
Unit		$dt\ ha^{-1}\ year^{-1}$
Spatial unit		Ecosystem
Value interpretation		<15.0 → very low potential; 15.1–20.0 → low; 20.1–30.0 → medium; 30.1–40.0 → high; >40 → very high
Source data		— Soil map (Institute of Soil Science and Plant Cultivation in Puławy) — Slope (relief) map based on digital elevation model (DEM) with a resolution of 33 m (General Staff of the Polish Army) — Map of groundwater conditions (Gawrysiak et al., 2004) — Statistical information on annual average yields (2010–14) in Podlaskie Voivodeship provided by Central Statistical Office (Local Data Bank) — Construction of the IAPSV indicator (Witek, 1994, 1981; Witek and Górski, 1977) — Regression equation (Filipiak, 2003)

edible plants used for human or animal nutrition. This basic provisioning service is usually quantified using harvested crops or fodder per area unit (dt ha^{-1}) (Burkhard et al., 2014; de Groot et al., 2010a; Haines-Young and Potschin, 2013; Koschke et al., 2013). However, other indicators such as area under crop cultivation (ha) (Mononen et al., 2016), net primary production (Kandziora et al., 2013), or protein content (Albizua et al., 2015) have also been used. Food production sourced from cultivated plants is commonly mapped across large areas using coarse-resolution land-use data in combination with agricultural statistics (Crossman et al., 2013). In most countries, respective statistics exist, and related scientific disciplines produce further knowledge and data. However, a small number of examples exist, where detailed commodity mapping has been completed (Bryan et al., 2009) by linking agricultural simulation process models to land use, soil, and climate variables. Data availability is still the most important limiting factor in quantification and mapping of this ES that acts on a rather local scale, with annual variations, e.g., due to crop rotation in agricultural areas (Kandziora et al., 2013). Therefore, the assessment of the potential supply of edible biomass of cultivated plants (the hypothetical maximum yield, Burkhard et al., 2012) is more adequate and viable than the actual supply capacity for this ES. ES potential depends strongly on ecosystem properties. Their analysis, predominantly driven by natural scientific methods, using analytical indicators, represents the starting point of the assessment (Bastian et al., 2013).

In this research, the equivalent of annual cereal yield (YIELD) provided by agroecosystems was estimated employing the indirect and complex indicator of the potential supply (Table 6.2). For the construction of the YIELD indicator, regression relationship between the indicator of agricultural production spatial valorization (IAPSV) (Witek, 1981, 1994) and cereal yielding was used. IAPSV has been widely applied in science and practice, for instance, to determine Less-Favored Areas in Agriculture (Czapiewski et al., 2008). It is based on the assessment of soil quality, climate, relief, and water conditions, which have the greatest influence on yields on the local scale (Witek and Górski, 1977). A similar approach has been applied in the assessment and mapping of production of agricultural plants in Germany (Rabe et al., 2016). The natural yield capability was quantified using data on soil type, soil condition, hydrological conditions, climate, yield period, relative humidity, and topography.

Assessment method

To calculate the IAPSV, data from soil maps at the scale 1:25,000, slope maps based on digital elevation model (DEM), maps of soil water retention, and statistical data on cereal yields in Podlaskie Voivodeship from the Local Data Bank of Polish Central Statistical Office were used. Each natural factor (soil quality, climate, relief, and water conditions) was defined using adequate parameter and was weighted reflecting its influence on agroecosystem productivity.

Soil quality was determined using complexes of agricultural suitability (basic units of soil maps), which cluster soil types with similar agricultural, physical, and chemical characteristics (soil texture, pH, buffer and retention abilities, etc.) and reflect their natural fertility and suitability for various crops. Ranking values (18—94) were taken from literature (Witek and Górski, 1977). They were obtained in the long-term field experiments on yields conducted throughout Poland (Table 6.3) and verified in present conditions using normalized difference vegetation index (NDVI) data (MRiRW, 2007). Very good wheat complex, for instance, includes the best soils in the country - rich in nutrients and organic matter, with regulated pH and water conditions, big retention abilities, and a thick humus horizon. In turn, very poor rye complex includes the worst soils formed from sands, poor in nutrients, and usually permanently too dry; even mineral fertilization does not amend their properties.

Agroclimate parameter corresponds with the so-called conversion yield obtained as multiyear average of uniform structure crops. It was based on measurements of weather parameters (temperature, precipitation, and insolation) conducted in 60 meteorological stations in Poland for 50 years. The possible values are 1—15 (Table 6.4). Conversion yield (P_p) for Podlaskie Voivodeship was calculated as

$$P_p = \frac{Z_{zb} + \dfrac{Z_{zi}}{7} + \dfrac{Z_{bc}}{12}}{P_{zb} + P_{zi} + P_{bc}},$$

where Z is the yields in dt, P is the cropland area in ha, zb is the cereals, zi is the potatoes, and bc is the sugar beets.

Table 6.3 Soil quality parameter for Poland.

Soil complexes of agricultural suitability		Parameter value
1	Very good wheat	94
2	Good wheat	80
3	Poor wheat	61
4	Very good rye	70
5	Good rye	52
6	Poor rye	30
7	Very poor rye	18
8	Strong grain and fodder	64
9	Poor grain and fodder	33
1z	Good and very good grassland	80
2z	Medium grassland	50
3z	Poor and very poor grassland	20

Based on Witek, T., Górski, T., 1977. Przyrodnicza bonitacja rolniczej przestrzeni produkcyjnej w Polsce. Wydawnictwa Geologiczne, Warszawa (in Polish).

Table 6.4 Agroclimate parameter values for Poland.

Parameter value	1	3	5	7	9	11	13	15
Cereal yields (dt ha^{-1})	28	29	30	31	32	33	34	35

Based on Witek, T., Górski, T., 1977. Przyrodnicza bonitacja rolniczej przestrzeni produkcyjnej w Polsce. Wydawnictwa Geologiczne, Warszawa (in Polish).

The average yield and cropland area in Podlaskie Voivodeship from years 2010−14[1] was taken into account in the calculations. The calculated conversion yield is 29.41 dt ha^{-1}, and it corresponds to the agroclimate parameter value 3.

Relief assessment was based on criteria concerning height above sea level, slope angle, and landform with distinction for eight types: plain, gently hilly, hilly, folded, moderately sloped, submountain, mountain, and high-mountain. The eight-degree scale was adopted (Table 6.5). The slope map based on DEM was used (the average value of the slope was assumed for each spatial unit).

Groundwater conditions were determined on the scale from 0.5 to 5 points using five soil-moisture categories, reflecting soil retention abilities and groundwater level. Data originated from the map of groundwater conditions developed in the Institute of Soil Science and Plant Cultivation,

[1] Data from the Local Data Bank of the Poland's Central Statistical Office - https://bdl. stat.gov.pl/BDL/start.

Table 6.5 Relief parameter values.

Landform	P	GH	H	F	MS	SM	M	HM
Denivelation (m)	0—3	3—7	7—20	20—40	40—75	75—200	200—400	>400
Slope angle (degree)	0—1	1—5	5—8	8—15	15—20	20—30	>30	>30
Parameter value	5	4	3.5	2.5	1	0.5	0.25	0

F, Folded; *GH*, Gently hilly; *H*, Hilly; *HM*, High-mountain; *M*, Mountain; *MS*, Moderately sloped; *P*, Plain; *SM*, Submountain.

Puławy (Gawrysiak et al., 2004). The water resources potentially available for plants of the 90—300 mm range were divided into five abovementioned categories (Table 6.6).

To calculate synthetic IAPSV, all component parameters (soil quality, agroclimate, relief, and soil water conditions) were summed up. In theory, IAPSV indicator can take 19.5—119 points (which means <30 very poor, 30—59 poor, 60—79 good, >80 very good arable fields). In the research, it was calculated for all spatial units distinguished using borders of complexes of agricultural suitability, slope angle, and soil-moisture categories (the value of the agroclimate parameter was adopted equal for the entire study area), whose boundaries overlap with agroecosystems verified with current orthophotomaps.

Taking advantage of the high correlation between the IAPSV and cereal yields (the IAPSV is known to explain the 87% variability of yields in some regions of Poland), the regression equation, $\text{YIELD} = -0.339 + 0.541\ \text{IAPSV}$, developed for regions with yields below the national average (e.g., Podlaskie Voivodeship) was applied (Filipiak, 2003). The YIELD indicator was calculated for all distinguished spatial units. Finally, after upscaling those

Table 6.6 Soil water conditions parameter.

Soil moisture categories	Parameter value
Permanently waterlogged (262—300 mm)	1.25
Periodically waterlogged (219—261 mm)	3.00
Optimally hydrated (176—218 mm)	5.00
Periodically too dry (133—175 mm)	2.00
Permanently too dry (90—132 mm)	0.50

units to agroecosystems, area weighted average YIELD indicator was calculated for each agroecosystem patch (Table 6.7).

Indicator values

In the study area, IAPSV takes values from 25 to 93. Over 55% of cropland belongs to very poor arable field category, related to soils formed from sands, poor in nutrients, and usually permanently too dry. Only 26% of cropland constitutes good and very good agroecosystems, related to fertile

Table 6.7 Potential of arable fields to produce edible plant biomass (1-very low; 5 - very high) indicated by annual cereal yield equivalent (dt ha^{-1} year^{-1}).

Ecosystem type	Acronym	YIELD indicator value	Ecosystem potential
Arable field on poor sandy soil	ARB1	13.5—39.6 (mean 17.7)	2
Arable field on fertile mesic soil	ARB2	13.5—45.7 (mean 37.7)	4
Arable field on fertile moist soil	ARB3	13.5—31.5 (mean 20.8)	3

soils with regulated water conditions and big retention abilities. Cereal yield equivalent (YIELD indicator) in the studied agroecosystems amounts to 13.5—45.7 dt ha^{-1} year^{-1} (Fig. 6.1). The highest values were recorded in agroecosystems with fertile mesic soil ARB2 (mean 37.7 dt ha^{-1} year^{-1}, SD 4.7), much lower in agroecosystems with fertile moist soil ARB3 (mean 20.8 dt ha^{-1} year^{-1}, SD 6.6), and poor sandy soil ARB1 (mean 17.7 dt ha^{-1} year^{-1}, SD 5.1).

The observed potential average yield from the studied agroecosystems (26.2 dt ha^{-1} year^{-1}, SD 10.9) was slightly lower than the mean actual value in Podlaskie Voivodeship (29.5 dt ha^{-1} year^{-1}, 2010—14). This difference (approximately 11%) may result from the fact that the calculations of potential yield comprise only natural conditions, and any improving agrotechnical measure such as fertilization, plant protection, or irrigation was not included.

6.1.1.2 Tree biomass for nonnutritional purposes
Ecosystem service description

This service belongs to *Provisioning* section and *Biomass* division in CICES V5.1 (Table 6.8). It can be further described as the ecological contribution

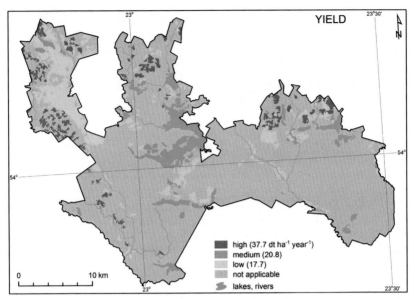

Figure 6.1 Potential of arable fields to produce edible plant biomass indicated by annual cereal yield equivalent (YIELD indicator; dt ha^{-1} year^{-1}).

to the production of standing tree biomass that can be harvested and used as raw material for nonnutritional purposes (e.g., construction, energy). The volume of timber resources has been proposed to indicate the potential of forest ecosystems to provide it.

Theoretical framework

Volume of timber resources was considered as an indicator of ecosystem potential adopting the definition of potential according to which "[…] ecosystems provide and ensure the necessary ecological structure and processes, which in turn determine the resource or potential to supply services regardless of whether the human being is currently using them" (Maes et al., 2012; Tallis et al., 2012).

Timber as raw material is used, amongst other, for the construction of buildings, roofing, as well as for the production of furniture, agricultural tools, paper, fabrics, ropes, etc. It should be emphasized that approximately 15% of the world energy consumption is provided by firewood and other plant biomass (in developing countries, almost 40%). The commercial value of wood is therefore huge, which consequently involves running such forest management, the primary objective of which is to increase timber

Table 6.8 Ecosystem service in Common International Classification of Ecosystem Services (CICES) V5.1 and indicator metadata.

Ecosystem service		Tree biomass for nonnutritional purposes
CICES V5.1	Section	Provisioning
	Division	Biomass
	Group(s)	Cultivated terrestrial plants for nutrition, materials, or energy; wild plants for nutrition, materials, or energy
	Class(es)	Fibers and other materials from cultivated plants for direct use or processing; cultivated plants grown as a source of energy; fibers and other materials from wild plants for direct use or processing; wild plants used as a source of energy
	Class code(s)	1.1.1.2; 1.1.1.3; 1.1.5.2; 1.1.5.3
Indicandum		**Harvestable tree biomass for nonnutritional purposes**
Indicator		**Volume of timber resources**
Acronym		**TIMBER**
Short description		Calculation of average volume of timber resources in different forest types and age classes using taxation data on timber resources on a subblock level; applicable only to forest ecosystems
Direct/indirect		Direct
Simple/compound/complex		Simple
Estimated/calculated		Calculated
Scale		Ratio
Value range		92−468
Unit		$m^3 \ ha^{-1}$
Spatial unit		Ecosystem
Value interpretation		$<100 \rightarrow$ very low potential; $100.0-200.0 \rightarrow$ low; $200.1-300.0 \rightarrow$ medium; $300.1-400.0 \rightarrow$ high; $>400 \rightarrow$ very high
Source data		— State Forests Information System (SILP); Forest Digital Map (LMN) of the four forest districts — Protection Plan of the Wigry National Park[a]

[a] Unpublished Protection Plan of the Wigry National Park (in Polish), developed in 2014, is available at the Directorate of WNP (Krzywe 82, 16−402 Suwałki).

production for economic benefits, and this in turn results in intervention in natural processes occurring in forest ecosystems (Duncker et al., 2012).

Assessment method

Volume of timber resources is defined as the volume (m^3) of all standing, living trees with diameter at breast height above 7 cm in a given area (ha). The input for the assessment of the volume of timber resources was the data collected by Forestry Offices in the four forest districts: Suwałki, Szczebra, Pomorze, and Głęboki Bród and in the Wigry National Park (WNP) (Fig. 6.2). Detail taxation data for individual forest subblocks were derived from the State Forests Information System (SILP), the Forest Digital Map (LMN) of the four forest districts, and Protection Plan of the WNP.

The database comprised information on timber volume (m^3 ha^{-1}) for approximately 16,000 forest subblocks. Characteristics of tree stand age and forest habitat type allowed to aggregate data into five age classes (0—40, 40—60, 60—80, 80—120, >120 years) and five types of forest communities corresponding to the applied ecosystem classification: (1) alder forests, (2) riparian forests, (3) oak-hornbeam forests, (4) coniferous and mixed forests, and (5) swamp coniferous forests. On this basis, the average timber volume was calculated separately for each age class and forest ecosystem type in four forest districts and in the WNP. Area-weighted means for the entire study area constitute the final TIMBER indicator values. Because of significant differences in timber volume between forest districts and the WNP we decided to show also the area-weighted means separately for managed and protected forests (Table 6.9).

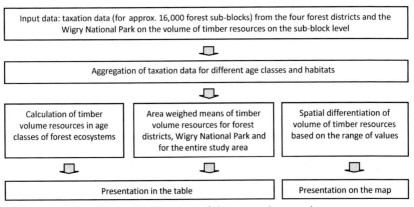

Figure 6.2 Basic stages of the research procedure.

Indicator values

The obtained values of timber volume in each age class of five forest ecosystems indicate that the lowest harvestable tree biomass characterize the youngest forests (<40 years), and it gradually increases with the age of the tree stand (Table 6.9). The maximum potential to supply timber characterizes forest ecosystems, whose tree stand reaches the age of 80−120 years. This trend is observed in all forest ecosystems in four forest districts and partially in the WNP. Forests within WNP have significantly larger timber volume in comparison with the four forest districts, resulting from different forest management. This shows that there are important factors other than habitat type and tree stand age that strongly affect timber volume, particularly the dominating forest function and the resulting forest management.

Area-weighted average timber volume for the entire study shows that the potential of the youngest (<40 years) oak-hornbeam and swamp coniferous forests to provide wood is very low (rank 1), while oak-hornbeam forests and coniferous and mixed forests with older tree stands (>80 years) rank highest in terms of potential wood supply.

The volume of timber resources in forest ecosystems with different tree stand age classes can be referred to the degree of forest maturity and this in turn to the stages of ecological succession of forests. During the succession, the structure changes as well as the species composition of forests. It should be noted that in our research the highest timber volume is observed in stands 80−120 years old in four forest ecosystems, and it is the optimal forest age category to maintain its maturity and therefore diversity. It may be related to the fact that species composition of forests over 120 years old is dominated by a small group of the so-called "ancient forest species" (Dzwonko and Loster, 2001; Hermy et al., 1999), while younger forests (up to 60 years old) are partially occupied by alien species, for example, from meadows, ruderal communities, and arable fields (Roo-Zielińska and Matuszkiewicz, 2016). These results are in line with the findings of Qing-fan and Song (2000), who analyzed the structure of 25 forest communities in two Forest Experimental Stations: Maoershan and Liangshui. These authors showed that the highest degree of diversity was achieved by forests in the middle stage of succession (and not in the oldest and in the youngest).

Actual volume of timber resources (TIMBER indicator) in managed forests is shaped first of all by forest management practices and depends on the applied treatments and cuts (Fig. 6.3). In turn, in protected forests, the

Table 6.9 Potential of forest ecosystems to produce tree biomass for nonnutritional purposes (1 - very low; 5 - very high) indicated by the volume of timber resources (m^3 ha^{-1}).

Ecosystem type	Ecosystem acronym	Area-weighted mean for four forest districts	Area-weighted mean for Wigry National Park	TIMBER indicator value – total area-weighted mean	Ecosystem potential
Alder forest <40 years	ALD1	95	235	166	2
Alder forest 40—60 years	ALD2	213	287	229	3
Alder forest 60—80 years	ALD3	258	305	288	3
Alder forest 80—120 years	ALD4	296	—	296	3
Alder forest >120 years	ALD5	—	364	364	4
Riparian forest <40 years	RIP1	134	—	134	2
Riparian forest 40—60 years	RIP2	235	328	278	3
Riparian forest 60—80 years	RIP3	288	—	288	3
Riparian forest 80—120 years	RIP4	358	381	367	4
Riparian forest >120 years	RIP5	354	330	333	4
Oak-hornbeam forest <40 years	OAK1	77	202	92	1
Oak-hornbeam forest 40—60 years	OAK2	285	378	319	4
Oak-hornbeam forest 60—80 years	OAK3	378	459	425	5
Oak-hornbeam forest 80—120 years	OAK4	434	497	468	5
Oak-hornbeam forest >120 years	OAK5	360	405	400	4
Coniferous and mixed forest <40 years	CON1	117	213	117	2
Coniferous and mixed forest 40—60 years	CON2	298	374	302	4
Coniferous and mixed forest 60—80 years	CON3	375	472	387	4

Coniferous and mixed forest 80–120 years	CON4	434	546	439	5
Coniferous and mixed forest >120 years	CON5	395	519	442	5
Swamp coniferous forest <40 years	SWP1	100	–	100	2
Swamp coniferous forest 40–60 years	SWP2	176	286	185	2
Swamp coniferous forest 60–80 years	SWP3	237	281	241	3
Swamp coniferous forest 80–120 years	SWP4	271	294	272	3
Swamp coniferous forest >120 years	SWP5	268	351	290	3

distribution of potential is determined by habitat conditions and the period for which the given area has already been protected. Besides, natural disturbances such as tornadoes, fires, or pest gradations play a significant role in shaping the potential of all kinds of forests to provide timber. As a result, the spatial pattern of potential is generally of mosaic character. Compared with patterns of other ES potentials, it is characterized by the second lowest number of patches, the highest mean patch size, the lowest largest patch index (LPI) value (the largest patch covers below 4% of the considered area), the smallest difference between area weighted mean shape index (AWMSI) and mean shape index (MSI), the highest MSI (which means that small patches are also irregular in shape but generally similar to bigger ones), and the longest distance to the nearest neighbor. Patches of the same level of potential are usually grouped in clusters. Patches with high and very high potential dominate covering 47% of forested area.

6.1.1.3 Honey
Ecosystem service description
This service belongs to *Provisioning* section and *Biomass* division in CICES V5.1 (Table 6.10). It can be further described as the ecological contribution

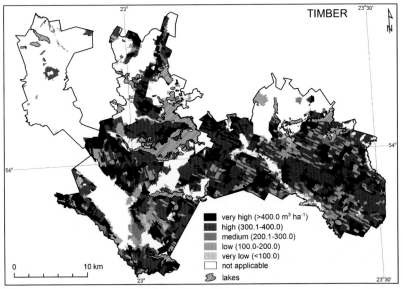

Figure 6.3 Potential of forest ecosystems to produce tree biomass for nonnutritional purposes indicated by the volume of timber resources (TIMBER indicator; m^3 ha^{-1}).

to the rearing of domesticated honeybees and the production of honey. The abundance of raw material for the production of honey (flower nectar, honeydew, indirectly also pollen) has been proposed to indicate the potential of ecosystems to provide it.

Theoretical framework

Apart from pollination services, managed honeybees support human well-being by way of several bee outputs, among which honey is the most important. Honey has a variety of positive nutritional and health effects on humans. As the only readily available natural sweetener, it was an important food for *Homo sapiens* from the very outset (Alvarez-Suarez et al., 2010). Honey mainly comprises carbohydrates, which constitute about 95% of its dry weight. In the long human tradition, it has been used not only as a nutrient but also as a medicine - various positive effects of honey consumption on human physiology and health have been proved (see Alvarez-Suarez et al., 2010 for review). All kinds of honey exhibit strong antibacterial activity. Depending on the honey source, significant antioxidant, antimutagenic, antitumor, and antiinflammatory activity is also manifested, while skin inflammation is reduced, wound healing promoted, scar size diminished, and tissue regeneration stimulated (Alvarez-Suarez et al., 2010; Molan, 2001). The annual world honey production as of 2013 was estimated to be 1.66 million tons (Food and Agriculture Organization of the United Nations, 2016), which is less than 1% of total sugar production (Alvarez-Suarez et al., 2010).

To construct an indicator and then map ecosystem potential for honey production, a more detailed definition of ecosystem potential was needed, so that ranges of values and the level of uncertainty might be reduced. In consequence, the new, more operational definition introduced holds that the potential of an ecosystem to produce honey is a maximum theoretical service supply in a given type of ecosystem and regional context, calculated for the environmental setting best suited for a given service (for example, as regards plant species composition, soil quality, water balance, etc.) (Affek, 2018; see also Section on pollination). This for instance means that in calculating the honey potential of cropland located on fertile soil, a selection needs to be made of a crop (cultivated in the region on such a soil) that has the highest honey potential, which is to say that a hectare of that crop can allow bees to produce a larger amount of honey than a hectare of any other crop. In turn, estimations of the honey potentials of given forest types entail selection of the most desirable plant composition that it is possible to

Table 6.10 Ecosystem service in Common International Classification of Ecosystem Services (CICES) V5.1 and indicator metadata.

Ecosystem service		Honey
CICES V5.1	Section	Provisioning
	Division	Biomass
	Group(s)	Reared animals for nutrition, materials, or energy
	Class(es)	Animals reared for nutritional purposes
	Class code(s)	1.1.3.1
Indicandum		**Ecosystem potential for honey production**
Indicator		**Abundance of raw material for the production of honey**
Acronym		**HONEY**
Short description		Calculation of the abundance of nectar, honeydew, and pollen using literature data on honey plant species combined with phytosociological relevés showing composition of plant communities; shows theoretical maximum annual honey yields per hectare.
Direct/indirect		Direct
Simple/compound/complex		Complex
Estimated/calculated		Calculated
Scale		Ratio
Value range		0–320
Unit		kg ha^{-1} year^{-1}
Spatial unit		Ecosystem
Value interpretation		0 → no relevant potential; 0.1–20.0 → very low potential; 20.1–60.0 → low; 60.1–120.0 → medium; 120.1–240.0 → high; >240.0 → very high
Source data		— Honey potential (nectar and pollen production) of single honey plants and whole plant communities (Demianowicz et al., 1960; Denisow, 2011; Kołtowski, 2006; Maksymiuk, 1960; Szklanowska, 1979, 1973) — Species composition and diversity of plant communities (Matuszkiewicz, 2001; phytosociological relevés, fieldwork) — Honeydew production (Crane and Walker, 1985; Haragsim, 1966; beekeeping literature)

encounter in the given region (within the given ecosystem type) - from the point of view of honey production. Taking environmental settings other than the optimal (for example, the average or most often occurring) would not reflect the full potential of the given ecosystem type. The potential understood in this way corresponds to the maximum possible annual honey yields per hectare of a given ecosystem, assuming that beekeepers were to ensure an optimum abundance of honeybees. It should be noted that, in the estimate made, no account was taken of interrelationships extending beyond a single growing season and capable of affecting long-term average yields (for example, the need for crop rotation or the process of succession in woodland).

Currently, there is practically no possibility of honey being obtained from wild-living European honey bee colonies (Krzysztofiak, 2001). Because of a number of bee diseases introduced from other continents (e.g., varroosis, nosemosis) and a scarcity of adequate natural nesting sites, a bee colony has little chance of overwintering without beekeepers' assistance and care. For these reasons, the contemporary production of honey requires a combination of natural and human capital.

The amount of honey produced within a growing season depends on several environmental factors, among which the most important are the quality of bee pasture and the weather (Crane, 1990). However, weather conditions, though playing a crucial role in determining honey yields (which can vary by 100% in any given region, from one year to another, in line with distinct weather conditions; Gerula et al., 2007; Semkiw and Ochal, 2009), were regarded as beyond the scope of the research, given that they are not ecosystem features. The assessment of ecosystem potential for honey production was thus confined to the quality of bee pasture, i.e., the abundance and accessibility of honey sources (nectar and honeydew) and pollen (necessary nutrition for a bee family to function properly) (Crane, 1990; Westrich, 1996). The amount of nectar and pollen potentially produced by an ecosystem (assuming favorable weather conditions) links up directly with the actual floral resources of a plant community and its melliferous potential. This is in turn determined by plant species composition and the honey potential of each individual species. However, it is not only the abundance but also the accessibility of sources from which bees can make honey that determine yields. Less-than-optimal use of existing resources arises, thanks to factors such as an unfavorable shape of a flower's corolla, limited flexibility of petals (e.g., due to low temperature), and also terrain obstacles hindering flight in bees.

About 16% of the world's flowering plant species are visited by honeybees with a view to nectar being collected (Crane, 1990). Moreover, most world honey is produced from nectar secreted by just 1.6% of those species. The assessment of an ecosystem's potential to produce honey thus depends crucially on the identification of important species and determination of their coverage. The honey potential of individual plant species in certain climatic conditions is usually estimated through direct measurement (Demianowicz et al., 1960).

In determining the potential honey production of an ecosystem, it is not enough to measure and sum up nectar secretion species by species because nectar secretion varies markedly, being dependent on habitat and plant community, not to mention seasonal differences stemming from changing weather conditions. Potential honey production calculated for a given species growing as a planted monoculture may vary considerably compared with the numbers obtained in home plant communities (Szklanowska, 1973). It was demonstrated, for instance, that flowers of bilberry (*Vaccinium myrtillus*), cowberry (*Vaccinium vitis-idaea*), and common cow-wheat (*Melampyrum pratense*) all secreted significantly more nectar (although less dense) in swamp coniferous forest than in coniferous and mixed forest, due to higher relative air humidity (Szklanowska, 1973).

The ecosystem potential to produce honey is not a widely estimated service, though some indirect indicators used to map pollination potential (such as floral resources) could be adapted for that purpose quite readily. Jarić et al. (2013) proposed a formula by which the melliferous potential of plant communities could be estimated, applying this subsequently in assessing selected forest and meadow ecosystems in Serbia (Jarić et al., 2013; Mačukanović-Jocić and Jarić, 2016). However, their calculations of honey potential did not account for honeydew as an underpinning raw material and for the impact of different environmental conditions in determining how varied actual production of nectar and pollen by individuals of the same plant species might be (Demianowicz et al., 1960; Szklanowska, 1973)

Szklanowska (1979, 1973) made multiseasonal direct measurements of total nectar secretion per ha in selected temperate forest communities. Although that author only collected nectar from undergrowth and understory plants (not trees), her valuable work offered a basis for the reliable assessment of floral resources within natural and seminatural ecosystems.

Assessment method

The abundance and availability of raw material for the production of honey (nectar and honeydew and pollen as an energy source) were taken into account to estimate in the indirect way how much honey can theoretically be produced from 1 ha of a given type of ecosystem. The indicator values were assigned to particular ecosystem types as a result of in-depth review of published research (entomological, botanical and ecological) conducted by various authors and own analysis of available vegetation data from the study area.

Potential production of nectar honey by the ecosystem is a resultant of the honey potential of individual melliferous plant species and their share in the plant cover. The honey potential of individual species is obtained from annual nectar secretion and its sugar content, assuming that mature honey is 80% sugar (Demianowicz et al., 1960). In turn, the total sugar production from 1 ha is calculated on the basis of a set of samples from several seasons with parameters such as (1) mean sugar content (the amount of sugar in nectar secreted per day per flower), (2) the average length of the blooming period, and (3) the average number of flowers per unit area.

In our estimates of ecosystem potential to produce honey, account was also taken of the abundance of floral pollen, this being a food for the brood that determines a colony's development and assures its ability to yield honey (Crane, 1990; Warakomska, 1972). Bees collect pollen not only from entomophilous but also from anemophilous species, which otherwise offer little of value for bees (Denisow, 2011; Warakomska, 1972). Pollen production and availability to bees are usually determined using the ether-ethanol method described in detail by Denisow (2011).

This study has made use of the results of extensive multiseasonal research on melliferous plants carried out in Poland since the 1950s, under the direction of Z. Demianowicz and B. Jabłoński. The fruit of this work - also providing some kind of a summary - is a *Great Atlas of Melliferous Plants* published (in Polish) in 2006 (Kołtowski, 2006). This provides information on nectar and pollen production in the cases of more than 250 honey plants growing in the temperate zone (the values for 32 most relevant honey species are also given after Kołtowski in Affek, 2018). Every species included in the Atlas was evaluated on a 1—4 scale in terms of its beekeeping value. The assessment took into account the length of the blooming period, blooming abundance, nectar and pollen production, frequency of insect visitation, and honey potential.

The recognition of key honey species within types of ecosystems was based on phytosociological relevés available for the study area (in which frequency of occurrence of species is established after Braun-Blanquet, 1964), as well as general typologies for plant communities that present species composition and dominance structure for Poland (Matuszkiewicz, 2001) (Table 6.11).

In the presented analysis, we differentiate the potential production of honey due to shading of the forest floor and the quality of the habitat because both factors were proven to affect the abundance and blooming of melliferous plants (Demianowicz et al., 1960; Szklanowska, 1979).

Besides nectar, bees also use honeydew in the production of honey. This sweet, sticky substance is excreted by aphids (plant-sucking insects - Hemiptera) and often deposited on leaves and stems (Crane and Walker, 1985; Haragsim, 1966). It appears irregularly, depending on weather conditions. Honeybees collect it and process it into a dark, strong "honeydew honey". Unique organoleptic properties and certain physico-chemical parameters are characteristic for this kind of honey (Rybak-Chmielewska et al., 2013), which is among the most desired and expensive kinds available on the market (Semkiw, 2015).

Globally, most honeydew flows are from trees in temperate-zone forests (Crane and Walker, 1985). Levels are particularly high in Central and Southeast Europe, where particular economic importance is therefore assumed. However, the intensity of honeydew flows is also known to vary greatly from year to year, and - given that temperature exerts most influence on aphid development and growth - it is this factor that also does much to affect the occurrence of honeydew. Aphid populations tend to peak when temperatures are in the $25-30°C$ range (Rybak-Chmielewska et al., 2013). On the other hand, heavy rains and sudden temperature drops reduce aphid populations markedly. Crane and Walker (1985), basing themselves on work by Haragsim (1966), listed trees constituting important sources of honeydew, along with information as to honey potentials in terms of yield per ha. Norway spruce, Scots pine, and English oak are among the most important sources of honeydew in the postglacial European landscape.

Indicator values

The highest potential to produce honey (approximately 300 kg ha^{-1}) was achieved by arable fields located on fertile soil, assuming that the selected crop would be most suited for honey production (e.g., phacelia - *Phacelia*

Table 6.11 The most important honey plants in different lowland ecosystems of Central Europe.

Ecosystem type	Honey plants
Settlement area	*Cerasus vulgaris* (sour cherry); *Malus domestica* (apple); *Prunus domestica* (wild plum); *Ribes nigrum* (blackcurrant); *Robinia pseudoacacia* (false acacia); *Tilia cordata* (small-leaved lime); *Solidago canadensis* (Canadian goldenrod)
Arable field	*Brassica napus* (rape); *Fagopyrum esculentum* (buckwheat); *Phacelia tanacaetifolia* (phacelia)
Dry grassland	*Centaurea jacea* (brown knapweed); *Echium vulgare* (viper's bugloss); *Knautia arvensis* (field scabious); *Melilotus alba* (white melilot); *Taraxacum officinale* (common dandelion); *Thymus serpyllum* (Breckland garden); *Trifolium pretense* (red clover)
Wet grassland	*Cirsium rivulare* (brook thistle); *Lamium album* (white dead-nettle); *Lychnis flos-cuculi* (Ragged-Robin); *Lythrum salicaria* (purple loosestrife); *Mentha longifolia* (horse mint)
Alder forest	*Salix alba* (white willow)
Riparian forest	*Eupatorium cannabinum* (hemp-agrimony); *Frangula alnus* (alder buckthorn); *Impatiens noli-tangere* (touch-me-not balsam)
Oak-hornbeam forest	*Ajuga reptans* (bugle); *Acer platanoides* (Norway maple); *Allium ursinum* (ramsons); *Corydalis solida* (bird-in-a-bush); *Galeobdolon luteum* (yellow archangel); *Lathyrus vernus* (spring vetchling); *Pulmonaria obscura* (Suffolk lungwort); *Tilia cordata* (small-leaved lime); *Viola sylvestris* (violet forest)
Coniferous and mixed forest	*Calluna vulgaris* (heather); *Cytisus nigricans* (black-rooted broom "Cyni"); *F. alnus* (alder buckthorn); *Melampyrum pratense* (common cow-wheat); *Rubus idaeus* (raspberry); *Sorbus aucuparia* (rowan); *Vaccinium myrtillus* (bilberry); *Vaccinium vitis-idaea* (cowberry)
Swamp coniferous forest	*C. vulgaris* (heather); *F. alnus* (alder buckthorn); *Ledum palustre* (Labrador tea); *Melampyrum pretense* (common cow-wheat);

Continued

Table 6.11 The most important honey plants in different lowland ecosystems of Central Europe. (cont'd)

Ecosystem type	Honey plants
Wetlands	*V. myrtillus* (bilberry); *Vaccinium uliginosum* (bog bilberry); *Vaccinium vitis-idaea* (cowberry) *Ledum palustre* (Labrador-tea); *Salix caprea* (goat willow); *Salix pentandra* (bay willow); *Vaccinium oxycoccos* (cranberry); *V. uliginosum* (bog bilberry)

Based on Maksymiuk, I., 1960. Nektarowanie lipy drobnolistnej *Tilia Cordata* Mill. w Rezerwacie Obrożyska koło Muszyny. Pszczelnicze Zeszyty Naukowe 4 (2), 105—125 (in Polish); Matuszkiewicz, W., 2001. Przewodnik do oznaczania zbiorowisk roślinnych Polski. Vademecum Geobotanicum, vol. 3, Państwowe Wydawnictwo Naukowe, Warszawa (in Polish); Szklanowska, K., 1979. Nektarowanie i wydajność miodowa ważniejszych roślin runa lasu liściastego. Pszczelnicze Zeszyty Naukowe 23, 123—130 (in Polish); Szklanowska, K., 1973. Bory jako baza pożytkowa pszczół. Pszczelnicze Zeszyty Naukowe 17, 51—85 (in Polish); phytosociological relevés from the study area.

tanacetifolia or buckwheat – *Fagopyrum esculentum*) (Table 6.12). Dry grassland along with young swamp pine forests also has one of the highest capacities to provide honey (up to $250 \, \text{kg ha}^{-1}$). The lowest capacity $(0–20 \, \text{kg ha}^{-1})$ was assigned to all alder forests, middle-aged riparian forests, and most of wetlands, as – apart from waterbodies and other non-vegetated areas – these also support the poorest bee pasture. The potential of forest ecosystems to produce honey is differentiated in relation to forest age in such a way that the youngest and oldest tree stands have higher values assigned to them than do middle-aged and mature forests. It was further assumed that, in forest communities with trees more than 120 years old, natural disturbances take place more often and result in a major improvement of light conditions on the forest floor.

A more detailed diversification of the potential of ecosystems for honey production, based on an eleven-point scale, along with maps and a wide discussion of results, can be found in the article recently published in *Ecological Indicators* (Affek, 2018).

Caution should be used in interpreting the obtained indicator values showing theoretical maximum annual honey yields per ha. The primary application here was in a ranking of ecosystems in relation to their capacity to provide the service, rather than in any provisioning of ultimate, precise figures. Differentiation of values in line with forest age sought to reflect a general regularity linking ecosystem potential with canopy closure and cover of melliferous plant species. Similar justifications led to a drawing of

Table 6.12 Ecosystem potential to produce honey (1 - very low; 5 - very high) indicated by the abundance of raw material for honey production (kg ha^{-1} year^{-1}).

Ecosystem type	Ecosystem acronym	HONEY indicator value	Ecosystem potential
Settlement area	SETTLE	60.1—120.0	3
Arable field on poor sandy soil	ARB1	120.1—240.0	4
Arable field on fertile mesic soil	ARB2	>240.0	5
Arable field on fertile moist soil	ARB3	120.1—240.0	4
Dry grassland	GRAS1	>240.0	5
Wet grassland on mineral soil	GRAS2	60.1—120.0	3
Peat grassland	GRAS3	60.1—120.0	3
Alder forest <40 years	ALD1	0.1—20.0	1
Alder forest 40—60 years	ALD2	0.1—20.0	1
Alder forest 60—80 years	ALD3	0.1—20.0	1
Alder forest 80—120 years	ALD4	0.1—20.0	1
Alder forest >120 years	ALD5	0.1—20.0	1
Riparian forest <40 years	RIP1	20.1—60.0	2
Riparian forest 40—60 years	RIP2	0.1—20.0	1
Riparian forest 60—80 years	RIP3	0.1—20.0	1
Riparian forest 80—120 years	RIP4	0.1—20.0	1
Riparian forest >120 years	RIP5	20.1—60.0	2
Oak-hornbeam forest <40 years	OAK1	20.1—60.0	2
Oak-hornbeam forest 40—60 years	OAK2	20.1—60.0	2
Oak-hornbeam forest 60—80 years	OAK3	20.1—60.0	2
Oak-hornbeam forest 80—120 years	OAK4	20.1—60.0	2
Oak-hornbeam forest >120 years	OAK5	20.1—60.0	2
Coniferous and mixed forest <40 years	CON1	120.1—240.0	4
Coniferous and mixed forest 40—60 years	CON2	60.1—120.0	3
Coniferous and mixed forest 60—80 years	CON3	60.1—120.0	3
Coniferous and mixed forest 80—120 years	CON4	60.1—120.0	3

Continued

Table 6.12 Ecosystem potential to produce honey (1 - very low; 5 - very high) indicated by the abundance of raw material for honey production (kg ha^{-1} year^{-1}). (cont'd)

Ecosystem type	Ecosystem acronym	HONEY indicator value	Ecosystem potential
Coniferous and mixed forest >120 years	CON5	60.1—120.0	3
Swamp coniferous forest <40 years	SWP1	>240.0	5
Swamp coniferous forest 40—60 years	SWP2	120.1—240.0	4
Swamp coniferous forest 60—80 years	SWP3	120.1—240.0	4
Swamp coniferous forest 80—120 years	SWP4	60.1—120.0	3
Swamp coniferous forest >120 years	SWP5	120.1—240.0	4
Wetlands - reed beds and sedge swamps	REED1	0.1—20.0	1
Wetlands - fens	FEN	0.1—20.0	1
Wetlands - bogs	BOG	20.1—60.0	2
Lakes (all types)	LAKE	0	0

distinctions between cropland and grassland capacity, with regard to soil conditions (fertility and humidity gradients). The intervals actually specified correspond with averages among maximum values that were, or possibly could be, recorded in the ecosystems considered. In reality, however, potential honey production in a particular ecosystem patch may differ significantly from the proposed ranges. Moreover, in actually existing ecosystems, environmental conditions (including plant cover) rarely allow dedicated potential to be realized. In some cases (e.g., heavily degraded grassland, arable fields with anemophilous crops, or dense planted pine forest), the estimated potential even very far exceeds the real supply.

More than 80% of the study area has been assessed as having at least medium potential to provide honey (Fig. 6.4). This shows that postglacial rural landscape with considerable share of pine forests, and open land (grassland and arable fields) may serve as good place to produce honey, despite climatic conditions not very favorable for beekeeping. The general spatial pattern of potential is dominated by a uniform matrix of average potential formed by a few large patches (the biggest one covers 18% of the study area). Small patches of other levels of potential are scattered randomly

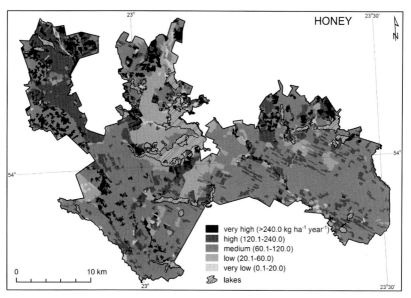

Figure 6.4 Ecosystem potential to produce honey indicated by the abundance of raw material for honey production (HONEY indicator; kg ha^{-1} year^{-1}).

on its background. In the northern part, the pattern changes into more mosaic-like with higher share of patches of high and low potential. Difference between area-weighted and nonweighted mean shape indices (AWMSI and MSI) shows that small patches are rather regular in shape (rectangular), while large and very large patches are irregular, with internal boundaries (holes). Small patches of the same level of potential are usually grouped in clusters (as it is shown by mean proximity index), and the interspersion of patches of different level of potential is of average level.

As HONEY indicator values were assigned to types of ecosystem, maps presenting the potential do not account for the spatial arrangement of ecosystem patches, which in case of highly fragmented landscape additionally increases the estimated honey yields (Affek, 2018). It should be noted that ecotone zones, field margins, and other small-scale habitats and the resulting high landscape complexity promote the abundance and diversity of honey-producing plants (Denisow and Wrzesień, 2015, 2007).

Unlike many relevant ES contributing greatly to human well-being (e.g., carbon sequestration, timber production, erosion control, and water retention), the potential supply of honey decreases with natural succession and biomass growth. The protected old-growth oak-hornbeam forest that covers extensive areas of the investigated WNP stands out as the largest compact area of low honey potential. In contrast, patches with young forest stands and clearings within extensive areas of managed pine forest (outside WNP) serve as potential hotspots for the production of honey. This apparent trade-off between services must thus be considered carefully as forest management practices are planned and when facing the dilemma between an active or passive approach to the conservation of open land.

6.1.1.4 Livestock and their outputs
Ecosystem service description

This service belongs to *Provisioning* section and *Biomass* division in CICES V5.1 (Table 6.13). It can be further described as the ecological contribution to the rearing of livestock and the production of their outputs that can be used as raw material for nutritional (e.g., meat, eggs) and nonnutritional purposes (e.g., clothing, energy). The maximum effective density of livestock units (LSU) has been proposed to indicate the potential of grassland to provide it.

Theoretical framework

Provision of fodder for domestic animals, whose products are used by man, is very important ES from grassland ecosystems (Burkhard et al., 2014; Haines-Young and Potschin, 2013). Animals are reared for nutritional purposes (e.g., meat, milk, fat, and eggs) and fibers and other materials (leather, fur, wool, feathers) for direct use or processing. They are also used as a tractive force, for riding on top, as pack animals or performing other utility functions. To quantify this ES, various indicators are used, e.g., not only the number of animals, respective animal products, or fodder plant harvest but also the simple area of pastures (Kandziora et al., 2013; Mononen et al., 2016). The assessment is usually based on data from land cover/land-use maps and national or other statistics. There are also animal production models (e.g., Common Agricultural Policy Regionalized Impact model or Agricultural Production Planning and Allocation model) accounting for the ecosystem's capacity, environmental effects, and human inputs to obtain more accurate results. Nevertheless, these models are time and data intensive. They are suitable if the aim is to better understand a certain production system or create animal production map for a certain location under a specific socioeconomic scenario or environmental constraint (Burkhard and Maes, 2017).

Table 6.13 Ecosystem service in Common International Classification of Ecosystem Services (CICES) V5.1 and indicator metadata.

Ecosystem service		Livestock and their outputs
CICES V5.1	Section	Provisioning
	Division	Biomass
	Group(s)	Reared animals for nutrition, materials, or energy
	Class(es)	Animals reared for nutritional purposes; fibers and other materials from reared animals for direct use or processing; animals reared to provide energy
	Class	1.1.3.1; 1.1.3.2; 1.1.3.3
	code(s)	
Indicandum		**Grassland capacity to feed livestock**
Indicator		**Maximum effective density of livestock units (LSU)**
Acronym		**MAXLSU**
Short description		Calculation of theoretical maximum effective density of reared animals (dairy cattle) using potential production of grassland and feed requirement of dairy cattle in the grazing season; applicable only to grassland ecosystems
Direct/indirect		Direct
Simple/compound/ complex		Complex
Estimated/calculated		Calculated
Scale		Ratio
Value range		1.6—1.8
Unit		LSU ha^{-1}
Spatial unit		Ecosystem
Value interpretation		1.6 → high potential; 1.8 → very high
Source data		— Production potential of grassland (Grzyb and Prończuk, 1995) — Feed requirement of dairy cattle in the grazing season (Preś and Rogalski, 1997)

In this research, the potential supply of livestock and their outputs was estimated employing the indirect and complex indicator of maximum effective density of LSU calculated according to the formula that takes into account the production potential of grassland and feed requirement of dairy cattle (the most common in the study area) in the grazing season (approximately 160 days per year). This approach combines two indicator types used to date and incorporates clearly environmental conditions that influence the level and quality of ES generated.

Assessment method

The production potential of three types of grassland was determined on the basis of grassland classification elaborated by Grzyb and Prończuk (1995). In general, for dry grassland, the average hay yield (dry matter in t ha^{-1}) was assumed to be 4.5 t ha^{-1}, for wet grassland on mineral soil 5 t ha^{-1}, and for peat grassland again 4.5 t ha^{-1}. Feed requirement of dairy cattle in the grazing season was adopted according to Preś and Rogalski (1997). The average demand of dairy cows for fodder was defined as 2.75 t of dry matter. The maximum effective density of LSU was calculated as

$$D_{DC} = \frac{P_{grass}}{FR_{DC}},$$

where D_{DC} is the livestock density (LSU ha^{-1})[2], P_{grass} is the production potential of grassland (t ha^{-1}), and FR_{DC} is the feed requirement of dairy cattle (t).

Indicator values

The maximum effective density of LSU for dry grassland and peat grassland takes value of 1.6 LSU ha^{-1}, while for wet grassland on mineral soil, characteristic of the higher production potential is 1.8 LSU ha^{-1} (Fig. 6.5,

---

[2] The livestock unit, abbreviated as LSU (or sometimes as LU), is a reference unit that facilitates the aggregation of livestock from various species and age as per convention, via the use of specific coefficients established initially on the basis of the nutritional or feed requirement of each type of animal. The reference unit used for the calculation of livestock units (1 LSU) is the grazing equivalent of one adult dairy cow producing 3000 kg of milk annually, without additional concentrated foodstuffs. Livestock unit coefficients: bovine animals: under 1-year-old 0.4 LSU, between 1 and 2 years old 0.7 LSU, male 2-year-old and over 1 LSU, heifers 2-year-old and over 0.8 LSU, dairy cows 1 LSU, other cows 2-year-old and over 0.8 LSU, sheep and goats 0.1 LSU, and equidae 0.8 LSU, pigs: piglets having a live weight of under 20 kg 0.027 LSU, breeding sows weighing 50 kg and over 0.5 LSU, and other pigs 0.3 LSU, poultry: broilers 0.07 LSU, laying hens 0.014 LSU, ostriches 0.35 LSU, other poultry (ducks, turkeys, geese, guinea fowls) 0.03 LSU, and rabbits (breeding females) 0.02 LSU. LSU is used to estimate the farm requirements on fodder (http://ec.europa.eu/eurostat/statistics-explained/index.php/Glossary:Livestock_unit (LSU)).

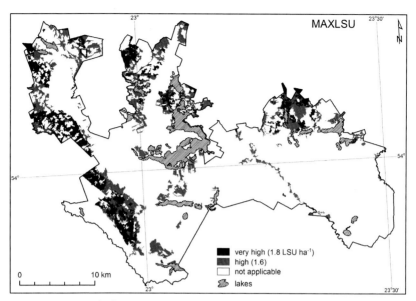

Figure 6.5 Potential of grassland to feed livestock indicated by maximum effective density of livestock units (MAXLSU indicator; LSU ha^{-1}).

Table 6.14). The results point to a very high potential of the studied grassland for cattle rearing. The received values are several times higher than the actual livestock density recorded in 2015 in Podlaskie Voivodeship and in the entire Poland (0.86 and 0.61 LSU ha^{-1}, respectively). The total maximum effective density of LSU for all studied grassland amounts to 19,782 LSU. This value is 24% higher than the livestock recorded in the study area (15,053 LSU) according to Agricultural Census 2010[3].

6.1.1.5 Edible wild berries
Ecosystem service description
This service belongs to *Provisioning* section and *Biomass* division in CICES V5.1 (Table 6.15). It can be further described as the ecological contribution to the production of noncultivated, edible fruits. The abundance of berry-producing plants has been proposed to indicate the potential of ecosystems to provide it.

---

[3] Data from the Local Data Bank of the Poland's Central Statistical Office - https://bdl.stat.gov.pl/BDL/start.

Table 6.14 Grassland potential to feed livestock (1 - very low; 5 - very high) indicated by maximum effective density of livestock units (LSU ha^{-1}).

Ecosystem type	Acronym	MAXLSU indicator value	Ecosystem capacity
Dry grassland	GRAS1	1.6	4
Wet grassland on mineral soil	GRAS2	1.8	5
Peat grassland	GRAS3	1.6	4

Theoretical framework

According to the terminology used in forestry, forest fruits belong to the nonwood forest products (NWFP). Based on the recommendations of an internal interdepartmental FAO meeting on definitions of NWFPs held in June 1999, the following new FAO working definition of NWFPs has been adopted: "non-wood forest products consist of goods of biological origin other than wood, derived from forests, other wooded land and trees outside forests" (Dembner and Perlis, 1999).

Edible wild berries are one of the most important provisioning ES and include bilberry (*V. myrtillus*), cranberry (*Oxycoccus palustris*), raspberry (*Rubus idaeus*), wild strawberry (*Fragaria vesca*), cowberry (*V. vitis-idaea*), bog bilberry (*Vaccinium uliginosum*), and blackcurrant (*Ribes nigrum*). It should be noted that in intensively used forests, where clear-cuts took place and soil is prepared before afforestation, the cover of berry-producing plants is reduced. On the other hand, the abundance of fruits decreases and the growth is weaker in old tree stands (Nestby et al., 2011). In selectively cut forests, the population of berry-producing plants is stable, as in natural forests (Mäkipää, 1999).

Because of the fact that berries contain large quantities of vitamins and nutrients valuable for humans and many of them (especially bilberry) can be called "natural" medicines that help with many diseases and ailments, individual acquisition of forest edible berries is very popular. The study conducted in Poland showed that for rural residents, nonwood products, mainly berries and mushrooms, constitute an important source of income (Barszcz, 2005). It is difficult to estimate the number of people harvesting berries, blackberries, and other fruits of the undergrowth. It can only be estimated that the amount of wild fruits delivered annually to collection points in Poland in recent years (2007—11) reached 8400—16,400 tons (Kuc et al., 2014). The deeply rooted traditions of forest gathering, as well

Table 6.15 Ecosystem service in Common International Classification of Ecosystem Services (CICES) V5.1 and indicator metadata.

Ecosystem service		Edible wild berries
CICES V5.1	Section	Provisioning
	Division	Biomass
	Group(s)	Wild plants for nutrition, materials, or energy
	Class(es)	Wild plants used for nutrition
	Class code(s)	1.1.5.1
Indicandum		**Harvestable volume of edible wild berries**
Indicator		**Cover of berry-producing plants**
Acronym		**BERRY**
Short description		Estimation of the cover of berry-producing plants in herb layer based on literature data and phytosociological relevés taken in the study area; shows indirectly the theoretical abundance of wild berries
Direct/indirect		Indirect
Simple/compound/complex		Compound
Estimated/calculated		Calculated
Scale		Ratio
Value range		0–100
Unit		%
Spatial unit		Ecosystem
Value interpretation		0 → no relevant potential; 0.1%–1.0% → very low; 1.1%–5.0% → low; 5.1%–20.0% → medium; 20.1%–50.0% → high; >50.0% → very high
Source data		— General habitat requirements of berry-producing species (Grau et al. (1983); Witkowska-Żuk (2008)) — Species distribution - over 400 unpublished phytosociological relevés, (A) made by the Authors, (B) taken from the Protection Plan of the Wigry National Park

as market needs, determine the dominance of fruits and mushrooms among all nonwoody forest benefits in Poland.

Because of the imperfect information system on the scale of forest use, available data, published by the Central Statistical Office in Poland, comprise only berries delivered to official collection points and do not include those collected for own needs, as well as those that reach the market in other ways (Staniszewski and Janeczko, 2012). Schulp et al. (2014) made the same observation on the European scale - a considerable part of collected wild berries goes to home consumption or informal market. Authors noted that for that reason, literature with statistics on the quantities of wild food collected (including berries) hardly exist and is scarce and scattered.

Therefore, it was only possible to indirectly estimate the potential of ecosystems to provide edible berries by analyzing the abundance of berry-producing plants. This approach is similar to the one proposed for Sweden (Snäll et al., 2015), according to which the potential production of berries is determined on the basis of bilberry (*V. myrtillus*) cover. As previously shown, the cover of *V. myrtillus* is strongly correlated with the annual production of its fruits (Miina et al., 2009). A slightly different approach was adopted in Finland (Mononen et al., 2016), where the ecosystem potential for berry supply is determined by two indicators: in a spatial perspective, an extent of habitat suitable for the occurrence of berries, and in the time perspective, as an average annual production of fruits.

Assessment method

To estimate the abundance of berry-producing plants (BERRY indicator), a total of seven species of shrubs and dwarfs that occurred in the study area were taken into account. The dwarfs are bog bilberry (*V. uliginosum*), cowberry (*V. vitis-idaea*), bilberry (*V. myrtillus*), cranberry (*O. palustris*), and wild strawberry (*F. vesca*) and shrubs are raspberry (*R. idaeus*) and blackcurrant (*R. nigrum*).

In the first stage, on the basis of literature data, the occurrence of individual species and their habitat requirements were determined (Grau et al., 1983; Witkowska-Żuk, 2008). Next, phytosociological relevés were collected, of the authors of the book and unpublished relevés from the Protection Plan of the WNP (434 phytosociological relevés in total), supplemented with literature data and expert evaluation. Phytosociological relevés described the composition, quantity, and cover of species in a given plant community.

The materials mentioned above made it possible to carry out an evaluation of ecosystem types in terms of their suitability for seven species of shrubs and dwarfs producing edible berries. An average number of berry-producing plant species in line with their average cover was estimated for each ecosystem type. Occurrence of accidental, single berry-producing plants corresponds to very low abundance of berries and thus very low ecosystem potential (1) to deliver this service. In turn, berry plant cover exceeding 50% means very high ecosystem potential (5).

Indicator values

The youngest oak–hornbeam forests (up to 40 years old) and the oldest coniferous and mixed forests (from 80 to over 120 years old) and all swamp coniferous forests show very high potential to provide edible berries (rank 5) (Table 6.16). Bilberry (*V. myrtillus*) appeared to be the primary berry species in oak–hornbeam and mixed coniferous forests, with a cover of more than 50%. Both *V. myrtillus* and *V. uliginosum* dominate in swamp coniferous forests. High potential to deliver wild berries was assigned to mature coniferous and mixed forests (60–80 years) as well as fens and bogs (rank 4). The main species are *V. myrtillus* in coniferous forests and *O. palustris* on fens and bogs, both with a cover of 20%–50%. The majority of ecosystems is characterized by the medium potential to provide berries (rank 3), these are the youngest alder forests and coniferous forests (up to 60 years old), all riparian forests, as well as oak–hornbeam forests (except for the youngest) and dry grassland. In alder and riparian forests, the main species are raspberry (*R. idaeus*) and blackcurrant (*R. nigrum*), in oak–hornbeam forests (40–60 years old) raspberry (*R. idaeus*), while in the older oak–hornbeam forests (60–80 years old) raspberry (*R. idaeus*) and bilberry (*V. myrtillus*). In the oldest oak–hornbeam forests and in the youngest coniferous forests, bilberry (*V. myrtillus*) is the prevailing species. Wild strawberry (*F. vesca*) was noted on grassland located on dry and mesic soil, covering up to 20% (rank 3). The oldest alder forests (from 60 to over 120 years) are characterized by low potential to provide wild berries (rank 2), with berry-producing plants covering less than 5% (mostly blackcurrant *R. nigrum*). The lowest potential (rank 1) was assigned to wet grassland on mineral soil where single raspberry shrubs (*R. idaeus*) were noted and to fens with isolated cranberry (*O. palustris*) dwarfs. The remaining ecosystem types do not supply edible wild berries, thus have no relevant potential to provide this service.

The potential of a typical postglacial landscape region to provide wild berries is strongly linked with the extent of forest cover, as the majority of

Table 6.16 Ecosystem potential to provide edible wild berries (1 - very low; 5 - very high) indicated by the cover of berry-producing plants (%).

Ecosystem type	Acronym	BERRY indicator value	Ecosystem potential	Main species
Settlement area	SETTLE	0	0	—
Arable field on poor sandy soil	ARB1	0	0	—
Arable field on fertile mesic soil	ARB2	0	0	—
Arable field on fertile moist soil	ARB3	0	0	—
Dry grassland	GRAS1	5.1—20.0	3	d
Wet grassland on mineral soil	GRAS2	0.1—1.0	1	e
Peat grassland	GRAS3	0	0	—
Alder forest <40 years	ALD1	5.1—20.0	3	g
Alder forest 40—60 years	ALD2	5.1—20.0	3	g
Alder forest 60—80 years	ALD3	1.1—5.0	2	h
Alder forest 80—120 years	ALD4	1.1—5.0	2	h
Alder forest >120 years	ALD5	1.1—5.0	2	h
Riparian forest <40 years	RIP1	5.1—20.0	3	g
Riparian forest 40—60 years	RIP2	5.1—20.0	3	g
Riparian forest 60—80 years	RIP3	5.1—20.0	3	g
Riparian forest 80—120 years	RIP4	5.1—20.0	3	g
Riparian forest >120 years	RIP5	5.1—20.0	3	g
Oak-hornbeam forest <40 years	OAK1	>50.0	5	c
Oak-hornbeam forest 40—60 years	OAK2	5.1—20.0	3	e
Oak-hornbeam forest 60—80 years	OAK3	5.1—20.0	3	f
Oak-hornbeam forest 80—120 years	OAK4	5.1—20.0	3	c
Oak-hornbeam forest >120 years	OAK5	5.1—20.0	3	c
Coniferous and mixed forest <40 years	CON1	5.1—20.0	3	c
Coniferous and mixed forest 40—60 years	CON2	5.1—20.0	3	c

Table 6.16 Ecosystem potential to provide edible wild berries (1 - very low; 5 - very high) indicated by the cover of berry-producing plants (%). (cont'd)

Ecosystem type	Acronym	BERRY indicator value	Ecosystem potential	Main species
Coniferous and mixed forest 60—80 years	CON3	20.1—50.0	4	c
Coniferous and mixed forest 80—120 years	CON4	>50.0	5	c
Coniferous and mixed forest >120 years	CON5	>50.0	5	c
Swamp coniferous forest <40 years	SWP1	>50.0	5	b
Swamp coniferous forest 40—60 years	SWP2	>50.0	5	b
Swamp coniferous forest 60—80 years	SWP3	>50.0	5	b
Swamp coniferous forest 80—120 years	SWP4	>50.0	5	b
Swamp coniferous forest >120 years	SWP5	>50.0	5	b
Wetlands - reed beds and sedge swamps	REED1	0	0	—
Wetlands - fens	FEN	0.1—1.0	1	a
Wetlands - bogs	BOG	20.1—50.0	4	a
Lakes (all types)	LAKE	0	0	—

a, *Oxycocus palustris;* b, *Vaccinium myrtillus and Vaccinium uliginosum;* c, *Vaccinium myrtillus;* d, *Fragaria vesca;* e, *Rubus idaeus;* f, *Rubus idaeus and Vaccinium myrtillus;* g, *Ribes nigrum and Rubus idaeus;* h, *Ribes nigrum.*

forest communities there have a considerable share of berry-producing species, while nonforest ecosystems have either no such plants or their abundance is low. Although eight types of the considered ecosystems (all of them forests) were ranked as having very high potential to produce wild berries, altogether covering more than 20% of the study area, in fact only pine and mixed forests >80 years old matter, with the coverage of 15% (Fig. 6.6). In turn, more than one-fourth of the study area (comprising lakes, arable fields, settlements, and selected wetlands and grassland) show no potential to produce wild berries.

Therefore, spatial pattern of potential is mosaic type with patches of various sizes. The largest patch covers only 6% of the study area. Areas of no

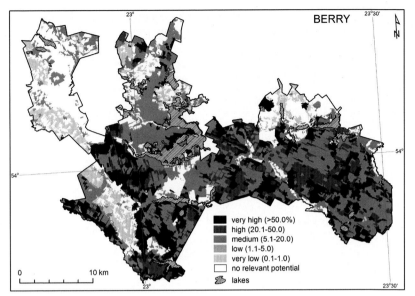

Figure 6.6 Ecosystem potential to provide edible wild berries indicated by the cover of berry-producing plants (BERRY indicator; %).

potential, mean potential, and very high potential codominate, covering together over 80% of the study area. Largest patches are generally more irregular in shape compared with small patches. Small patches of the same level of potential are randomly distributed, and the interspersion of patches of different level of potential is of average level compared with other ES potentials.

6.1.1.6 Edible biomass of wild animals
Ecosystem service description
This service belongs to *Provisioning* section and *Biomass* division in CICES V5.1 (Table 6.17). It can be further described as nondomesticated, wild animal species that can be used as raw material for the production of food. Two measures have been proposed to indicate the potential supply of this service: total biomass of large game animals within hunting units and annual net productivity of fish of commercial meaning within lake ecosystems.

GAME indicator: Total biomass of large game animals
Theoretical framework
Meat from wild animals is widely acknowledged as a provisioning service in ES classifications (de Groot et al., 2010b; Maes et al., 2013). This kind of

Table 6.17 Ecosystem service in Common International Classification of Ecosystem Services (CICES) V5.1 and indicator metadata.

Ecosystem service		Edible biomass of wild animals	
CICES V5.1	Section	Provisioning	
	Division	Biomass	
	Group(s)	Wild animals for nutrition, materials, or energy	
	Class(es)	Wild animals used for nutritional purposes	
	Class code(s)	1.1.6.1	
Indicandum		**Harvestable biomass of wild animals for nutritional purposes**	**Harvestable biomass of fish for nutritional purposes**
Indicator		**Total biomass of large game animals**	**Annual net productivity of fish of commercial meaning**
Acronym		**GAME**	**FISH**
Short description		Calculation of total biomass per km^2 of four large ungulate species (elk, roe deer, red deer, wild boar) using hunting data on their abundance and literature data on average unit weight	Estimation of annual net productivity of fish of commercial meaning based on literature data and abiotic conditions of lakes; applicable only to aquatic ecosystems
Direct/indirect		Direct	Direct
Simple/compound/complex		Compound	Simple
Estimated/calculated		Calculated	Estimated

Continued

Table 6.17 Ecosystem service in Common International Classification of Ecosystem Services (CICES) V5.1 and indicator metadata. (cont'd)

Ecosystem service		Edible biomass of wild animals
Scale	Ratio	Ratio
Value range	40–600	5–40
Unit	$kg\ km^{-2}$	$kg\ ha^{-1}\ year^{-1}$
Spatial unit	Hunting unit	Ecosystem
Value interpretation	<100.0 → very low ecosystem potential; 100.0–300.0 → low; 300.1–400.0 → medium; 400.1–500.0 → high; >500.0 → very high	5–20 → very low ecosystem potential; 20–30 → low; 25–33 → medium; 30–35 → high; 35–40 → very high
Source data	Annual hunting reports for the years 2011–14 from the four forest districts and the Wigry National Park	– Bathymetry, morphometry, physical, chemical, and biotic characteristics of lakes: Wigry National Park Protection Plan (Jańczak, 1999) – Fishery classification of lakes (Rudnicki et al., 1971; Skrzypczak et al., 2011) – Comparative data on fish species and density in selected lakes from the Wigry National Park Protection Plan

food constitutes a relatively large part of the diet in Poland compared with other European countries (Schulp et al., 2014). Ungulate mammals are among the most popular game meat providers (Sandalj et al., 2016; Schulp et al., 2014).

The abundance of wild animals can serve well as an indirect indicator of nature potential to deliver harvestable animal biomass for nutritional purposes. The potential of the area (country, region) to provide this service can be described in various ways and most simply as the number of species or - in a more complex way - as a density of animals or their biomass. The spatial variability of animal densities is mainly related to the availability of food resources (Dzięciołowski, 1970; Grabińska, 2011; Okarma and Tomek, 2008). The presence of herbivores (European bison, elk, red deer, roe deer) depends more on the share of deciduous forests, while the distribution of omnivorous wild boars is associated primarily with the occurrence of old trees.

It should be noted that game animals contribute to many other ES belonging to all three sections, provisioning (e.g., ornaments: antlers, skins), regulation and maintenance (e.g., pest control) and cultural (e.g., wildlife observation, symbolic meaning), and play significant role in maintaining human well-being on many different dimensions.

Assessment method

The data on the number of wild animals were used to assess the total biomass of large game animals per km^2, indicating the potential of nature to provide edible biomass from wild animals.

They were derived from annual hunting reports from 2011 to 2014 prepared by the four Forest District Offices separately for each hunting unit (14 units in total in the study area) and from the literature on WNP (Jamrozy, 2008; Misiukiewicz, 2014). Four species of ungulates were taken into account: elk, red deer, roe deer, and wild boar.

The analyzed hunting units differ in terms of land use. Open land, formed by meadows and pastures, prevails in units 47, 48, 64, 65, and 70. Units 46, 96, and the WNP have complex land use pattern, while in the remaining ones, forests predominate.

The GAME indicator (total biomass of large game animals) is measured in kg per 1 km^2 separately for each hunting unit and for the WNP. The density of total biomass was calculated according to the following formula: Σ (species density $\times$ unit weight). The unit weight was the simple

Table 6.18 Average weight (in kg) of large game animals included in calculating GAME indicator.

Species	Male	Female
Elk	400	300
Red deer	120	80
Roe deer	25	25
Wild boar	120	90

Based on Czyżyk, P., Żurkowski, M., Ciepluch, Z., Struziński, T., Czajka, W., 2007. Parametry populacyjne jelenia szlachetnego (Cervus elaphus L.) w Leśnym Kompleksie Promocyjnym "Lasy Mazurskie". Część I. Ocena masy poroża i masy tuszy byków pozyskanych w wyniku odstrzałów selekcyjnych. Sylwan 9, 41—50 (in Polish); Komosińska, H., Podsiadło, E., 2002. Ssaki kopytne. Wydawnictwo Naukowe PWN, Warszawa (in Polish); Okarma, H., Tomek, A., 2008. Łowiectwo. Wydawnictwo Edukacyjno-Naukowe H2O, Kraków (in Polish); Pielowski, Z., 1999. Sarna. Oficyna Edytorska "Wydawnictwo Świat", Warszawa (in Polish).

arithmetic mean of average male and female weight, with an assumption made that each population is equally divided on males and females (Table 6.18).

Indicator values

The estimated total biomass of large wild animals (GAME indicator) ranges from approximately 40 kg to approximately 600 kg per 1 km^2 (Fig. 6.7). The highest values were reported for the hunting units with the largest forest cover, such as hunting unit number 77, forested in 72%, and forming the Animal Breeding Center. Very low potential to provide edible biomass

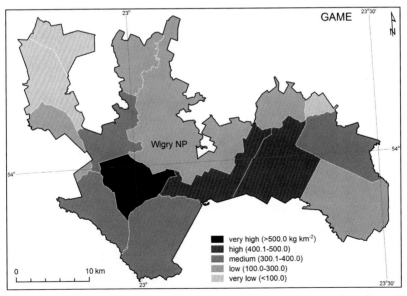

Figure 6.7 Potential of nature to provide edible biomass of wild animals indicated by the total biomass of large game animals within hunting units and Wigry National Park (GAME indicator; kg km^{-2}).

of wild animals (<100 kg km^{-2}) was noted in hunting units comprising mostly open land with forest cover not exceeding 9%.

FISH indicator: Annual net productivity of fish of commercial meaning
Theoretical framework

Lakes provide a wide range of ES. In general, we can divide all services of lakes into two groups: those that are a result of water retention and those that are a result of lake ecological functioning (Turkowski, 2011).

In the metaanalysis conducted by Reynaud and Lanzanova (2017) on worldwide data set of 699 observations drawn from (Table 6.4) 133 studies, the ES delivered by lakes were grouped into the following categories: provisioning (water for drinking and nondrinking purposes); regulation and maintenance (maintaining populations and habitats, flood protection, erosion prevention); cultural (amenity, fishing, boating, swimming, camping, sightseeing, and unspecified recreational services). According to this metaanalysis, fishery is placed in recreational services (265 observations). Commercial fishing is not mentioned among provisioning services, which are generally poorly represented in the database (Reynaud and Lanzanova, 2017). On the other hand, lake-induced spatial changes of climate variables (precipitation, mean minimum and mean maximum temperatures, cloud cover, vapor pressure), so important in some regions (Scott and Huff, 1996), are also not mentioned in the analysis.

Catching fish is classified differently in particular ES classifications. According to Schallenberg et al. (2013), ES provided by lakes and important for New Zealand belong to four groups:

a) Fulfillment of responsibilities under global conventions (biodiversity, climate change mitigation);

b) Provisioning (drinking water, fisheries, waterfowl, recreation, and tourism including catch-and-consume angling);

c) Support and regulation (nutrient and sediment processing, sequestration, hydrological regulation);

d) Cultural (recreation and tourism, fisheries, waterfowl).

In this approach, fish catching and fishery are located in three different places: twice as two separate services in the provisioning section, and once in the cultural section. This distinction results from the different goals of human activities.

According to different definitions, the term "fishing" refers to extracting wild fish from natural waters, as opposed to "aquaculture", i.e., rearing of aquatic animals in controlled environments and with human intervention (stocking, feeding, etc.) (European Commission, 2011).

Traditionally, catching fish is divided into two categories: fishery (commercial) and angling (recreational). Recently, the division into three categories has been proposed, both in the United States (Final Ecosystem Goods and Services Classification System [FEGS-CS], elaborated by the US Environmental Protection Agency, see Landers and Nahlik, 2013) and in Europe (the European Inland Fisheries Advisory Commission EIFAC - see European Commission, 2011). In both approaches, the following categories of activities (EIFAC) or of beneficiaries connected to fish as ES (FEGS-CS) are distinguished:

a) Commercial fisheries (European Union) or food extractors (USA), concerning the fishery that utilizes the natural abundance of edible fish (i.e., noncultivated or bred) for commercial use or sale on domestic and export markets;

b) Subsistence fisheries (European Union) or food subsisters (USA), concerning the utilization of the natural abundance of edible fish as a major supplement to the personal existence (individual's nutritional needs). Fishing products are not traded on formal domestic or export markets but are consumed personally or within a close network of family and friends. Pure subsistence fisheries sustain a basic level of livelihood and constitute a culturally significant food-producing and food-distributing activity;

c) Recreational fishing (European Union) or anglers (USA), concerning catching fish recreationally (i.e., neither for survival nor for the market) and including catch-and-release or catch-and-consume activities. According to US classification, stocked fish are not a service, as they are considered a human input.

The first and the second categories in both classifications correspond to the 1.1.6.1 class (wild animals used for nutritional purposes) in CICES V5.1, while the third matches the 3.1.1.1 class (physical and experiential interactions with natural environment), but, in fact, precise separation between them is not always fully possible.

Inland (lake) fishery of commercial character is present in 21 countries out of 27 belonging to the European Union. Only in six Member States typical commercial fishery is absent, and these are Belgium, Cyprus, Luxemburg, Malta, Slovakia, and Slovenia. In Denmark and the Czech Republic, fishing is practiced on a very small scale. In non-EU countries, economic catches in inland waters are also carried out. Thus, it can be said that inland fisheries of economic importance still prevail in most of the European countries. According to data for the years 2000—09, catches in European Union member states exceeded 35,000 tons per year. The largest catch was in Finland (about 4500 tons), while the smallest in the Czech Republic (24 tons). In relation to total catches (inland, maritime, aquaculture) in Europe, inland fishery provided 1% of fish mass on markets (Mitchell et al., 2010).

Assessment method

Direct measurement/assessment of lake ecosystem potential to provide edible biomass of fish is very difficult and often questionable. That is why indirect approaches are in use. In this study, the assessment method is based on the scheme presented on Fig. 6.8.

According to this scheme, total fish productivity depends on ecological parameters of a given lake, as well as on an actual biomass of all fish. In the lakes in Poland, as in the entire European lowland, the size of the total fish biomass in eutrophic and mesotrophic lakes of medium and large size usually lies between 120 and 150 kg per hectare (Krzysztofiak, 2012). Many works show (see Krzywosz and Kamiński, 2012; Randall et al., 1995) that for those lakes the ratio of annual production to total biomass is within 0.3—0.7. Thus, the total annual production of fish in lakes ranges between 35 and 105 kg per hectare, depending on the ecological status of a lake.

Harvestable productivity is a subset of total productivity determined by fish species and the size of specimen having the market meaning and allowing maintaining further production capacities of fish populations.

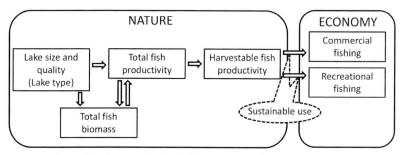

Figure 6.8 General scheme of the indirect approach to the assessment of lake ecosystem potential to provide edible biomass of fish.

According to Mann (1969) and Zawisza (1973), harvestable productivity should not exceed 25% of total production.

Taking into account the above relationships, it is reasonable to accept the harvestable fish productivity as an indirect indicator of lake ecosystem potential to provide edible biomass of fish.

The evaluation of the harvestable fish productivity was based on literature data showing the relationship between ecological status of the lake and the amount and productivity of fish.

Bathymetry, morphometry, physical, chemical, and biotic characteristics of lakes were taken from Jańczak (1999) and WNP Protection Plan. On this basis, all lakes were ascribed to classes of fishery use, distinguished by Rudnicki et al. (1971) and Skrzypczak et al. (2011). Detailed data on fish species composition, size structure, and their density in selected lakes within WNP were taken from the WNP Protection Plan.

Indicator values

The highest potential of lake ecosystems to provide edible biomass of fish was assigned to medium-sized eutrophic lakes of medium depth ($35-40$ kg ha^{-1} year^{-1}), while the lowest is characteristic for small dystrophic lakes ($5-20$ kg ha^{-1} year^{-1}). The potential of remaining types of lake is of intermediate character (Table 6.19).

Spatial distribution of lakes of different categories is shown on Fig. 6.9. Different levels of potential are represented by differentiated number of lakes. Dystrophic lakes with the lowest potential are represented by 20 lakes, with total area of only 75.5 ha. These lakes (together with several smallest waterbodies of the same type, not presented on the map) are not the subject of regular fishing, only control catches for scientific purposes are allowed.

The next category, small shallow eutrophic lakes, with low potential of $20-30$ kg ha^{-1} year^{-1} is represented by 14 lakes, covering together 78.9 ha.

The following two categories with medium ($25-33$) and high potential ($30-35$ kg ha^{-1} year^{-1}) dominate in the study area (6 lakes, 2611.4 ha and 16 lakes, 1559.4 ha, respectively). They represent very similar types of lake, differing slightly in terms of trophy. Sixteen lakes, mostly medium-sized and eutrophic, have the highest potential ($35-40$ kg ha^{-1} year^{-1}) and cover 396.2 ha.

The observed spatial distribution of lakes with different potential is typical for the most of European lake lands, with the general rule of

Table 6.19 Potential of lake ecosystems to provide edible biomass of fish (1 - very low; 5 - very high) indicated by the annual net productivity of fish of commercial meaning (kg ha^{-1} year^{-1}).

Ecosystem type	Ecosystem acronym	FISH indicator value	Ecosystem potential
Large, deep mesotrophic lakes	LAKE1	25−33	3
Large, deep eutrophic lakes	LAKE2	30−35	4
Big eutrophic lakes of medium depth	LAKE3	30−35	4
Medium-sized eutrophic lakes of medium depth	LAKE4	35−40	5
Small shallow eutrophic lakes	LAKE5	20−30	2
Small dystrophic lakes	LAKE6	5−20	1
Artificial lakes	LAKE7	30−35	4

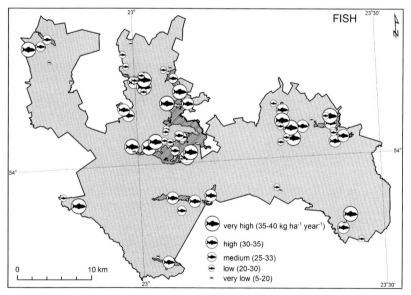

Figure 6.9 Potential of lake ecosystems to provide edible biomass of fish indicated by the annual net productivity of fish of commercial meaning (FISH indicator; kg ha^{-1} year^{-1}).

decreasing trophy (and potential to provide edible biomass of fish) northwards.

The estimated values of potential of lakes to provide edible biomass of fish are generally utilized by commercial catches and angling fishery. For the lakes of the entire northeast Poland, the commercial catches in years 2007−14 oscillated around 8.34 kg ha^{-1} year^{-1} (Wołos et al., 2015a) and were 2.4 times lower than angling catches. These values are concordant with the pan-European data, according to which recreational fisheries are responsible for around 70%−80% of the human exploitation of natural inland fish stocks (Mitchell et al., 2010; Wołos et al., 2015b). Taking account of the above data and adding the role of other consumers (such as cormorant *Phalacrocorax carbo*), it may be stated that all the harvestable fish potential is utilized (see Krzywosz and Traczuk, 2011).

6.1.2 Regulating ecosystem services
6.1.2.1 Erosion control
Ecosystem service description
This service belongs to *Regulation and Maintenance* section and *Regulation of physical, chemical, biological conditions* division in CICES V5.1 (Table 6.20). It can be further described as the reduction in the loss of material by virtue of the stabilizing effects of the presence of vegetation that mitigates or prevents potential damage to human use of the environment or human health and safety. The vegetation capacity to prevent erosion has been proposed to indicate the capacity of ecosystems to provide it.

Theoretical framework
Erosion processes are complex natural and geological phenomena that destroy the surface of the Earth's crust (Ziemnicki, 1978). One of them is soil erosion consisting on removal of parts of the soil cover under the influence of various mechanical factors, both natural and anthropogenic. Erosion degrades soil quality in each ecosystem type, thereby reducing the land productivity. As a result, biodiversity of eroded area is reduced and stability of entire ecosystem threatened (Pimentel et al., 1995). The occurrence and intensity of soil water erosion depends on many factors (Laflen and Moldehauer, 2003; Renard et al., 1997; Wischmeier and Smith, 1978): atmospheric precipitation and surface runoff, soil susceptibility to erosion, especially grain size, terrain features (relief, slope steepness, length, and the shape of slopes), type and condition of plant cover, and antierosion treatments. One of the most common phenomena is the occurrence of surface erosion caused by water runoff with soil molecules along the slope. Undulated relief, high precipitation, including intensive rains, lithology, lack of plant cover, and intensification of agricultural crops or inappropriate land use favor the intensification of erosion

Table 6.20 Ecosystem service in Common International Classification of Ecosystem Services (CICES) V5.1 and indicator metadata.

Ecosystem service		Erosion control
CICES V5.1	Section	Regulation and Maintenance
	Division	Regulation of physical, chemical, biological conditions
	Group(s)	Regulation of baseline flows and extreme events
	Class(es)	Control of erosion rates
	Class code(s)	2.2.1.1
Indicandum		**Ecosystem capacity to control erosion**
Indicator		**Vegetation capacity to prevent erosion**
Acronym		**EROSION**
Short description		Calculation of soil mass that will not be eroded due to vegetation, calculated by means of USLE model as the difference between mass of eroded soil in case of presence and absence of vegetation
Direct/indirect		Direct
Simple/compound/ complex		Complex
Estimated/calculated		Calculated
Scale		Ratio
Value range		<0.6->1.0
Unit		$t\ ha^{-1}\ year^{-1}$
Spatial unit		Ecosystem
Value interpretation		<0.60 → very low potential; 0.60–0.75 → low; 0.76–0.90 → medium; 0.91–1.00 → high; >1.00 → very high
Source data		Drzewiecki and Mularz (2005); Molnár and Julien (1998); Pistocchi et al. (2002); Święchowicz (2013); Wischmeier and Smith (1978)

processes including mainly water erosion. This process often results in significant transformations of the relief and, as a result, changes in soil properties. About 33% of area of Poland is exposed to this process and its intensity varies and refers to hypsometric conditions, rainfall, land cover, and lithology (Nowocień, 2007). Among the lowland areas, lake districts (including the research area) are most vulnerable to erosion, due to the significant share of slopes steeper than nine degrees (Chudecki and Niedźwiecki, 1983; Chudecki et al., 1993; Churska, 1973; Koćmit et al., 2006; Niewiadomski and Skrodzki, 1964; Smolska, 2002; Uggla et al., 1968). Grassland with dense vegetation cover is among the most resistant to erosion, and even heavy rainfall does not affect them significantly (Lipski and Kostuch, 2005). The same applies to deciduous and mixed forests with close canopies. Vegetation prevents the occurrence of soil erosion by effectively protecting the soil against direct impacts of raindrops, which destroy the soil structure and rinse away floatable fraction (Gerlach, 1976). The vegetation ability to prevent erosion is a very important factor in the context of ecosystem functioning, but also human activity, mainly agriculture. In addition to field research, to determine the threats related to erosion, probabilistic models are used, e.g., USLE, RUSLE, and LISEM.

Assessment method

The assessment of ecosystem potential to prevent erosion was based on USLE erosion model (Wischmeier and Smith, 1978), literature data on the resistance of various types of vegetated land to water erosion, and soil cover characteristics. The six major input parameters used in the study are vegetation cover factor (C), slope length factor (L), slope steepness factor (S), soil erodibility factor (K), rainfall erosivity factor (R), and erosion control factor (P). The analysis was carried out on individual ecosystem patches. C values were set as follows according to the observed herb layer density in particular ecosystem types: arable field - 0.2, grassland - 0.001−0.003, wetlands - 0.001, coniferous forest (>40 years) - 0.003, coniferous and mixed forest (<40 years) - 0.005, and deciduous and swamp coniferous forest - 0.002. It was assumed that in forest ecosystems, in which deciduous species prevail, as well as on grassland and wetlands, vegetation strongly protects the terrain surface against erosion. In the case of coniferous and mixed forests, a lower erosion resistance index was used. Assuming no vegetation cover in arable fields, these ecosystems were considered the least resistant.

L values (slope length factor) and S values (slope steepness factor) were assumed constant almost over the entire study area and equal to 19.5 and 2.5, respectively. Other values were applied only for ecosytems on peat (L = 9, S = 1), as they are inherently flat. Another assumption was that mineral soils are characterized by loamy sand/sandy loam texture, typical for lowland postglacial landscape, thus K values were in the range

0.001—0.0195, after Drzewiecki and Mularz (2005), Molnár and Julien (1998), and Pistocchi et al. (2002). The other USLE parameters were assumed constant: $P = 1$ (antierosive treatments not performed in the study area) and $R = 55.3$, after Święchowicz (2013). In this way, the main factor differentiating ecosystems in terms of potential to prevent erosion was land cover.

After calculating the average annual mass of eroded soil from the study area, the same values were calculated assuming that the research area is not covered by vegetation. The soil mass that is retained in each ecosystem type due to vegetation (EROSION indicator) is the difference between the obtained results.

Indicator values

The most resistant to water erosion are ecosystems with dense vegetation, both forests and meadows, especially on wet soil - wetlands, both wet grassland on mineral soil and peat grassland, and all categories of deciduous and swamp coniferous forests (Table 6.21). On the basis of field research, it has been found that coniferous and mixed forests under 40 years old have lower potential to prevent erosion than older stands, due to less dense herb layer. Agroecosystems appear to have the lowest potential for preventing erosion due to only temporary presence of vegetation cover. The threat of erosion there is additionally intensified by agricultural human activity resulting in tillage erosion. Ecosystems with similar soil texture, but permanently covered with vegetation (meadows, forests), are moderately or highly resistant to erosion. Among the forest ecosystems, deciduous forests are characterized by high potential to prevent erosion.

In rural landscape, the often occurring mosaic of arable fields and pastures result in the mosaic of ecosystem potential to prevent erosion, where very high capacity of wet grassland contrasts with very low potential of arable fields (Fig. 6.10). More than 40% of the study area is characterized by medium vegetation potential to prevent erosion, although only dry grassland and older pine and mixed forests were evaluated this way. Within vast forested areas sufficiently protected against erosion, only patches of young coniferous and mixed forests stands out as areas with low capacity to prevent erosion processes.

Spatial pattern of EROSION indicator values is dominated by a uniform matrix of medium potential consisting of a few large patches (the largest one covers over 22% of the study area). Small patches of other levels of potential form clusters on its background. In the northern part of the study area, the pattern changes into matrix of very low potential with patches of high

Table 6.21 Ecosystem capacity to control erosion (1 - very low; 5 - very high) indicated by vegetation capacity to prevent erosion (t ha^{-1} year^{-1}).

Ecosystem type	Acronym	EROSION indicator value	Ecosystem potential
Settlement area	SETTLE	Not considered	
Arable field on poor sandy soil	ARB1	<0.60	1
Arable field on fertile mesic soil	ARB2	<0.60	1
Arable field on fertile moist soil	ARB3	<0.60	1
Dry grassland	GRAS1	0.76—0.90	3
Wet grassland on mineral soil	GRAS2	>1.00	5
Peat grassland	GRAS3	>1.00	5
Alder forest <40 years	ALD1	0.91—1.00	4
Alder forest 40—60 years	ALD2	0.91—1.00	4
Alder forest 60—80 years	ALD3	0.91—1.00	4
Alder forest 80—120 years	ALD4	0.91—1.00	4
Alder forest >120 years	ALD5	0.91—1.00	4
Riparian forest <40 years	RIP1	0.91—1.00	4
Riparian forest 40—60 years	RIP2	0.91—1.00	4
Riparian forest 60—80 years	RIP3	0.91—1.00	4
Riparian forest 80—120 years	RIP4	0.91—1.00	4
Riparian forest >120 years	RIP5	0.91—1.00	4
Oak-hornbeam forest <40 years	OAK1	0.91—1.00	4
Oak-hornbeam forest 40—60 years	OAK2	0.91—1.00	4
Oak-hornbeam forest 60—80 years	OAK3	0.91—1.00	4
Oak-hornbeam forest 80—120 years	OAK4	0.91—1.00	4
Oak-hornbeam forest >120 years	OAK5	0.91—1.00	4
Coniferous and mixed forest <40 years	CON1	0.61—0.75	2
Coniferous and mixed forest 40—60 years	CON2	0.76—0.90	3
Coniferous and mixed forest 60—80 years	CON3	0.76—0.90	3
Coniferous and mixed forest 80—120 years	CON4	0.76—0.90	3
Coniferous and mixed forest >120 years	CON5	0.76—0.90	3

Table 6.21 Ecosystem capacity to control erosion (1 - very low; 5 - very high) indicated by vegetation capacity to prevent erosion (t ha^{-1} year^{-1}). (cont'd)

Ecosystem type	Acronym	EROSION indicator value	Ecosystem potential
Swamp coniferous forest <40 years	SWP1	0.91–1.00	4
Swamp coniferous forest 40 –60 years	SWP2	0.91–1.00	4
Swamp coniferous forest 60 –80 years	SWP3	0.91–1.00	4
Swamp coniferous forest 80 –120 years	SWP4	0.91–1.00	4
Swamp coniferous forest >120 years	SWP5	0.91–1.00	4
Wetlands - reed beds and sedge swamps	REED1	>1.00	5
Wetlands - fens	FEN	>1.00	5
Wetlands - bogs	BOG	>1.00	5
Lakes (all types)	LAKE	Not considered	

potential. The analyzed potential belongs to the group of ES potentials with the lowest mean patch size and high patch size variance. Large patches are generally more irregular in shape compared with small patches, but differences are generally little compared with other ES potentials. Small patches of the same potential are clustered, and the interspersion of patches of different level of potential is quite low compared with other types ES potentials.

6.1.2.2 Water retention
Ecosystem service description
This service belongs to *Regulation and Maintenance* section and *Regulation of physical, chemical, biological conditions* division in CICES V5.1 (Table 6.22). It can be further described as the regulation of water flows by virtue of the chemical and physical properties or characteristics of ecosystems, which assists people in managing and using hydrological systems and mitigates or prevents potential damage to human use, health, or safety. The soil water holding capacity (field capacity - FC) has been proposed to indicate the capacity of ecosystems to provide it.

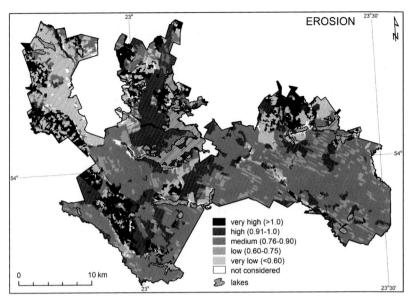

Figure 6.10 Ecosystem potential to control erosion indicated by vegetation capacity to prevent erosion (EROSION indicator; t ha^{-1} year^{-1}).

Theoretical framework

The ability of soil to retain water is an important factor in ecosystem water management, enabling mitigation of the effects of both periods of water regime, that is semiarid and arid, as well as excessive water during the heavy rain. One of the measures enabling the assessment of the ecosystem potential to retain water in soil is the amount of water stored at the given soil pedon (the smallest homogeneous soil unit) in conditions of FC. The physical definition of FC is the bulk water content retained in soil at -33 J kg^{-1} (or -0.33 bar) of hydraulic head or suction pressure. This stock corresponds to the ability of a given type of soil to store water under Earth's gravity. To a large extent, it depends on the granulometric composition of the substrate, the type of humus, and the forms of land use. Mathematical formulas (Puchalski and Prusinkiewicz, 1975) are used to determine it, which on the basis of the measurable features of soil in a defined ecosystem allow for a fairly precise determination of soil retention properties.

Assessment method

Data obtained as a result of field and laboratory research were the basis for determining ecosystem potential to retain water in soil. Standard soil science methods were used to determine soil properties in samples with an

Table 6.22 Ecosystem service in Common International Classification of Ecosystem Services (CICES) V5.1 and indicator metadata.

Ecosystem service		Water retention
CICES V5.1	Section	Regulation and Maintenance
	Division	Regulation of physical, chemical, biological conditions
	Group(s)	Regulation of baseline flows and extreme events
	Class(es)	Hydrological cycle and water flow regulation
	Class code(s)	2.2.1.3
Indicandum		**Ecosystem capacity to retain water**
Indicator		**Soil water holding capacity (field capacity)**
Acronym		**SOILH2O**
Short description		Calculation of potential water stock in 10,000 m^3 of soil (1 ha $\times$ 1 m) as the sum of potential stocks in organic and mineral horizons based on reference field data and laboratory analysis
Direct/indirect		Direct
Simple/compound/complex		Complex
Estimated/calculated		Calculated
Scale		Ratio
Value range		13,000—66,000
Unit		hl ha^{-1}
Spatial unit		Ecosystem
Value interpretation		<20,000 $\rightarrow$ very low potential; 20,000—30,000 $\rightarrow$ low; 30,001—40,000 $\rightarrow$ medium; 40,001—50,000 $\rightarrow$ high; >50,000 $\rightarrow$ very high
Source data		— Field and laboratory studies at selected points representing particular types of terrestrial ecosystem — Soil map (Institute of Soil Science and Plant Cultivation in Puławy)

intact structure. The mathematical formula used for calculating soil water content at field capacity (WCFC) is

$$WCFC = SMFC\ D\ h\ 10\ q^{-1},$$

where SMFC is the soil moisture at field capacity, D is the soil volumetric density, h is the soil level thickness, and q is the water density. The obtained results characterizing retention properties of particular soil horizons at FC were summed up to a depth of 1 m.

Indicator values

The largest values of soil water content at FC ($>50,000$ hL ha^{-1}) were noted in ecosystems with semihydrate and autogenic soils, formed in heavier lithologic material (Table 6.23). These include the soils of both fens and bogs, swamp forests, and middle-age riparian forests. In peat wetlands, soil is able to store even $65,000-66,000$ hL per ha to the depth of 1 m. In turn, ecosystems of pine forests, mixed pine forests, and all arable fields have very low potential to retain water in soil ($<20,000$ hL ha^{-1}). High potential ($40,001-50,000$ hL ha^{-1}) was assigned to ecosystems of alder and riparian forests with younger stands and grassland on peat substratum.

Values of SOILH20 indicator are related to the trophic and edaphic conditions of the soil cover (Fig. 6.11). Apart from waterbodies, which

Table 6.23 Ecosystem potential to retain water (1 - very low; 5 - very high) indicated by soil water holding capacity (field capacity- FC) (hL ha^{-1}).

Ecosystem type	Ecosystem acronym	SOILH2O indicator value	Ecosystem potential
Settlement area	SETTLE	Not considered	
Arable field on poor sandy soil	ARB1	17,000	1
Arable field on fertile mesic soil	ARB2	17,000	1
Arable field on fertile moist soil	ARB3	19,000	1
Dry grassland	GRAS1	13,000	1
Wet grassland on mineral soil	GRAS2	19,000	1
Peat grassland	GRAS3	47,000	4
Alder forest <40 years	ALD1	26,000	2
Alder forest 40–60 years	ALD2	41,000	4
Alder forest 60–80 years	ALD3	39,000	3
Alder forest 80–120 years	ALD4	27,000	2
Alder forest >120 years	ALD5	27,000	2
Riparian forest <40 years	RIP1	48,000	4
Riparian forest 40–60 years	RIP2	49,000	4
Riparian forest 60–80 years	RIP3	52,000	5
Riparian forest 80–120 years	RIP4	50,000	5

Table 6.23 Ecosystem potential to retain water (1 - very low; 5 - very high) indicated by soil water holding capacity (field capacity- FC) (hL ha^{-1}). (cont'd)

Ecosystem type	Ecosystem acronym	SOILH2O indicator value	Ecosystem potential
Riparian forest >120 years	RIP5	27,000	2
Oak-hornbeam forest <40 years	OAK1	22,000	2
Oak-hornbeam forest 40 −60 years	OAK2	21,000	2
Oak-hornbeam forest 60 −80 years	OAK3	19,000	1
Oak-hornbeam forest 80 −120 years	OAK4	20,000	2
Oak-hornbeam forest >120 years	OAK5	19,000	1
Coniferous and mixed forest <40 years	CON1	17,000	1
Coniferous and mixed forest 40−60 years	CON2	19,000	1
Coniferous and mixed forest 60−80 years	CON3	22,000	2
Coniferous and mixed forest 80−120 years	CON4	19,000	1
Coniferous and mixed forest >120 years	CON5	19,000	1
Swamp coniferous forest <40 years	SWP1	34,000	3
Swamp coniferous forest 40 −60 years	SWP2	52,000	5
Swamp coniferous forest 60 −80 years	SWP3	55,000	5
Swamp coniferous forest 80 −120 years	SWP4	54,000	5
Swamp coniferous forest >120 years	SWP5	51,000	5
Wetlands - reed beds and sedge swamps	REED1	37,000	3
Wetlands - fens	FEN	65,000	5
Wetlands - bogs	BOG	66,000	5
Lakes (all types)	LAKE	Not considered	

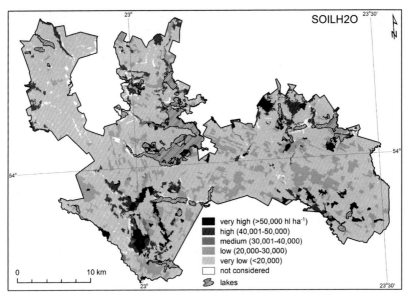

Figure 6.11 Ecosystem potential to retain water indicated by soil water holding capacity (field capacity - FC) (SOILH2O indicator; hl ha^{-1}).

were not included in the analysis, only riparian forests, swamp forests, and wetlands exhibit very high potential to hold water, which anyway cover only 3.5% of the study area. Therefore, spatial pattern of ecosystem potential to retain water is dominated by a uniform matrix of very low potential (65% of the study area, with the single largest patch covering 40%). Small patches of other levels of potential are randomly scattered on its background. Difference between AWMSI and MSI shows that small patches are rather regular in shape (rectangular), while large and very large patches are extremely irregular, with internal boundaries (holes). Small patches of very high and high potential are quite clustered, while other patches are rather randomly distributed. Interspersion of patches of different level of potential is on a relatively low level.

6.1.2.3 Pollination
Ecosystem service description
This service belongs to *Regulation and Maintenance* section and *Regulation of physical, chemical, biological conditions* division in CICES V5.1 (Table 6.24). It can be further described as the fertilization of plants by wild pollinators that maintains or increases the abundance and/or diversity of species that people

Table 6.24 Ecosystem service in Common International Classification of Ecosystem Services (CICES) V5.1 and indicator metadata.

Ecosystem service		Pollination
CICES V5.1	Section	Regulation and Maintenance
	Division	Regulation of physical, chemical, biological conditions
	Group(s)	Lifecycle maintenance, habitat, and gene pool protection
	Class(es)	Pollination
	Class code(s)	2.2.2.1
Indicandum		**Ecosystem potential to provide pollination service**
Indicator		**Abundance of nesting wild bees**
Acronym		**POLLIN**
Short description		Estimation of potential maximum abundance of wild bees (*Apoidea*) in growing season based on availability of food resources and nesting sites. Shows theoretical maximum number of nesting wild bees per hectare and—indirectly—pollination potential
Direct/indirect		Indirect
Simple/compound/complex		Complex
Estimated/calculated		Estimated
Scale		Ratio
Value range		0—2000
Unit		Individuals ha^{-1}
Spatial unit		Ecosystem
Value interpretation		0 → no relevant potential; 0.1—100.0 → very low ecosystem potential; 100.1—200.0 → low; 200.1—300.0 → medium; 300.1—600.0 → high; >600.0 → very high
Source data		— Honey potential (nectar and pollen production) of single honey plants and whole plant communities (Demianowicz et al., 1960; Denisow, 2011; Kołtowski, 2006; Maksymiuk, 1960; Szklanowska, 1979, 1973)

Continued

Table 6.24 Ecosystem service in Common International Classification of Ecosystem Services (CICES) V5.1 and indicator metadata. (cont'd)

Ecosystem service	Pollination
	— Species composition and diversity of plant communities (Matuszkiewicz, 2001, phytosociological relevés, fieldwork)
	— Honeydew production (Crane and Walker, 1985; Haragsim, 1966; beekeeping literature)
	— Habitat requirements for key wild pollinators (Banaszak, 1983; Denisow and Wrzesień, 2015; Westrich, 1996)
	— Wild bee abundance and species richness in different lowland ecosystems (Banaszak, 2010, 1983; Banaszak and Cierzniak, 2000; Banaszak and Jaroszewicz, 2009; Banaszak and Krzysztofiak, 1996, 1992; Banaszak and Szefer, 2013; Krzysztofiak, 2001; Pawlikowski, 2010)

use or enjoy. The abundance of nesting wild bees has been proposed to indicate the capacity of ecosystems to provide it.

Theoretical framework

Only bees (superfamily *Apoidea* within order *Hymenoptera*) were taken into account when the potentials of ecosystems to deliver pollination services were estimated. This reflects the status of these insects as the key pollinator taxon all over the world (Greenleaf et al., 2007). However, other taxa including butterflies, flies, beetles, wasps, bats, birds, lizards, and even mammals can all prove to be important pollinators in certain habitats and in the case of particular plants. Nevertheless, none of these achieve the numerical dominance as visitors to flowers that bees are capable of (Winfree, 2010). Of the more than 16,000 bee species described worldwide (Michener, 2007), honeybees (*Apis mellifera*), bumblebees (*Bombus*), leafcutting bees (*Megachile*), and mason bees (*Osmia*) have been recognized as the most efficient pollinators of a wide variety of crops (Nogué et al., 2016). However, managed honeybees were excluded from the estimates of pollination potentials, as their abundance is not driven by natural ecosystem conditions but by the decisions beekeepers take. In turn, the latter decisions are driven, not only by the quality of pasture for bees but also by economic conditions and local traditions as regards rearing (Gerula et al., 2007).

It is worth noting here that wild bees are as effective pollinators as honeybees in the cases of many crops, and in fact more effective in some cases (Winfree, 2010). Unmanaged bees alone can fully pollinate crops in some agricultural contexts and are frequent flower visitors in others (Ricketts et al., 2008). Therefore, only wild bees were considered when estimating the ecosystem potential for providing pollination, and their potential abundance was considered the best indicator.

To construct indicator and then map ecosystem potential to provide pollination service, a more detailed definition of potential was needed so that ranges of values and the level of uncertainty might be reduced. In consequence, the new, more operational definition introduced holds that the potential of an ecosystem to deliver pollination service is a maximum theoretical service supply in a given type of ecosystem and regional context, calculated for the environmental setting best suited for a given service (for example, as regards plant species composition, soil quality, water balance, etc.) (Affek, 2018, see also section on honey). This, for instance, means that, in calculating potential of cropland located on fertile soil, a selection needs to be made of a crop (cultivated in the region on such a soil) that can nest and feed the number of wild bees larger than a hectare of any other crop. In turn, estimations of the pollination potentials of given forest types entail selection of the most desirable plant composition and nesting resources that it is possible to encounter in the given region (within the given ecosystem type) – from the point of view of bees. Taking environmental settings other than the optimal (for example, the average or most often occurring) would not reflect the full potential of the given ecosystem type. In reality, however, the potential number of wild pollinators in a particular ecosystem patch may differ significantly from the proposed ranges. Moreover, in actually existing ecosystems, environmental conditions (including plant cover) rarely allow dedicated potential to be realized. In some cases (e.g., heavily degraded grassland, arable fields with anemophilous crops, or dense planted pine forest), the estimated potential even very far exceeds the real supply.

Assessment method

It was decided based on the literature review that the best indicator of an ecosystem capacity to deliver pollination service is the number of pollinating insects (in particular, the number of wild bees) capable of nesting and foraging within a defined ecosystem area. The abundance of wild pollinators within nesting habitats was estimated (after Lonsdorf et al., 2009) by reference to the availability of the resources bees require to survive and reproduce. These can be roughly divided into those relating to nesting (e.g., appropriate substratum, such as bare soil, stems, or cavities, as well as [leaf or resinous] materials needed by some species to create a nest interior) and those relating to nutrition (mainly floral resources providing nectar and pollen) (Kremen et al., 2007; Westrich, 1996; Winfree, 2010). Similar to managed honeybees, non-Apis bees are also able to exploit honeydew produced by aphids as a carbohydrate source (Banaszak, 1983; Konrad et al., 2009).

In reality, it is often the case that favorable nesting conditions are offered in one ecosystem, while necessary floral resources are present in another, neighboring one. In such circumstances, it becomes difficult to evaluate the potentials of ecosystems independently, with spatial analysis at the landscape level thus looking like a more informative approach, providing that account can be taken of proportions of occurring ecosystem types, as well as the spatial arrangement of particular patches (see Banaszak, 1983). In this work, however, assessment of the potential represented by different ecosystem types rather than the local landscape structure eschews the spatial aspect, instead basing itself around a theoretical assumption that a bee is within flying distance of just one ecosystem type (hence with the food base and nesting sites accessible within the given ecosystem only).

The large body of entomological and ecological literature relating to the nesting and food preferences of wild bees (e.g., Kremen et al., 2007; Westrich, 1996; Winfree, 2010) was set against data from multiseasonal observations of wild bees carried out in WNP by Krzysztofiak (2001) and Banaszak and Krzysztofiak (1996). Other research reports presenting direct calculations of the abundance of wild bees were also drawn on (Banaszak, 2010, 1983; Banaszak and Cierzniak, 2000; Banaszak and Jaroszewicz, 2009; Banaszak and Krzysztofiak, 1992; Banaszak and Szefer, 2013; Pawlikowski, 2010), as were data on melliferous plants (Denisow and Wrzesień, 2015, 2007) in the different lowland ecosystems of Central Europe. Data on the number of bees per unit area (densities) obtained by these authors using the belt method developed by Banaszak (1980) served as a starting point for estimating the potential numbers of wild bees nesting in 1 ha of a given ecosystem type.

The results of entomological research suggest there is a less-diverse bee fauna in woodlands than in open habitats (Banaszak and Jaroszewicz, 2009). In Central Europe, unimproved meadows are among the most important habitats for bees (Westrich, 1996). In Poland, the highest noted densities of wild pollinators (more than 8000 individuals per ha) were recorded on dry grassland in Masuria (Banaszak, 2010, 2009). In another study (Banaszak and Szefer, 2013), the authors reported much greater abundance and species richness of wild bees on fallow land, around rural settlements, and on clearings than in mature forest (be this either of pine or oak-hornbeam) or on moist meadows. Interestingly, all of the habitats with markedly greater abundance of bees were created as a result of human activity. In turn, in a study conducted in northeast Poland's Białowieża National Park - which is renowned for its "primaeval" character - the smallest numbers of bee species are recorded in old-growth oak-hornbeam and riparian forest, while forest clearings in coniferous forests accommodated the greatest numbers of wild bees (Banaszak and Jaroszewicz, 2009).

Banaszak and Krzysztofiak (1992) conducted a metaanalysis regarding abundances of wild bees in different forest communities of Poland and concluded that there were much greater total abundances of bees in coniferous forests (75—406 individuals per ha) than in broad-leaved de-ciduous forests (69—88). Research also shows a significant inverse corre-lation between bee abundance and forest age, be this in naturally regenerated or planted forest (Krzysztofiak, 2001; Pawlikowski, 2010; Taki et al., 2013) (Table 6.25). A slight increase in numbers of nesting bees can only be expected in the oldest forests retaining certain "primaeval" features. Here, the presence of a large amount of standing and lying dead wood increases numbers of nesting sites, while natural disturbances give rise to canopy gaps that appear to let in more light to the forest floor and thus to create favorable conditions for honey plants (see Banaszak and Jaroszewicz, 2009).

Indicator values
The highest potential to provide pollination service was achieved by dry grassland together with clearings in coniferous forests with high honey plant coverage (Table 6.26). They were considered the best habitats for

Table 6.25 Values for abundance and species richness among wild bees inhabiting mixed forest at different successional stages, in Wigry National Park (northeast Poland). Values from three studied seasons.

Forest age	Maximum abundance (wild bees ha^{-1})	Mean abundance (wild bees ha^{-1})	Species richness	Shannon-Weaver diversity index (H)
7—9	1550	532	106	3.52
25	550	194	66	3.13
50	750	239	27	2.63
100	450	71	33	2.59

Based on Krzysztofiak, A., 2001. Struktura zgrupowań pszczół (Apoidea, Hymenoptera) w różnowiekowych drzewostanach świerkowo-sosnowych Wigierskiego Parku Narodowego. Zeszyty Naukowe Akademii Bydgoskiej im. Kazimierza Wielkiego w Bydgoszczy. Studia Przyrodnicze 15, 113—215 (in Polish).

wild pollinators (supporting 1000–2000 or even 8000 nesting individuals per ha) in European postglacial landscape. The potential of forest ecosystems was differentiated in relation to forest age. Forest clearings and young forests were regarded as better wild bee habitats than mature stands with dense canopies. In turn, cropland, notwithstanding the highest recorded densities of pollinators when covered with honey crop plants (Banaszak, 1983), only serves as an averagely good habitat for wild bees (with up to 100–200 nesting individuals per ha) due to the short blooming period in the monoculture and the potentially disturbed soil nesting properties (thanks to tillage or pesticide and fertilizer application). On the basis of other studies (e.g., Banaszak, 1983; Banaszak and Cierzniak, 2000), it was assumed that extremely high but ephemeral abundance results from the migration of bees from neighboring refuge habitats in search of food, with only a small fraction of individuals actually nesting directly in the crop.

Rural settlement areas were also considered moderately good bee habitat (with up to 300–600 nesting individuals per ha) because, apart from vegetated (garden and orchard) areas rich in melliferous species, as well as constructions made of natural materials (such as wood and clay) convenient for nesting, much of the space is paved and completely unsuitable for bees.

A more detailed diversification of the potential of ecosystems to provide pollination, based on an eleven-point scale, along with maps and a wide discussion of results, can be found in Affek (2018).

Spatial pattern of ecosystem potential to provide pollination in the central and eastern part of the study area is formed by a uniform matrix of high values with small patches of other values clustered on its background (Fig. 6.12). In the northern and western parts, the pattern changes into matrix of low potential with bigger number of patches of high and medium potential. The largest patch covers over 15% of the study area. Areas of low and high potential codominate, covering together over 67% of the total area. The largest patches are generally more irregular in shape compared with small patches, but differences are generally little compared with other ES potentials. Small patches of the same level of potential are quite clustered, and interspersion of patches of different level of potential is on an average level compared with other ES potentials.

As POLLIN indicator values were assigned to ecosystem types, maps presenting potential service supply do not account for the spatial arrangement of ecosystem patches. There are several proven landscape-level effects modifying wild bee abundance and pollination efficiency (Banaszak, 1992;

Table 6.26 Ecosystem potential to provide pollination service (1 - very low; 5 - very high) indicated by the abundance of nesting wild bees (individuals ha^{-1}).

Ecosystem type	Ecosystem acronym	POLLIN indicator value	Ecosystem potential
Settlement area	SETTLE	300.1−600.0	4
Arable field on poor sandy soil	ARB1	100.1−200.0	2
Arable field on fertile mesic soil	ARB2	100.1−200.0	2
Arable field on fertile moist soil	ARB3	0.1−100.0	1
Dry grassland	GRAS1	>600.0	5
Wet grassland on mineral soil	GRAS2	200.1−300.0	3
Peat grassland	GRAS3	100.1−200.0	2
Alder forest <40 years	ALD1	0.1−100.0	1
Alder forest 40−60 years	ALD2	0.1−100.0	1
Alder forest 60−80 years	ALD3	0.1−100.0	1
Alder forest 80−120 years	ALD4	0.1−100.0	1
Alder forest >120 years	ALD5	0.1−100.0	1
Riparian forest <40 years	RIP1	100.1−200.0	2
Riparian forest 40−60 years	RIP2	0.1−100.0	1
Riparian forest 60−80 years	RIP3	0.1−100.0	1
Riparian forest 80−120 years	RIP4	0.1−100.0	1
Riparian forest >120 years	RIP5	0.1−100.0	1
Oak-hornbeam forest <40 years	OAK1	200.1−300.0	3
Oak-hornbeam forest 40−60 years	OAK2	100.1−200.0	2
Oak-hornbeam forest 60−80 years	OAK3	100.1−200.0	2
Oak-hornbeam forest 80−120 years	OAK4	0.1−100.0	1
Oak-hornbeam forest >120 years	OAK5	100.1−200.0	2
Coniferous and mixed forest <40 years	CON1	>600.0	5
Coniferous and mixed forest 40−60 years	CON2	300.1−600.0	4
Coniferous and mixed forest 60−80 years	CON3	300.1−600.0	4
Coniferous and mixed forest 80−120 years	CON4	300.1−600.0	4

Continued

Table 6.26 Ecosystem potential to provide pollination service (1 - very low; 5 - very high) indicated by the abundance of nesting wild bees (individuals ha^{-1}). (cont'd)

Ecosystem type	Ecosystem acronym	POLLIN indicator value	Ecosystem potential
Coniferous and mixed forest >120 years	CON5	300.1−600.0	4
Swamp coniferous forest <40 years	SWP1	300.1−600.0	4
Swamp coniferous forest 40 −60 years	SWP2	200.1−300.0	3
Swamp coniferous forest 60 −80 years	SWP3	200.1−300.0	3
Swamp coniferous forest 80 −120 years	SWP4	100.1−200.0	2
Swamp coniferous forest >120 years	SWP5	200.1−300.0	3
Wetlands - reed beds and sedge swamps	REED1	0.1−100.0	1
Wetlands - fens	FEN	0.1−100.0	1
Wetlands - bogs	BOG	0.1−100.0	1
Lakes (all types)	LAKE	0	0

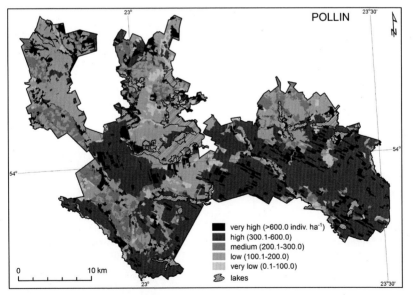

Figure 6.12 Ecosystem potential to provide pollination service indicated by the abundance of nesting wild bees (POLLIN indicator; individuals ha^{-1}).

Ricketts et al., 2008). The works of different authors have demonstrated that landscape structure, i.e., the sizes and shapes of patches (Banaszak, 1992; Winfree, 2010), proximity and neighborhood (Denisow and Wrzesień, 2015), and connectivity (Diekötter et al., 2008) play a crucial role in shaping the abundance of pollinators, as well as the distribution and richness of the melliferous flora (Denisow and Wrzesień, 2015, 2007). The presence of green linear elements is also considered to have a positive impact on wild bee populations (Schulp et al., 2014).

Unlike many relevant ES contributing greatly to human well-being (e.g., carbon sequestration, timber production, erosion control, and water retention), the potential to provide pollination service (such as the potential to produce honey) decreases with natural succession and biomass growth. Preforest ecosystems are not particularly stable, as both canopy gaps, but also unmanaged grassland, are subject to succession, thereby losing their main relevance as refuge habitats and bee pastures. This apparent trade-off between services must thus be considered carefully as forest management practices are planned and when facing the dilemma between an active or passive approach to the conservation of open land.

6.1.2.4 Nursery habitat maintenance
Ecosystem service description

This service belongs to *Regulation and Maintenance* section and *Regulation of physical, chemical, biological conditions* division in CICES V5.1 (Table 6.27). It can be further described as the presence of ecological habitats necessary for sustaining populations of plant, animal, and fungi species that people use or enjoy. Six measures have been proposed to indicate the potential supply of this service: two of them were calculated within ecosystem types (occurrence of rare and endangered habitats that are considered to be of European interest, species richness of herb layer vascular plants), the other two within hunting units (species diversity of game mammals and hazel grouse population density), and the last two within landscape units (ecosystem diversity and abundance of small-scale habitats).

ES *maintaining nursery populations and habitats (including gene pool protection)* classified as a regulating service in CICES V5.1 is the one that raises the most controversy among researchers. The main reason for various interpretations of this service is the fact that it can be treated not only as final ES but also as an intermediate ES underpinning several other services, including provisioning. Broad literature analysis carried out by Liquete et al. (2016) shows that ES related to the presence of habitats are treated by

Table 6.27 Ecosystem service in Common International Classification of Ecosystem Services (CICES) V5.1 and indicator metadata.

Ecosystem service		Nursery habitats maintenance					
CICES V5.1	Section	Regulation and Maintenance					
	Division	Regulation of physical, chemical, biological conditions					
	Group(s)	Lifecycle maintenance, habitat, and gene pool protection					
	Class(es)	Maintaining nursery populations and habitats					
	Class code(s)	2.2.2.3					
Indicandum		**Potential of nature to maintain nursery habitats**					
Indicator		Occurrence of rare and endangered habitats that are considered to be of European interest	Species richness of herb layer vascular plants	Species diversity of game mammals	Hazel grouse population density	Ecosystem diversity	Abundance of small-scale habitats
Acronym		NATURA	RICHNESS	MAMMAL	GROUSE	DIVERSITY	HABITAT
Short description		Assessment of ecosystem rarity and maturity based on the correspondence with Natura 2000 Habitat	Calculation of 90th percentile of the number of vascular plant species from the a large set of phytosociological relevés	Calculation of game mammal diversity using Shannon–Weaver diversity	Calculation of the number of individuals of hazel grouse per km²; as rare umbrella	Calculation of ecosystem diversity using Shannon–Weaver diversity index: $H = -\sum p_i \log_2 p_i$,	Calculation of the number of small-scale habitats (below 3 ha) per km²; includes refuge habitats such as hedges,

types and age of tree stands		index: H = $-\sum p_i \log_2 p_i$, where p_i is the number of animals of the ith species in the total number of analyzed animals; 15 mammal species considered	species, its abundance reflects well the level of biodiversity of the entire ecosystem, both fauna and flora	where p_i is the proportion of ith type of ecosystem in the total area	clumps, and belts of trees, scrub, wet and swampy areas, boulders and boulder fields, large single trees, etc.
Direct/indirect	Indirect	Indirect	Indirect	Indirect	Indirect	Indirect
Simple/compound/complex	Simple	Simple	Compound	Simple	Compound	Compound
Estimated/calculated	Estimated	Calculated	Calculated	Calculated	Calculated	Calculated
Scale	Rank	Ratio	Ratio	Ratio	Ratio	Ratio
Value range	1–5	10–60	1.8–3.3	0–14	0.1–4.1	0.3–103.4
Unit	–	–	–	Individuals km^{-2}	–	Objects km^{-2}
Spatial unit	Ecosystem	Ecosystem	Hunting unit	Hunting unit	Landscape unit	Landscape unit
Value interpretation	1 → very low potential;	<20.0 → very low potential;	<2.50 → very low	0 → no relevant	<0.75 → very low	<5.0 → very low potential;

Continued

Table 6.27 Ecosystem service in Common International Classification of Ecosystem Services (CICES) V5.1 and indicator metadata. (cont'd)

Ecosystem service	Nursery habitats maintenance				
2 → low; 3 → medium; 4 → high; 5 → very high	20.0–30.0 → low; 30.1–40.0 → medium; 40.1 −50.0 → high; >50.0 → very high	potential; 2.50–3.00 → low; 3.01 −3.10 → medium; 3.11–3.20 → high; >3.2 → very high	potential; 0.1 −1.0 → very low potential; 1.1 −2.5 → low; 2.6 −5.0 → medium; 5.1 −10.0 → high; >10.0 → very high	potential; 0.75 −1.50 → low; 1.51 −2.25 → medium; 2.26 −3.00 → high; >3.00 → very high	5.0–25.0 → low; 25.1 −50.0 → medium; 50.1 −75.0 → high; >75.0 → very high
Source data	The list of Natura 2000 Habitats included in the Habitats Directive (Council Directive 92/ 43/EEC) and the Regulation of the Minister of the Environment (2014)	Phytosociological relevés: over 400 from the study area (made by the Authors and from Wigry National Park Protection Plan) and other 100 from the region	Annual hunting reports for the years 2011−14 from the four forest districts and the Wigry National Park	Map of ecosystems of the study area	Spatial database of topographic objects (BDOT10k), aerial photos, vegetation map of the Wigry National Park

different authors in various ways, which is reflected in specific names, e.g., habitat for species, maintenance of habitats and reproductive populations, maintenance of life cycles, maintenance of habitats of biological origin, and maintenance of reproductive areas, refuge, and resting areas. It should be noted, however, that the differences in names also reflect differences in conceptual approaches.

Different authors include the role of ecosystems and/or landscapes in enabling species populations to survive, grow, breed, and expand into different ES groups (see Liquete et al., 2016): (a) regulating, (b) supporting, and (c) a separate group of habitat services. The scope of these services is also different, as they may relate to habitats necessary for (a) all life processes of all species (plants, animals, and fungi), (b) reproductive functions and raised young populations but only for species - direct producers of provisioning services, and (c) reproduction and resting places of migratory species (Fig. 6.13).

A different approach, which does not directly link indicators with the abovementioned categories of areas and species, was used by Maes et al. (2014). They proposed to quantify "maintenance of nursery populations and habitats" service by means of proxy indicators of ecosystem condition (such as conservation investments, habitat landscape protection, biodiversity value, and ecological status or habitat diversity) developed under the *European Union Biodiversity Strategy*. These indicators can be also treated as ES surrogates (Maes et al., 2016, 2014).

Different concepts have resulted in the development of a whole range of different indicators, some of which are presented in Table 6.28. The presented selection is representative for the majority of European approaches used on a European, national, or regional scale. In most approaches, separate indicators are proposed for forest areas and for agricultural ones.

Based on the considerations summarized above, the concept of relatively broad meaning of the service was adopted for our study. This service applies mainly to breeding and nursery areas and does not include resting areas for migratory species that are not identifiable in the case study scale. This approach is valid for all species from all taxonomic groups but undoubtedly is focused on vascular plants and vertebrate animals for which the links to selected indicators are much better documented.

Because of the complexity of the concept of this service, a set of six various indicators was used (NATURA, RICHNESS, MAMMAL, GROUSE, DIVERSITY, HABITAT). This set meets the following conditions simultaneously:

a) adjustment to the spatial scale of the study;

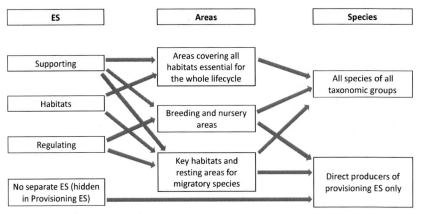

Figure 6.13 Location and scope of the habitat maintenance ecosystem service. *(After Liquete, C., Cid, N., Lanzanova, D., Grizzetti, B., Reynaud, A., 2016. Perspectives on the link between ecosystem services and biodiversity: The assessment of the nursery function. Ecological Indicators 63, 249–257; Maes, J., Zulian, G., Thijssen, M., Castell, C., Baró, F., Ferreira, A.M., Melo, J., Garrett, C.P., David, N., Alzetta, C., Geneletti, D., Cortinovis, C., Zwierzchowska, I., Louro Alves, F., Souto Cruz, C., Blasi, C., Alós Ortí, M.M., Attorre, F., Azzella, M.M., Capotorti, G., Copiz, R., Fusaro, L., Manes, F., Marando, F., Marchetti, M., Mollo, B., Salvatori, E., Zavattero, L., Zingari, P.C., Giarratano, M.C., Bianchi, E., Duprè, E., Barton, D., Stange, E., Perez-Soba, M., van Eupen, M., Verweij, P., de Vries, A., Kruse, H., Polce, C., Cugny-Seguin, M., Erhard, M., Nicolau, R., Fonseca, A., Fritz, M., Teller, A., 2016. Mapping and Assessment of Ecosystems and their Services. Urban Ecosystems (fourth Report — Final, May 2016). Technical Report — 2016 –102, European Union, Luxembourg, modified.)*

b) the possibility to use existing data collected in a standard manner;

c) adoption of proxy indicators whose diagnostic value results from statistically confirmed relationships between landscape structure or ecosystem properties and functions performed for specific groups of species;

d) representation by indicators of species, ecosystem, and landscape levels in such a way that at least some of them concern areas of any land cover type, without limitation only to forest or agricultural areas.

All of the above indicators refer in a complex manner to habitats for plants, animals, and fungi, but – because of their specificity – two first are more focused on habitats for plants, while the other four better indicate habitats for animals from different taxonomic groups.

Table 6.28 Selected measures showing the potential of nature to maintain habitats for plants, animals, and fungi at various levels of biosphere organization.

Level	General	Forests and similar	Agroecosystems and similar	Source
Landscape level		Spatial structure, connectivity, ecological corridors		Saastamoinen et al. (2014)
		Investments in forest protection		Maes et al. (2014)
			Trees in agricultural areas, share of High Nature Value farmland	Maes et al. (2014)
	The share of wetlands suitable for amphibians			Turkelboom et al. (2014)
Habitat level			Homogeneity, fragmentation, intensive agriculture/ extensive agriculture	Helfenstein and Kienast (2014)
		Coarse dead wood, protected areas, nitrogen deposit		Helfenstein and Kienast (2014)
		Habitats of common species, protected habitats (various forms of protection), endangered habitats, old-growth forests		Saastamoinen et al. (2014)

Continued

Table 6.28 Selected measures showing the potential of nature to maintain habitats for plants, animals, and fungi at various levels of biosphere organization. (cont'd)

Level	General	Forests and similar	Agroecosystems and similar	Source
	Quality of habitats (common birds index)			Maes et al. (2015)
		Protected forests (Natura, 2000 and similar), protected refugee of selected animal species		Maes et al. (2014)
Species level		Total number of species, endangered species, invasive species	Total number of species, endangered species, invasive species	Helfenstein and Kienast (2014)
		Structure and distribution of tree species		Maes et al. (2014)
The level of intraspecific genetic diversity			Races of cultivated plants, races of farm animals	Helfenstein and Kienast (2014)
		In situ and ex situ protection		Saastamoinen et al. (2014)
			Traditional orchards	Maes et al. (2014)

NATURA indicator: Occurrence of rare and endangered habitats that are considered to be of European interest
Theoretical framework

The first indicator used is "occurrence of rare and endangered habitats that are considered to be of European interest" (NATURA). The Habitats Directive (Council Directive 92/43/EEC; Regulation (EU) No 1143/2014) introduced the concept of Natural Habitats of Community interest, also called NATURA 2000 Habitats[4]. These are in fact ecosystems that have to (a) be in danger of disappearance in their natural range; (b) have a small natural range following their regression or by reason of their intrinsically restricted area; or (c) present outstanding examples of typical characteristics of one or more of the five following biogeographical regions: Alpine, Atlantic, Continental, Macaronesian, and Mediterranean. The best-developed Natura 2000 Habitats are characterized by high species richness, compound vertical structure, and complex relationships between species occupying different ecological niches. In turn, common synanthropic habitats are characterized by more variable (unstable) and often nonspecific species composition with simplified relationships between species.

NATURA 2000 Habitat types of priority conservation status in EU were listed in the Appendix to the Habitats Directive and then adapted to the Polish law, together with a more precise characteristic of vegetation corresponding to each NATURA 2000 Habitat type. The aim of activities related to these habitats/ecosystems, in individual countries, is to preserve them in a proper state of conservation or to restore them, which is to ensure maintaining natural biodiversity of specific areas. According to the above assumptions, the occurrence of such habitats/ecosystems in the given area directly testifies, on the one hand, the presence of rare and endangered species, and on the other hand on the presence of sites that have the capacity to nurture and maintain rare and endangered species. Therefore, the occurrence of NATURA 2000 Habitats and their level of maturity can be considered as a direct and simple indicator describing one aspect of the potential of a region to maintain nursery habitats.

---

[4] Natura 2000 Habitats - "terrestrial or aquatic areas distinguished by geographic, abiotic, and biotic features, whether entirely natural or seminatural" (Council Directive 92/43/EEC). In most cases, a single individual Natura 2000 Habitat is equivalent to one ecosystem or a group of similar ecosystems (in a narrow sense).

Assessment method

All 42 types of ecosystem were evaluated on a 1—5 rank scale in terms of their maturity and correspondence with NATURA 2000 Habitat types listed in the EU Habitats Directive. Ecosystem abiotic conditions, plant species composition, and succession stage were taken into account (Table 6.29).

Indicator values

The ecosystems with the highest rank (the highest potential) include riparian, oak-hornbeam, and swamp coniferous forests with tree stands over 80 years old (Fig. 6.14, Table 6.30). Peat grassland, bogs, and dystrophic lakes received the same top value. In turn, the same forest types but with middle age trees (60—80 years old) were ranked as having high potential, such as dry grassland, fens, and both mesotrophic and eutrophic lakes.

The medium potential is characteristic for riparian, oak-hornbeam, and swamp coniferous forests younger than 60 years old, as well as for alder forests and coniferous and mixed forests over 80 years old. This group also comprises wet grassland on mineral soil. Alder forests, as well as coniferous and mixed forests aged 60—80, were assessed as having low potential. All other ecosystems received the lowest rank.

Table 6.29 Characteristics of ecosystems in relation to their estimated potential (on the 1—5 rank scale) to maintain nursery habitats.

Ecosystem category	Rank
Best-developed and mature ecosystems corresponding to Natura 2000 Habitats: (A) forest ecosystems - trees older than 80 years old, (B) nonforest ecosystems - types rare in Poland and with well-established characteristic combination of species	5
Premature ecosystems corresponding to Natura 2000 Habitats: (A) forest ecosystems - middle age trees (60—80 years old), (B) nonforest ecosystems - types more common in Poland and/or under visible human pressure	4
A. Remaining ecosystems corresponding to Natura 2000 Habitats - only young forest ecosystems with tree stands below 60 years old B. Mature forest ecosystems (with tree stands above 80 years old) not representing Natura 2000 Habitats	3
Remaining forest ecosystems not representing Natura 2000 Habitats	2
Remaining nonforest ecosystems not representing Natura 2000 Habitats	1

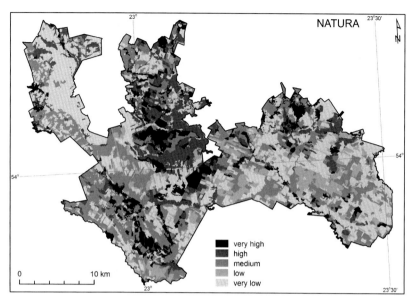

Figure 6.14 Ecosystem potential to maintain nursery habitats indicated by the occurrence of rare and endangered habitats that are considered to be of European interest (NATURA indicator).

The spatial pattern of NATURA indicator values is determined by land cover diversity on the one hand and by forest management in individual forest districts on the other hand. In general, the spatial pattern is mosaic type with patches of various sizes, but – compared with other ES potentials – the size variance is the lowest. The largest patch covers only 5% of the study area. Patches of different level of potential occupy similar areas, which results in the highest value of Shannon evenness index (SEI) compared with other ES potentials studied. The largest complex of ecosystems with the highest potential is located within the WNP and comprises mainly forests. In addition to this large complex, also valleys and depressions with valuable nonforest ecosystems have the highest potential to maintain nursery habitats. The largest patches are generally more irregular in shape compared with small patches, but differences are not very high. Small patches of the same level of potential are quite evenly distributed in the local scale, and the interspersion of patches of different level of potential is high compared with other ES potentials.

Table 6.30 Ecosystem potential to maintain nursery habitats (1 - very low; 5 - very high) indicated by the occurrence of rare and endangered habitats that are considered to be of European interest.

Ecosystem type	Ecosystem acronym	NATURA indicator value/ ecosystem potential
Settlement area	SETTLE	1
Arable field on poor sandy soil	ARB1	1
Arable field on fertile mesic soil	ARB2	1
Arable field on fertile moist soil	ARB3	1
Dry grassland	GRAS1	4
Wet grassland on mineral soil	GRAS2	3
Peat grassland	GRAS3	5
Alder forest <40 years	ALD1	1
Alder forest 40–60 years	ALD2	1
Alder forest 60–80 years	ALD3	2
Alder forest 80–120 years	ALD4	3
Alder forest >120 years	ALD5	3
Riparian forest <40 years	RIP1	3
Riparian forest 40–60 years	RIP2	3
Riparian forest 60–80 years	RIP3	4
Riparian forest 80–120 years	RIP4	5
Riparian forest >120 years	RIP5	5
Oak-hornbeam forest <40 years	OAK1	3
Oak-hornbeam forest 40 –60 years	OAK2	3
Oak-hornbeam forest 60 –80 years	OAK3	4
Oak-hornbeam forest 80 –120 years	OAK4	5
Oak-hornbeam forest >120 years	OAK5	5
Coniferous and mixed forest <40 years	CON1	1
Coniferous and mixed forest 40 –60 years	CON2	1
Coniferous and mixed forest 60 –80 years	CON3	2
Coniferous and mixed forest 80 –120 years	CON4	3
Coniferous and mixed forest >120 years	CON5	3
Swamp coniferous forest <40 years	SWP1	3

Table 6.30 Ecosystem potential to maintain nursery habitats (1 - very low; 5 - very high) indicated by the occurrence of rare and endangered habitats that are considered to be of European interest. (cont'd)

Ecosystem type	Ecosystem acronym	NATURA indicator value/ ecosystem potential
Swamp coniferous forest 40 −60 years	SWP2	3
Swamp coniferous forest 60 −80 years	SWP3	4
Swamp coniferous forest 80 −120 years	SWP4	5
Swamp coniferous forest >120 years	SWP5	5
Wetlands - reed beds and sedge swamps	REED1	1
Wetlands - fens	FEN	4
Wetlands - bogs	BOG	5
Large, deep mesotrophic lakes	LAKE1	4
Large, deep eutrophic lakes	LAKE2	4
Big eutrophic lakes of medium depth	LAKE3	4
Medium-sized eutrophic lakes of medium depth	LAKE4	4
Small shallow eutrophic lakes	LAKE5	4
Small dystrophic lakes	LAKE6	5
Artificial lakes	LAKE7	1

RICHNESS indicator: Species richness of herb layer
Theoretical framework

The number of vascular plant species in plant communities depends on a number of different factors on a regional and local scale, the most important of which are:

- habitat diversity (quality and quantity of resources – including fertility and moisture, as well as the presence of barriers);
- possible pool of species (biogeographical elements);
- diversity of land uses;
- biology of specific groups of organism;
- level of anthropogenic impacts;
- succession stage;
- size of plant community patch;
- neighborhood and distance to the nearest neighbor.

The combination of these conditions causes that each patch of vegetation has largely individual character and is characterized by a specific species composition, which additionally changes over time. Therefore, for different studies, evaluation, and model needs, the ideal solution would be to determine the number of species in each patch separately. However, it is hardly possible due to practical reasons. Therefore, for the needs of indicator systems, appropriate for regional or wider scales, most commonly the average number of species in particular ecosystem types is determined. However, it should be emphasized that for such a measure to be operational, it should be referred to narrow typological units, preferably distinguished through phytosociological analysis. The use of indicators employing species richness (not only plants) to broader categories (e.g., Corine Land Cover or ecosystem types in the MAES approach) does not make much sense.

The use of species richness as an indicator of the potential provision of some ES, despite some theoretical reservations (see e.g., Vos et al., 2014), is relatively widely accepted. It is based on a known and generally occurring mechanism, in that the more producer species there are, the more species on each higher trophic level find their food base. However, in such conditions, there is a higher probability of occurrence of highly competitive species, and there are less free ecological niches (Hassan et al., 2005). These dependencies have been demonstrated many times, e.g., in relation to plant–butterfly relationships (Dolek and Geyer, 2002) or plants–birds and plants–invertebrates (Billeter et al., 2008). In different assessment systems (regarding issues other than ES), plant species richness is commonly used as proxy indicator for the richness of other systematic groups or general diversity (see Britton and Fischer, 2007; Magurran, 2004).

Taking account of the above considerations, it can be assumed that the number of vascular plant species in the ecosystem, especially the species of herb layer, is a reasonable (although "proxy") indicator of ecosystem potential to maintain habitats for plants, animals, and fungi. Such an indicator was also included in a set developed for Switzerland (Helfenstein and Kienast, 2014) and in Swedish ES assessments in agricultural areas - although based on a slightly different theoretical approach (Andersson et al., 2015).

Assessment method

The input data were phytosociological relevés from various sources. Most of them are authors' own data and unpublished documentation from the Protection Plan of the WNP. They were supplemented with phytosociological materials from publications related to the analyzed area and its closest neighborhood. In total, about 500 relevés were analyzed. In the majority of them, the reference area for the list of species was in the range of 100–400 m^2. Regardless of the tabular and syntaxonomic approaches introduced by the authors of the source materials, individual phytosociological relevés were combined into groups corresponding to the types of ecosystem distinguished for the needs of the current study. Then, for each of these groups, relevés were sorted according to the increasing number of species. Because of the fact that the number of species in the relevés relating to at least some ecosystem types was quite diverse, and there were some relevés with an exceptionally high or small number of species, neither the value of the average number of species nor the minimum–maximum range would adequately reflect the average species richness. Hence, the values of 10th and 90th percentile were taken as accurate measures of variability, which were then rounded to the nearest five species. Finally, the 90th percentile was selected as RICHNESS indicator to show ecosystem potential to maintain nursery habitats.

Indicator values

Ecosystems with the highest indicator value (often more than 50 species revealed in phytosociological relevés) cover only some types of forest in specific age ranges, i.e., alder forests over 60 years old, riparian forests, and swamp coniferous forests over 80 years (Table 6.31). Oak-hornbeam forests in younger age classes (less than 60 years) also belong to that group. In this case, the high species richness results from the coexistence of many non-forest species, which disappear during the further vegetation succession.

The second group, that is, ecosystems with high values of RICHNESS indicator (the number of species is usually around 40), covers the majority of the remaining forest communities, riparian forests (less than 80 years), oak-hornbeam forests (over 60 years), and coniferous and mixed forests with pine and spruce stands (40–60 years old and over 80 years old).

The third group, which includes ecosystems with an average potential to maintain nursery habitats (the number of species in the relevé is often close to 30), includes a few forest communities (alder forests below 60 years, riparian forests below 40 years, and some mixed coniferous forests). Non-forest communities constitute large part of that group, including all agro-ecosystems and the majority of meadows and pastures (dry grassland and peat grassland). This group also includes fens.

The group characterized by a low value of the indicator (the number of species in the relevé rarely exceeds 20) includes swamp coniferous forests with a tree stand of up to 80 years old and wet grassland on mineral soil.

Table 6.31 Ecosystem potential to maintain nursery habitats (1 - very low; 5 - very high) indicated by species richness of herb layer vascular plants.

Ecosystem type	Ecosystem acronym	Species richness - 10 percentile	RICHNESS indicator value - 90 percentile	Ecosystem potential
Settlement area	SETTLE	Not considered		
Arable field on poor sandy soil	ARB1	20	30	3
Arable field on fertile mesic soil	ARB2	20	30	3
Arable field on fertile moist soil	ARB3	20	30	3
Dry grassland	GRAS1	20	30	3
Wet grassland on mineral soil	GRAS2	10	20	2
Peat grassland	GRAS3	20	30	3
Alder forest <40 years	ALD1	25	35	3
Alder forest 40 −60 years	ALD2	25	35	3
Alder forest 60 −80 years	ALD3	35	55	5
Alder forest 80 −120 years	ALD4	35	55	5
Alder forest >120 years	ALD5	35	55	5
Riparian forest <40 years	RIP1	25	35	3
Riparian forest 40 −60 years	RIP2	35	40	4
Riparian forest 60 −80 years	RIP3	35	40	4
Riparian forest 80 −120 years	RIP4	40	50	5
Riparian forest >120 years	RIP5	40	50	5
Oak-hornbeam forest <40 years	OAK1	50	60	5
Oak-hornbeam forest 40−60 years	OAK2	40	50	5
Oak-hornbeam forest 60−80 years	OAK3	35	45	4

Table 6.31 Ecosystem potential to maintain nursery habitats (1 - very low; 5 - very high) indicated by species richness of herb layer vascular plants. (cont'd)

Ecosystem type	Ecosystem acronym	Species richness - 10 percentile	RICHNESS indicator value - 90 percentile	Ecosystem potential
Oak-hornbeam forest 80 −120 years	OAK4	35	45	4
Oak-hornbeam forest >120 years	OAK5	35	45	4
Coniferous and mixed forest <40 years	CON1	15	30	3
Coniferous and mixed forest 40 −60 years	CON2	20	40	4
Coniferous and mixed forest 60 −80 years	CON3	15	30	3
Coniferous and mixed forest 80 −120 years	CON4	25	40	4
Coniferous and mixed forest >120 years	CON5	25	40	4
Swamp coniferous forest <40 years	SWP1	15	25	2
Swamp coniferous forest 40−60 years	SWP2	15	25	2
Swamp coniferous forest 60−80 years	SWP3	15	25	2
Swamp coniferous forest 80 −120 years	SWP4	25	50	5
Swamp coniferous forest >120 years	SWP5	25	50	5
Wetlands - reed beds and sedge swamps	REED1	10	15	1
Wetlands - fens	FEN	20	35	3
Wetlands - bogs	BOG	10	15	1

Continued

Table 6.31 Ecosystem potential to maintain nursery habitats (1 - very low; 5 - very high) indicated by species richness of herb layer vascular plants. (cont'd)

Ecosystem type	Ecosystem acronym	Species richness - 10 percentile	RICHNESS indicator value - 90 percentile	Ecosystem potential
Large, deep mesotrophic lakes	LAKE1	10	15	1
Large, deep eutrophic lakes	LAKE2	10	15	1
Big eutrophic lakes of medium depth	LAKE3	10	15	1
Medium-sized eutrophic lakes of medium depth	LAKE4	10	15	1
Small shallow eutrophic lakes	LAKE5	10	15	1
Small dystrophic lakes	LAKE6	5	10	1
Artificial lakes	LAKE7	5	10	1

The last group consists of ecosystems with a very low indicator value (the number of species in the relevé rarely exceeds 10). This group includes bogs, which are inherently poor in vascular plant species, even on extended patches, as well as reed beds and sedge swamps, and littoral zone plant communities of all types of lake.

It should be emphasized once again that certain numbers of species are taken from relatively small test areas. Considering the general richness of vascular plant species that can occur within large patches of specific communities, the hierarchy of ecosystems may be slightly different. In particular, the littoral communities of different types of lake, meadow communities, and some fens could be clearly richer.

Spatial pattern of RICHNESS indicator values is mosaic type with predominance of large patches with complex structure (Fig. 6.15). It is determined mostly by soil fertility and land use. Agricultural areas are generally characterized by medium and low values, while forests by high or very high values. The largest patch covers over 8% of the study area. Patches with low and average potential dominate, as they cover 70% of the total area. The largest patches are generally more irregular in shape compared with

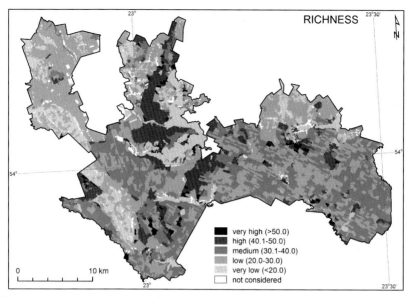

Figure 6.15 Ecosystem potential to maintain nursery habitats indicated by species richness of herb layer vascular plants (RICHNESS indicator).

small patches, but differences are not very high. Small patches of the same level of potential are quite evenly distributed in the local scale, and the interspersion of patches of different level of potential is medium compared with other ES potentials. The WNP is characterized by the highest share of ecosystems with high and very high number of vascular plant species.

MAMMAL indicator: Species diversity of mammals
Theoretical framework

Wild fauna with diverse species composition (herbivores and predators) is as necessary for the proper functioning of ecosystems as floristic diversity (herbaceous and woody plants) (Jędrzejewska and Jędrzejewski, 2001; Kamieniarz and Panek, 2008; Okarma and Tomek, 2008). The species richness and animal density reflect the potential of nature in a given region to provide various regulating services such as life cycle maintenance and habitat and gene pool protection.

Potential of nature to maintain nursery habitats for animals can be estimated directly by identifying suitable habitats for species or indirectly by specifying different types of diversity indices. The latter approach results from the well-known relationship between the diversity of abiotic factors and vegetation (Riera et al., 1998), the diversity of habitats and vegetation, and the diversity of flora and fauna (Barthlott et al., 1999; Larsen and Jensen, 2000). Because it is rather impossible to include all species occurring in a given area in the diversity index, the question arises as to how many groups of species provide information on the biodiversity of spatial units (Faith and Walker, 1996; Fleishman et al., 2005; Lawton and Gaston, 2001). One of the possible approach to this issue is to assume that proxy complex indicators should take into account various functional groups of organism, e.g., multihabitat species indicating the potential for many habitats, species of special interest (endangered, protected, game) due to their actual indicating value, as well as species dependent on specific resources, e.g., food or shelter (Carignan and Villard, 2002; Dale and Beyeler, 2001). The MAMMAL indicator applied was developed in accordance with the above criteria.

Assessment method

Species diversity of game mammals (MAMMAL indicator) was used as an indirect "proxy" measure of the potential of nature in a given region to maintain nursery habitats for animals (Table 6.27). The data on the abundance of 15 small and big mammal game species (Table 6.32) were used as input. They were derived from annual hunting reports from 2011 to 2014 prepared by the four Forest District Offices separately for each hunting unit (14 units in total in the study area) and from the WNP inventory (Jamrozy, 2008; Misiukiewicz, 2014).

The Shannon-Weaver diversity index (H) was used to calculate game mammal diversity, according to the formula (Shannon and Weaver, 1949, after Odum, 1953):

$$H = - \sum p_i \log_2 p_i,$$

where p_i is the number of animals of the ith species in the total number of analyzed animals.

Indicator values

Diversity of game mammals (MAMMAL indicator) ranges from 1.8 to 3.3 in the study area (Fig. 6.16). The highest value was reported for hunting unit 75 (Animal Breeding Center) covered in 80% with mostly coniferous and mixed forests. The WNP is characterized by equally high species diversity, where oak-hornbeam forests prevail. The lowest potential to maintain nursery habitats for animals indicated by MAMMAL indicator was assigned to hunting unit 70, which is mostly grassland with only 19% forest cover.

Table 6.32 Game mammals selected for analysis.

Order	Species
Big game	
Ungulate	Red deer, *Cervus elaphus* Linnaeus, 1758 Roe deer, *Capreolus Capreolus* Linnaeus, 1758 Elk, *Alces alces* Linnaeus, 1758 Wild boar, *Sus scrofa* Linnaeus, 1758
Small game	
Carnivorous	Red fox, *Vulpes vulpes* Linnaeus, 1758 Raccoon dog, *Nyctereutes procyonoides* Gray, 1834 Badger, *Meles meles* Linnaeus, 1758 Otter, *Lutra lutra* Linnaeus, 1758 Pine marten, *Martes martes* Linnaeus, 1758 Beech marten, *Martes foina* Erxleben, 1777 European polecat, *Mustela putorius* Linnaeus, 1758 American mink, *Mustela vison* Schreber, 1777
Lagomorphs Rodents	Brown hare, *Lepus europaeus* Pallas, 1778 European beaver, *Castor fiber* Linnaeus, 1758 Muskrat, *Ondatra zibethicus* Linnaeus, 1766

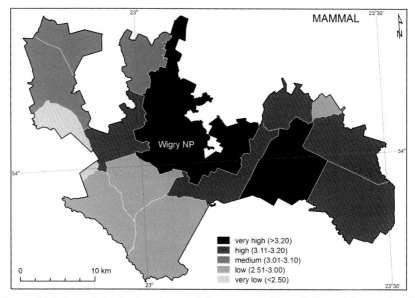

Figure 6.16 Potential of nature to maintain nursery habitats indicated by species diversity of game mammals within hunting units and Wigry National Park (MAMMAL indicator).

GROUSE indicator: Hazel grouse population density
Theoretical framework

Three species of grouse, Western capercaillie *Tetrao urogallus*, black grouse *Tetrao tetrix*, and hazel grouse *Bonasa bonasia*, are often mentioned as examples of umbrella species of European forests, selected for making conservation-related decisions (protecting umbrella species typically indirectly protects the many other species that make up the ecological community of its habitat). Their occurrence is related not only with large complexes of old-growth forests and specialized biocoenosis of vertebrates (rare and endangered species) but also with close-to-natural forest structure and extensive forest management (Zawadzka and Zawadzki, 2006). The three species mentioned differ in details as to habitat preferences, but in each case, their presence and abundance depend to a large extent on landscape heterogeneity and the presence of rowans, willows, nesting sites, and forest islands (Müller et al., 2009). For instance, in the case of hazel grouse, the presence of young deciduous stands within old forest complexes is essential (Swenson and Angelstam, 1993).

The conducted human-induced restitution of Western capercaillie and black grouse populations in the study area partly disrupt the natural pattern of relationships between those species and the environment, therefore only the density of hazel grouse has been taken into account in the assessments, treating their abundance as an indirect indicator of the potential of nature in a given region to maintain nursery habitats (Table 6.27).

Hazel grouse is a typical forest species. It lives mainly in coniferous forests, but also inhabits mixed and deciduous forests, preferring old-growth forests with dense tree canopy of tall conifers (Nüßlein, 1988). Its nests are usually located in shallow hollow in the ground under dense vegetation (dense tree cover, shrub, or thick undergrowth) (Bonczar, 2004; Weisner et al., 1977). Well-developed and rich undergrowth is particularly important in mixed forests. The abundance of hazel grouse is steadily decreasing, which is caused by the reduction of forest cover and the implementation of single-species and even-aged stands with poor shrub and herb layers. As hazel grouse prefers large forest complexes with little human interference (Okarma and Tomek, 2008), its abundance can serve as an indirect measure of forest habitat quality (naturalness) and the intensity of forest exploitation.

Assessment method

The density of hazel grouse (GROUSE indicator) was used to indicate the potential of nature to maintain nursery habitats, in particular for rare and endangered forest animals. The data on the abundance of hazel grouse were derived from annual hunting reports from 2011 to 2014 prepared by the

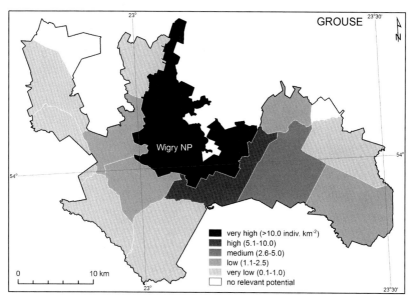

Figure 6.17 Potential of nature to maintain nursery habitats indicated by hazel grouse (*Bonasa bonasia* L. 1758) population density within hunting units and Wigry National Park (GROUSE indicator; individuals km^{-2}).

four Forest District Offices separately for each hunting unit (14 units in total in the study area) and from the literature on the WNP (Jamrozy, 2008; Misiukiewicz, 2014). The density is calculated in individuals per 1 km^2 of the reference area (hunting unit or WNP).

Indicator values

Density of hazel grouse (GROUSE indicator) ranges from 0 to 14 individuals per km^2 in the study area (Fig. 6.17). The highest value was reported for the WNP, rich in old-growth oak-hornbeam, coniferous, and mixed forests - habitats preferred by this species. In highly forested hunting units (70%-80% forest cover) in the vicinity of WNP, hazel grouse densities reach 5 per km^2, which means that the natural potential to maintain nursery animal habitats there is high. In turn, densities in forested hunting units located further from WNP do not exceed one per km^2.

DIVERSITY indicator: Ecosystem diversity
Theoretical framework

The use of DIVERSITY indicator (Table 6.27) as an indirect measure of nature potential to maintain nursery habitats results from several theoretical assumptions. Firstly, it has been proven that the highest values of ecosystem diversity, when removing the effect of confounding variables, occur in the situation of average anthropogenic pressure on the landscape, which is associated on the one hand with the occurrence of natural ecosystems and on the other with the functioning of ecosystems differently transformed by human activity. In such conditions, species may react in different ways to human pressure (see Richling and Solon, 2011 and the literature cited there).

Secondly, the applied Shannon-Weaver diversity index, although referring to the actual land cover, also indirectly covers the abiotic conditions, which determine the possibility of occurrence of species with different ecological requirements (see, for example, factors shaping species richness of bryophytes and lichens in forest communities - Lõhmus et al., 2007).

Thirdly, it has been shown that species richness of different systematic groups of organism is strongly correlated with spatial diversity of ecosystems in the landscape. This regularity applies not only to species generalists but also to various groups of species specialists and has been demonstrated many times in different landscapes and geographical zones, and in relation to different groups of species (e.g., vascular plants - Gould and Walker, 1999; plants, butterflies, and spiders - Záhlavová et al., 2009). It results from the fact that on a local scale, higher heterogeneity - manifested by higher diversity of ecosystems - is associated with the occurrence of more diverse resources and a greater number of separate ecological niches enabling the coexistence of a larger number of species.

Fourthly, the importance of ecosystem diversity in the landscape is emphasized for the possibility of occurrence of species that use different environments during life cycle (multiecosystem species). This is particularly important for numerous invertebrates and birds (Konvicka et al., 2006; Söderström et al., 2001; Zimmermann et al., 2005).

However, taking account also other spatial conditions, including the nature of existing ecosystems, fragmentation, distance to the nearest neighbor, barriers in the landscape, and other characteristics of composition and configuration, ecosystem diversity cannot be treated as the only and sufficient indicator of species richness (cf. Amarasekare, 2003; Chesson, 2000). Therefore, we treat this measure as a surrogate indicator of nature potential to provide this service, similarly as proposed by Maes et al. (2014).

It should also be noted that the use of diversity measure as an indicator of nature potential to maintain nursery habitats makes sense only with relatively narrow definition of ecosystem types, i.e., as in our study. When using Corine Land Cover units or the MAES ecosystem categories, the obtained results do not provide any valuable or useful information.

Assessment method

The layer with landscape units was intersected with the map of ecosystems, and for each of the landscapes, the share of each type of ecosystem was calculated. Then, for the calculation of the diversity index, the following Shannon-Weaver formula was used:

$$H = -\sum p_i \log_2 p_i,$$

where p_i is the fraction of the ith type of the ecosystem in the given landscape (McGarigal and Marks, 1995; Odum, 1953). The results were reclassified into five-grade scale (Table 6.27).

Indicator values

The range of DIVERSITY indicator values is relatively wide and extends from 0.1 to 4.1. Landscapes with the lowest values are dominated by large lakes. Low potential (H from 0.75 to 1.50) was assigned to landscapes including large arable fields and built-up areas, while landscapes of medium potential (H from 1.51 to 2.25) are either mosaic-like with the dominance of meadows, pastures, and arable fields or are dominated by homogeneous matrix of pine (spruce) forests. Landscapes of high potential (H from 2.26 to 3.00) are composed mainly by the mosaic of different forest ecosystems, while landscapes with the highest potential (H > 3.00) by complex mosaic of different ecosystems, with a significant share of moist and wet habitats (Fig. 6.18).

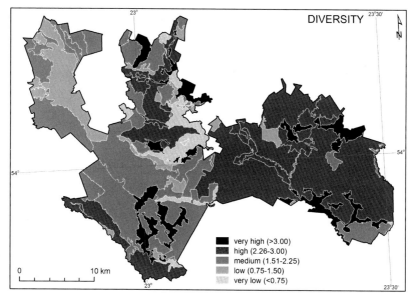

Figure 6.18 Potential of nature to maintain nursery habitats indicated by ecosystem diversity within landscape units (DIVERSITY indicator).

HABITAT indicator: Abundance of small-scale habitats
Theoretical framework

The existence of potential breeding sites for rare species associated with specific small habitats is highly related to potential of a region to provide regulating ES: nursery habitat maintenance. Small-scale ecosystems occur in all types of landscape, but they play the most important role in maintaining nursery habitats in agricultural landscapes. Small habitats (small-scale ecosystems) include clumps and belts of trees and scrub, small water reservoirs and swamps, grassy belts (connected with boundary strips or fire fighting belts), and boulder fields. The above spatial elements (apart from woody and shrub vegetation) also play an important, though not always recognized, biocenotic role within forest landscapes.

Research conducted in many regions of Europe and in many types of landscape used for farming clearly showed the decisive influence of small landscape elements on the preservation and functioning of complex ecological systems. Also in Poland, such relations were documented many times.

The presence of hedges, tree rows or clumps, and other small-scale elements affects the abundance and distribution of all groups of animals. For many species of insect, essential nourishment necessary for proper reproduction is nectar or pollen of specific plant species that grow mainly in the ecotone zones. Their presence decides, for example, about the possibility of occurring in the landscape of many species of these arthropods. Midfield trees and woods and forest islands are preferred by families such as *Phoridae, Lonchaeidae, Asteridae, Mycetophilidae* (*Diptera*), *Ichneumonidae* (*Hymenoptera*), *Scolytidae* (*Coleoptera*), and *Chrysopidae* (*Neuroptera*). The mentioned families achieve the highest values of density and biomass in these environments (Karg, 1989).

A significant part of plant species (even 40%—50%) in hedges, tree rows, or clumps produces fruits that are food for animals, mainly birds. Midfield woodlands are the nesting place for about 70 bird species, and the number of breeding pairs can reach up to 420 per hectare. These numbers are much larger than the densities given to other environments. The most common and most numerous species in Central Europe that use small woods and forest islands in midfield for nesting include finch (16% of the entire group of birds of the small woods), yellowhammer (13%), sparrow (7%), and bunting (5%). In periods of migration, mainly in winter, midfield small woods and forest islands serve as places of food acquisition by many bird species, shelter (more species overwinter in small woods than in forests), as well as a route between larger forest complexes (Cieślak and Dombrowski, 1993; Ryszkowski and Bałazy, 1994).

Small swamps, waterbodies, and watercourses together with accompanying vegetation play an important role for different plant and animal groups. Especially on large, deforested areas, they constitute the only possible habitats for breeding, foraging, or wintering as well as stepping stones of ecological corridors for numerous plants, insects, small mammals, amphibians, and reptiles (Dajdok and Wuczyński, 2005; Karg, 2004).

Small ecosystems are a part of intralandscape ecological corridors whose presence in the agricultural landscape is particularly important for species of low mobility, for example, lacking flight capacity. Ecological corridors are therefore very important for amphibians and reptiles, as well as for small mammals, some birds, and groups of insects.

Different types of hedge and tree shelterbelt link the bigger ecosystems dispersed in the agricultural landscape (forests, grassland, water reservoirs, wetlands, etc.). For many species, they are the only relatively safe route of movement in the area. For example, for amphibians penetrating agroecosystems, midfield tree rows are primarily communication routes and a place where it is easy to find convenient hiding places during periods of reduced activity (Ryszkowski and Bałazy, 1994).

Assessment method

Primary data used in the construction of the HABITAT indicator come mainly from the Spatial Database of Topographic Objects in scale 1:10,000 (BDOT10k) (Table 6.27). Complementary sources were aerial photos and a detailed map of the actual vegetation of the WNP. The following categories of BDOT10k objects have been interpreted as small-scale elements in the landscape (with an area of less than 3 ha) (Regulation of the Minister of Internal Affairs and Administration, 2011):

a) Class: Wetlands - areas periodically or permanently swamped, flooded, or covered with water, areas with shallow groundwater level;

b) Class: Rushes - areas covered with high grassland vegetation, occurring both in coastal waters and on land;

c) Class: Land cover - the following subclasses were included: forest or wooded area, shrub vegetation, and water;

d) Class: Natural objects (objects that are elements of the natural environment and were not included into land cover Class) - the following objects were included: tree or group of trees, erratic boulder or group of boulders, clump of shrubs, row of trees, hedges.

The newly created shapefile containing the abovementioned spatial objects has been verified and corrected with the help of available aerial photos and cartographic data from the Protection Plan of the WNP.

The map of small objects (small-scale ecosystems) was then intersected with the layer with landscape units, and the total number of these ecosystems was counted in individual landscapes and their density was calculated, expressed in number per km^2.

Indicator values

The HABITAT indicator values are highly scattered, ranging from 0.3 objects per $1 \, km^2$ to over 103 (Fig. 6.19). Thirty landscapes are characterized by very low or low potential (<25 objects per km^2), while other 50 were evaluated as demonstrating high or very high potential (>75 objects per km^2) (Fig. 6.20). According to the typology of landscapes presented in Table 5.6, low potential to maintain nursery habitats was assigned to forest landscapes (group B) except landscapes with the domination of alder forests – B2 subgroup – and lake-dominated landscapes (group C), while high values are generally characteristic for agriculture landscapes of different type.

6.1.2.5 Invasive plant species control
Ecosystem service description

This service belongs to *Regulation and Maintenance* section and *Regulation of physical, chemical, biological conditions* division in CICES V5.1 (Table 6.33). It can be further described as the reduction by biological interactions of the incidence of plant species that prevent or reduce the output of food, material, or energy from ecosystems, or their cultural importance, by consumption of biomass or competition. The biotic resistance to invasive plant species has been proposed to indicate the capacity of ecosystems to provide it.

Theoretical framework

Invasion of alien plant species is a kind of territorial expansion of species outside the area of their natural distribution, proceeding rapidly and massively as a result of indirect or direct human pressure (Falińska, 2004). Invasive species threaten local biodiversity and cause economic losses, therefore the problem of biological invasions is not only the subject of scientific research but it is also important for legislators at various levels.

EU regulation No 1143/2014 of October 22, 2014, on the prevention and management of the introduction and spread of invasive alien species states that "invasive alien species are one of the main threats to biodiversity

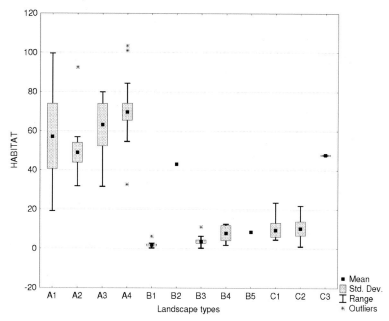

Figure 6.19 HABITAT indicator values showing nature potential to maintain nursery habitats within landscape types (landscape typology - see Table 5.6).

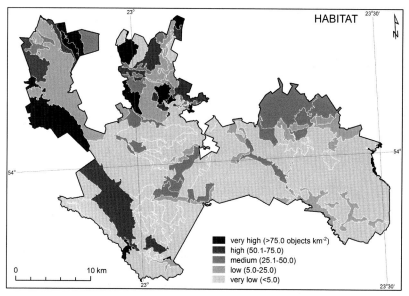

Figure 6.20 Potential of nature to maintain nursery habitats indicated by the abundance of small-scale habitats within landscape units (HABITAT indicator; objects km^{-2}).

Table 6.33 Ecosystem service in Common International Classification of Ecosystem Services (CICES) V5.1 and indicator metadata.

Ecosystem service		Invasive plant species control
CICES V5.1	Section	Regulation and Maintenance
	Division	Regulation of physical, chemical, biological conditions
	Group(s)	Pest and disease control
	Class(es)	Pest control (including invasive species)
	Class code(s)	2.2.3.1
Indicandum		**Ecosystem capacity to control invasive plant species**
Indicator		**Biotic resistance to invasive plant species**
Acronym		**INVAS**
Short description		Estimation of potential theoretical resistance to invasions of alien plant species resulting from the level of competition of native plant species and abiotic limitations on plant growth and development; resistance values obtained by adding normalized B, N, C, S, and A, where B is the herb layer biomass, N is the number of vascular plant species, C is the share of species of C-type life strategy, S is the share of species of S-type life strategy, A is the tree stand crown density
Direct/indirect		Indirect
Simple/compound/ complex		Complex
Estimated/ calculated		Estimated
Scale		Ratio
Value range		0.806−2.339
Unit		−
Spatial unit		Ecosystem
Value interpretation		$<1.027 \rightarrow$ very low ecosystem potential; $1.027-1.486 \rightarrow$ low; $1.487-1.873 \rightarrow$ medium; $1.874-2.078 \rightarrow$ high; $>2.078 \rightarrow$ very high
Source data		− Total organic carbon content (t ha^{-1}) in the herb layer (calculated as part of the "carbon stock in herb layer" indicator) − Species richness (phytosociological relevés, fieldwork)

Table 6.33 Ecosystem service in Common International Classification of Ecosystem Services (CICES) V5.1 and indicator metadata. (cont'd)

Ecosystem service	Invasive plant species control
	— Tree crowns density (taxation descriptions of forests in Suwałki, Pomorze, Głęboki Bród, and Szczebra Forest Districts) — Plant life strategies (Grime, 1979) the study on plant communities (Roo-Zielińska, 2014)

and related ecosystem services […]. The risk posed by these species may be greater due to the increase in global trade, transport, tourism and climate change". A list of 54 invasive species for Poland (about 1.5% of flora) was developed by Tokarska–Guzik (2005). Some of them are included in the Regulation of the Minister of the Environment of September 9, 2011, on the list of nonnative species of plants and animals, which in the case of release into the environment can threaten native species or natural habitats.

The predominance of invasive species over noninvasive ones results from various reasons. Faliński (2004) emphasizes such properties of invasive species, as frequent dioecy, polygamy, facultative autogamy, accelerated disintegration of the parent individual to the progeny, ability to compete with other species by growth, speed of regeneration of damaged organs, production of allelopathic substances, phenotypic variability, ability to create mutants, polyploids, crossbreeding with related native species. Although these features do not occur simultaneously in a given species, their multiplicity shows that invasive species can interact in many different ways with the existing ecosystem.

Large-scale plant invasions are also dependent on ecosystem properties. One of the most important hypotheses related to the concept of environmental vulnerability to invasions is that the resistance of the environment increases with the increase of its biodiversity or with the increase of species richness (Elton, 1958). Rejmánek et al. (2005) suggest that biodiversity at other than plant trophic level may be even more important. They point out factors conducive to invasions, e.g., presence of nonspecific symbiotic species, such as pollinators, species facilitating seed dispersal, microbial species coexisting with plant roots (Richardson et al., 2000a,b; Simberloff and Von Holle, 1999), lack of natural enemies, especially specific herbivores and pathogens (DeWalt et al., 2004; Elton, 1958; Keane and Crawley, 2002; Wolfe, 2002), and ecosystem disturbances (Di Castri, 1990; Fox and Fox, 1986; Kornaś, 1990; Lake and Leishman, 2004; Planty-Tabacchi et al., 1996).

Thus, the final success of invasive species consists of many factors related both to its biology and ecology, as well as to the broadly understood environment in which it spreads. Chytrý et al. (2008) emphasize the combined role of environmental susceptibility to invasions, climatic factors, and supply intensification of alien species propagules.

Referring to the above considerations, the quantification of invasive plant species control is very difficult. To date, no universal method for determining the ecosystem resistance to invasions has been developed. Our attempt takes account of only a few possible properties of ecosystems that impact their resilience. Therefore, the results of the analysis should not be equated with the actual level of invasion pressure in the study area but only with the relative invasion resistance resulting from ecosystem characteristics that were taken into account in the indicator construction.

Assessment method

To estimate the potential of ecosystems to provide invasive plant species control, available data on a theoretical level of competition from native plant species and on theoretical abiotic limitations for plant growth and development were used. The value of biotic resistance to invasive plant species indicator was calculated for natural and seminatural ecosystems, thus excluding settlement areas and agroecosystems.

Based on the analysis of scientific literature and the technical possibilities of specific feature estimation, it was assumed that the following factors influence the level of resistance to invasions: plant biomass determined as organic carbon stock in the herb layer (B), plant species richness (N), the share of species in the competitive life strategy type (C - competitors), and average density of tree crowns (A - for forests).

The abiotic limitations were determined using the share of S - life strategy species (S - stress tolerant) - particularly resistant to stress resulting, for example, from extreme moisture or temperature conditions as well as from extreme contents of certain soil elements.

We used C-S-R life strategies identified by Grime (1979) and based on plant characteristics including morphological features, resource allocation, phenology, and response to stress. The competitive strategy (C - competitor species) prevails in productive, relatively undisturbed ecosystems, the stress-tolerant strategy (S - stress-tolerant species) is associated with continuously unproductive conditions, and the ruderal strategy (R - ruderal species) is characteristic of severely disturbed but potentially productive habitats (Table 6.34).

The methods of calculating the carbon stock in the herb layer and the number of vascular plant species are presented in Sections 6.1.2.4 and 6.1.2.10. Data on plant life strategies were taken from the research on plant communities (Roo-Zielińska, 2014), assigned as close as possible to ecosystem types (Table 6.35). Additionally, for forest ecosystems, to diversify the level of species competition depending on the stand age, the average density of tree crowns was taken into account on the basis of forest taxation descriptions. Then, to obtain one value expressing the species competition in the forests, the share of C-strategists and the density of tree crowns were multiplied.

The values of all factors building up the indicator have been normalized to fit in the range of 0—1 and summed up. The obtained values were then

Table 6.34 Types of plant life strategies.

Life strategy type		Description
C	Competitors	Plant species living in areas of low stress and disturbance intensity and excel in biological competition; mainly trees and shrubs
S	Stress tolerant	Plant species living in areas of high stress intensity and low disturbance intensity; often found in stressful environments (deep shade, nutrient deficient soils, and extreme pH levels)
R	Ruderal species	Plant species living in high disturbance intensity and low stress intensity; fast-growing and complete rapidly their life cycles and generally produce large amounts of seeds. Plants that have adapted this strategy are often found to colonize recently disturbed land and are often annuals
CR	Mixed type	Plant species living in areas of low stress intensity; their competition is limited by disturbance
SR	Mixed type	Plant species living in low disturbance areas
CS	Mixed type	Plant species living in relatively not disturbed areas with moderate stress intensity
CSR	Mixed type	Plant species living in areas with limited competition level influenced by moderate stress and disturbances

Based on Grime, J.P., 1979. Plant Strategies and Vegetation Processes. John Wiley & Sons, Chichester—New York—Brisbane—Toronto.

classified, and five classes of the ecosystem resistance to invasions of alien plant species were distinguished.

Indicator values

Relatively low resistance to invasions of alien plant species is observed in alder, riparian, and oak–hornbeam forests, as well as in reed beds and sedge swamps. These all ecosystems cover about 13% of the study area. Small patches of very low potential were frequently observed in river valleys. Open areas such as wet grassland on mineral soil, peat grassland, and bogs and among forests - coniferous and mixed forests up to 80 years old and the oldest as well as swamp coniferous forests up to 120 years old - are more resistant.

The highest resistance is found in fens, dry grassland, coniferous and mixed forests (80–120 years old), and the oldest swamp coniferous forests (Table 6.36, Fig. 6.21). Ecosystems of medium, high, and very high potential form quite extensive and mosaic-like pattern with patches of various sizes, but - compared to other potentials - the size variance is very low (the second lowest among the studied ES potentials). The largest patch covers less than 5% of the study area. Patches of different level of potential occupy similar areas (except of very low potential - poorly represented in the study area), which results in the second highest SEI value. The largest patches are generally more irregular in shape compared with small patches, but differences are not very high. Small patches of the same level of potential are quite evenly distributed, and the interspersion of patches of different level of potential is medium compared with other ES potentials.

6.1.2.6 Pest rodent control
Ecosystem service description

This service belongs to *Regulation and Maintenance* section and *Regulation of physical, chemical, biological conditions* division in CICES V5.1 (Table 6.37). It can be further described as the reduction by biological interactions of the incidence of pest rodents that prevent or reduce the output of food, material, or energy from ecosystems, or their cultural importance, by consumption of biomass or competition. The number of small rodents eaten by predators within hunting units has been proposed to indicate the capacity of nature to provide it.

Table 6.35 Percentage shares of plant species by types of life strategy (C, competitive type, R, ruderal type, S, stress type, CR, CS, CSR, SR, mixed types according to Grime, 1979). Calculation based on shares of characteristic species of different life strategy types in plant communities (belonging to different phytosociological alliances).

Ecosystem type	Life strategy type						Plant community type
	C	CR	CS	CSR	S	SR	
Settlement area	Not considered						
Alder forest	50.8	0.00	49.2	0.00	0.00	0.00	*Alnion glutinosae* – group of alder forest communities
Riparian forest	25.0	0.00	29.1	43.5	1.2	1.2	*Alno-Ulmion* – lowland riparian forest communities
Oak-hornbeam forest	23.6	0.5	26.1	46.7	1.5	1.5	*Carpinion betuli* – group of oak-hornbeam forests
Coniferous and mixed forest	31.3	0.00	19.8	31.3	17.6	0.00	*Dicrano-Pinion* – group of coniferous and mixed forests
Swamp coniferous forest	14.1	0.00	50.3	11.9	17.6	0.00	*Dicrano-Pinion* – group of swamp coniferous forests
Dry grassland	51.2	2.4	12.2	34.1	0.00	0.00	*Arrhenatherion elatioris*
Wet grassland on mineral soil	42.0	1.7	14.3	40.4	0.00	1.7	*Molinion caerulae*
Peat grassland	51.5	3.7	13.2	29.8	0.00	1.8	*Calthion palustris*
Arable field on poor sandy soil	Not considered						
Arable field on fertile mesic soil							
Arable field on fertile moist soil							
Wetlands – reed beds and sedge swamps	10.4	0.00	64.7	24.0	0.9	0.00	*Magnocaricion*
Wetlands – fens	0.00	0.00	36.8	42.1	21.1	0.00	*Caricion nigrae*
Wetlands – bogs	14.3	0.00	54.8	0.00	31.0	0.00	*Sphagnion magellanici* – group of forest mires
Wetlands – reed beds and sedge swamps	0.9	0.00	85.2	14.0	0.00	0.00	*Phragmition*

After Roo-Zielińska, E., 2014. Wskaźniki ekologiczne zespołów roślinnych Polski. Wydawnictwo Akademickie Sedno, Warszawa (in Polish).

Table 6.36 Ecosystem potential to control plant invasive species (1 - very low; 5 - very high) indicated by biotic resistance to invasive plant species.

Ecosystem type	Ecosystem acronym	INVAS indicator value	Ecosystem potential
Settlement area	SETTLE	Not considered	
Arable field on poor sandy soil	ARB1		
Arable field on fertile mesic soil	ARB2		
Arable field on fertile moist soil	ARB3		
Dry grassland	GRAS1	2.139	5
Wet grassland on mineral soil	GRAS2	1.656	3
Peat grassland	GRAS3	1.872	3
Alder forest <40 years	ALD1	1.009	1
Alder forest 40—60 years	ALD2	1.153	2
Alder forest 60—80 years	ALD3	1.396	2
Alder forest 80—120 years	ALD4	1.427	2
Alder forest >120 years	ALD5	1.427	2
Riparian forest <40 years	RIP1	0.806	1
Riparian forest 40—60 years	RIP2	1.026	1
Riparian forest 60—80 years	RIP3	0.999	1
Riparian forest 80—120 years	RIP4	1.014	1
Riparian forest >120 years	RIP5	1.283	2
Oak-hornbeam forest <40 years	OAK1	1.305	2
Oak-hornbeam forest 40—60 years	OAK2	1.234	2
Oak-hornbeam forest 60—80 years	OAK3	1.354	2
Oak-hornbeam forest 80—120 years	OAK4	1.413	2
Oak-hornbeam forest >120 years	OAK5	1.374	2
Coniferous and mixed forest <40 years	CON1	1.973	4
Coniferous and mixed forest 40—60 years	CON2	1.756	3
Coniferous and mixed forest 60—80 years	CON3	2.064	4
Coniferous and mixed forest 80—120 years	CON4	2.101	5

Table 6.36 Ecosystem potential to control plant invasive species (1 - very low; 5 - very high) indicated by biotic resistance to invasive plant species. (cont'd)

Ecosystem type	Ecosystem acronym	INVAS indicator value	Ecosystem potential
Coniferous and mixed forest >120 years	CON5	2.073	4
Swamp coniferous forest <40 years	SWP1	2.028	4
Swamp coniferous forest 40 −60 years	SWP2	2.030	4
Swamp coniferous forest 60 −80 years	SWP3	1.834	3
Swamp coniferous forest 80 −120 years	SWP4	2.067	4
Swamp coniferous forest >120 years	SWP5	2.339	5
Wetlands - reed beds and sedge swamps	REED1	1.485	2
Wetlands - fens	FEN	2.158	5
Wetlands - bogs	BOG	2.077	4
Lakes (all types)	LAKE	Not considered	

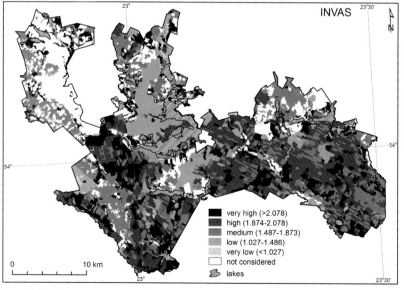

Figure 6.21 Ecosystem potential to control plant invasive species indicated by biotic resistance to invasive plant species (INVAS indicator).

Theoretical framework

On all continents, rodents are the cause of damage to plant crops, yields, and in storage locations (Elliot, 1988; Hopf et al., 1976; Jackson, 1977). In central Europe, losses in cereal crops are caused primarily by voles (*Microtus* sp.), but also by other rodents, e.g., mice (*Apodemus* spp.) (Heroldová and Tkadlec, 2011). In temperate forest, rodents are an important link in the food chain. Small rodents constitute the largest group of prey for small- and medium-sized predators: mammals and birds (Bunevič and Dackevič, 1985; Jędrzejewska and Jędrzejewski, 2001; Jędrzejewski and Jędrzejewska, 1993; Jędrzejewski et al., 1993a,b) (Table 6.38). Predators play an important role in maintaining a constant level of rodent population during annual cycles but probably have no significant effect during the plagues of pests. A substantial increase in the number of predators usually indicates an increase in the number of rodents.

Assessment method

The data on the abundance of small predators were used to assess the number of small rodents eaten daily by predators indicating the potential of nature to control populations of pest rodents. They were derived from annual hunting reports from 2011 to 2014 prepared by the four Forest District Offices separately for each hunting unit (14 units in total in the study area) and from the literature on the WNP (Jamrozy, 2008; Misiu-kiewicz, 2014). Six species of small predators were taken into account: red fox *Vulpes vulpes*, badger *Meles meles*, raccoon dog *Nyctereutes pro-cyonoides*, American mink *Mustela vison*, pine marten *Martes martes*, and European polecat *Mustela putorius*. Based on literature review, it was assumed that the daily diet of each predator considered comprises a certain number of rodents. The average number of rodents eaten daily by one representative of each species was then estimated (Table 6.39). The total number of rodents eaten daily per 1 km^2 was calculated according to the formula:

$$T = \sum (D_i \times a_i),$$

where T is the total number of rodents eaten daily by predators (rodents km^{-2} day^{-1}), D_i is the density of ith predator species (predators km^{-2}), and a_i is the number of rodents eaten daily by one individual of ith predator species.

Table 6.37 Ecosystem service in Common International Classification of Ecosystem Services (CICES) V5.1 and indicator metadata.

Ecosystem service		Pest rodent control
CICES V5.1	Section	Regulation and Maintenance
	Division	Regulation of physical, chemical, biological conditions
	Group(s)	Pest and disease control
	Class(es)	Pest control (including invasive species)
	Class code(s)	2.2.3.1
Indicandum		**Potential of nature to control populations of pest rodents**
Indicator		**Number of small rodents eaten by predators**
Acronym		**RODENT**
Short description		Calculation of total number of rodents eaten daily by small predatory mammals (T) using the following formula: $T = \Sigma\ (D_i \times a_i)$, where D_i is the density of ith predator species (predators km^{-2}), a_i is the number of rodents eaten daily by one individual of ith predator species; shows regulatory role of small predators controlling the size of pest rodent population
Direct/indirect		Indirect
Simple/compound/ complex		Compound
Estimated/calculated		Calculated
Scale		Ratio
Value range		2–27
Unit		Individuals $km^{-2}\ day^{-1}$
Spatial unit		Hunting unit
Value interpretation		<5.0 → very low potential; 5.0–7.5 → low; 7.6 –10.0 → medium; 10.1–20.0 → high; >20.0 → very high
Source data		— Annual hunting reports for the years 2011–14 from the four forest districts and the Wigry National Park — Literature on the structure of the predators diet

Table 6.38 Primary and secondary predator prey in Białowieża Primeval Forest.

Predator species	Primary prey	Secondary prey (ranked by importance)		
		1	2	3
Raccoon dog	Ungulate carrion	Amphibians *Rana*	Rodents *Microtus*	Invertebrates
Pine marten	Forest rodents *Clethrionomys*	Hole-nesting birds *Turdus*	Insectivores	Ungulate carrion
Red fox	Voles	Ungulate carrion	Brown hare	Forest rodents *Clethrionomys*
Polecat	Amphibians *Rana*	Forest rodents *Apodemus*	Shrews	
American mink	Amphibians *Rana*	Forest rodents *Microtus*	Fish	
Badger	Earthworms	Amphibians *Bufo bufo*		

Based on Jędrzejewska, B., Jędrzejewski, W., 2001. Ekologia Zwierząt Drapieżnych Puszczy Białowieskiej. Wydawnictwo Naukowe PWN, Warszawa (in Polish).

Indicator values

The number of small rodents eaten daily by predators (RODENT indicator) ranges from 2 to 27 per km^2 in the study area (Fig. 6.22). The highest values were reported for hunting units with high forest cover (60%—80%), where coniferous and mixed forests prevail. Small predators are mainly associated with forests, and rodents are largely their food in these environments, that is why the predation pressure from the analyzed six species is probably the highest in these ecosystems (Grabińska, 2011; Ostler and Roper, 1998; Panek and Bresiński, 2002; Sidorovich et al.,

Table 6.39 Average number of rodents eaten daily by predators.

Predator species	Number of rodents
Pine marten	7.64
Red fox	7.00
European polecat	0.92
Raccoon dog	0.38
American mink	0.16
Badger	0.05

Based on Jędrzejewska, (B), Jędrzejewski, W., 2001. Ekologia Zwierząt Drapieżnych Puszczy Białowieskiej. Wydawnictwo Naukowe PWN, Warszawa (in Polish); Jędrzejewski, W., Jędrzejewska, (B), 1993. Predation on rodents in Białowieża primeval forest, Poland. Ecography 16 (1), 47—64; Ward, O.G., Wurster-Hill, D.H., 1990. Nyctereutes procyonoides. The American Society of Mammologists, Mammalian species 358, 1—5.

1996). The lowest potential of nature to control pest rodent population was achieved in forest and open land hunting units in the southwestern part of the study area (70, 95, 96), which shows that also other than land cover—related factors impact RODENT indicator.

6.1.2.7 Soil formation
Ecosystem service description
This service belongs to *Regulation and Maintenance* section and *Regulation of physical, chemical, biological conditions* division in CICES V5.1 (Table 6.40). It can be further described as the biological decomposition of minerals that maintain fertility or conditions necessary for human use. The degree of base saturation has been proposed to indicate the capacity of ecosystems to provide it.

Theoretical framework
The degree of base saturation is considered an important biochemical indicator of ecosystem performance (Degórski, 2002). Base saturation is defined as the percentage of the soil exchange sites occupied by basic cations, such as potassium (K), magnesium (Mg), calcium (Ca), and sodium

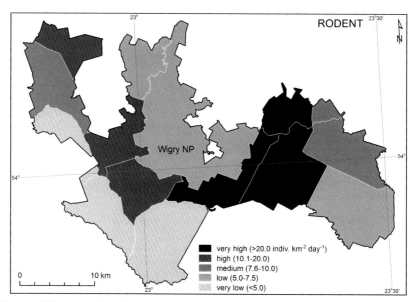

Figure 6.22 Potential of nature to control pest rodent population indicated by the number of small rodents eaten by predators within hunting units and Wigry National Park (RODENT indicator; individuals km^{-2} day^{-1}).

Table 6.40 Ecosystem service in Common International Classification of Ecosystem Services (CICES) V5.1 and indicator metadata.

Ecosystem service		Soil formation
CICES V5.1	Section	Regulation and Maintenance
	Division	Regulation of physical, chemical, biological conditions
	Group(s)	Regulation of soil quality
	Class(es)	Weathering processes and their effect on soil quality
	Class code(s)	2.2.4.1
Indicandum		**Ecosystem capacity to form productive soil**
Indicator		**Degree of base saturation**
Acronym		**SATURATION**
Short description		Calculation of the degree of base saturation (V) using the following formula: $V = S\,T^{-1}\,100$, where $S = Ca^{2+} + K^+ + Mg^{2+} + Na^+$ and $T = S + Hh$; where Hh is the hydrolytic acidity; the content of exchangeable cations in total sorption capacity was calculated based on reference field data and laboratory analysis; base saturation shows the availability of nutrients to plants and – indirectly – soil fertility
Direct/indirect		Indirect
Simple/compound/complex		Complex
Estimated/calculated		Calculated
Scale		Ratio
Value range		16–96
Unit		%
Spatial unit		Ecosystem
Value interpretation		<25.0 → very low potential; 25.0–50.0 → low; 50.1–70.0 → medium; 70.1–80.0 → high; >80.0 → very high
Source data		Field and laboratory studies at selected points representing particular types of terrestrial ecosystem

(Na). The content of exchangeable cations in total sorption capacity is also an important feature of soil fertility. In addition, the degree of base saturation is an important indicator of agronomic quality of cultivated soils. The base saturation depends on many factors, including lithologic and mineralogical conditions, soil acidity, humus type, etc. The higher the value of this indicator, the higher the ecosystem resistance to external factors, e.g., anthropogenic pollutants (Ulrich et al., 1984).

Assessment method

The soil material was collected during field studies from the organic (O) and humus (A) soil horizons. The mineral samples collected after grinding with a rubber stopper were screened through a sieve with a 2 mm mesh diameter to remove the skeleton fractions. Further analyzes were performed in earthy parts. The organic samples were precrushed and separated into two parts, one of which was ground in a knife mill made of tungsten carbide.

The sorption properties were determined in the crushed samples. The content of exchangeable cations was analyzed by emission of spectrometry with microwave-excited plasma after extraction of samples in a $1 \, mol \, L^{-1}$ ammonium acetate solution buffered to pH $= 7.0$. Hydrolytic acidity (Hh) was determined using the Kappen method.

On the basis of the obtained results, the sum of base (S) was calculated as the sum of alkaline exchangeable cations: Ca^{2+}, Mg^{2+}, K^+, and Na^+, absorbed by the soil, expressed in cmol(+)/kg. Next, the sorption capacity (T) was calculated as S + Hh, which made it possible to determine the soil base saturation as $S \, T^{-1} \, 100\%$.

Indicator values

The highest values of base saturation occur in arable soils, where fertilization is applied, i.e., artificial reinforcement of soil fertility potential. The best natural biochemical properties are characterized by ecosystems with moist soil, moder or mull type of humus, and high trophy. The highest saturation values were found in the ecosystems with the oldest stand of alder and riparian forests growing on fen peat soil, fens with fiber peat soils, as well as with fibrous sap peat soils, which is consistent with the natural content of these chemical components in studied soil types, resulting from their trophy. The lowest fertility potential is characterized by podzolic soils overgrown with coniferous and mixed forest (Table 6.41, Fig. 6.23).

Spatial pattern of SATURATION indicator values is determined mostly by soil properties and land use (Fig. 6.23). In central and eastern part of the study area, the uniform matrix of very low potential dominates, with a few randomly scattered small patches of higher levels of potential. In the northern and western parts, the pattern changes into more mosaic-like with high share of patches of high and medium potential. The largest patch covers over 18% of the total area. Areas of very low potential dominate, covering together over 42%. Large patches are generally more irregular in shape compared with small patches, but differences are rather small

Table 6.41 Ecosystem potential to form productive soil (1 - very low; 5 - very high) indicated by the degree of base saturation (%).

Ecosystem type	Ecosystem acronym	SATURATION indicator value	Ecosystem potential
Settlement area	SETTLE	Not considered	
Arable field on poor sandy soil	ARB1	81	5
Arable field on fertile mesic soil	ARB2	96	5
Arable field on fertile moist soil	ARB3	61	2
Dry grassland	GRAS1	31	2
Wet grassland on mineral soil	GRAS2	59	2
Peat grassland	GRAS3	78	4
Alder forest <40 years	ALD1	56	2
Alder forest 40−60 years	ALD2	67	2
Alder forest 60−80 years	ALD3	67	2
Alder forest 80−120 years	ALD4	68	2
Alder forest >120 years	ALD5	68	2
Riparian forest <40 years	RIP1	68	2
Riparian forest 40−60 years	RIP2	73	4
Riparian forest 60−80 years	RIP3	86	5
Riparian forest 80 −120 years	RIP4	88	5
Riparian forest >120 years	RIP5	88	5
Oak-hornbeam forest <40 years	OAK1	30	2
Oak-hornbeam forest 40 −60 years	OAK2	33	2
Oak-hornbeam forest 60 −80 years	OAK3	36	2
Oak-hornbeam forest 80 −120 years	OAK4	43	2
Oak-hornbeam forest >120 years	OAK5	43	2

Table 6.41 Ecosystem potential to form productive soil (1 - very low; 5 - very high) indicated by the degree of base saturation (%). (cont'd)

Ecosystem type	Ecosystem acronym	SATURATION indicator value	Ecosystem potential
Coniferous and mixed forest <40 years	CON1	16	1
Coniferous and mixed forest 40–60 years	CON2	18	1
Coniferous and mixed forest 60–80 years	CON3	24	1
Coniferous and mixed forest 80–120 years	CON4	17	1
Coniferous and mixed forest >120 years	CON5	20	1
Swamp coniferous forest <40 years	SWP1	44	2
Swamp coniferous forest 40–60 years	SWP2	43	2
Swamp coniferous forest 60–80 years	SWP3	51	3
Swamp coniferous forest 80–120 years	SWP4	69	3
Swamp coniferous forest >120 years	SWP5	82	5
Wetlands – reed beds and sedge swamps	REED1	68	3
Wetlands – fens	FEN	91	5
Wetlands – bogs	BOG	35	2
Lakes (all types)	LAKE	Not considered	

compared with other ES potentials. Small patches of the same level of potential are quite clustered, and the interspersion of patches of different level of potential is the highest compared with other ES potentials.

6.1.2.8 Organic matter decomposition
Ecosystem service description

This service belongs to *Regulation and Maintenance* section and *Regulation of physical, chemical, biological conditions* division in CICES V5.1 (Table 6.42). It can be further described as the decomposition of biological materials and their incorporation in soils that maintain their characteristics necessary for human use. The ratio of organic carbon to total nitrogen in soil has been proposed to indicate the capacity of ecosystems to provide it.

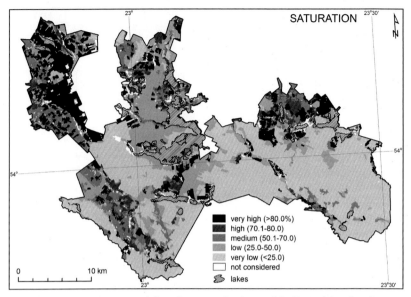

Figure 6.23 Ecosystem potential to form productive soil indicated by the degree of base saturation (SATURATION indicator; %).

Theoretical framework

One of the important regulating ES is maintenance of soil biogeochemical properties. The correct soil-forming processes determine the efficiency of the entire soil system. One measure of the biological activity in soil is the ratio of organic carbon to total nitrogen. Different plant residues contain varying quantities and availability of carbon (energy), nitrogen, and nutrients. Nitrogen availability for plants released during decomposition of plant debris depends very much on the C/N ratio of the decomposing organic matter. When the ratio is too wide (above 32:1), the organic matter mineralization slows down and the nitrogen potentially available for plants is absorbed by microorganisms (Thompson and Troeh, 1973). Generally, the lower the C/N ratio in soils, the more nitrogen in the mineral form (NH_4^+, NO_3^-) available for plants (Wójcik, 2004), and the soil biological activity is higher.

Assessment method

Soil material was collected during field studies from the organic (O) and humus (A) horizons of the analyzed soils. Mineral samples after grinding with a rubber stopper were screened through a sieve with a 2 mm mesh

Table 6.42 Ecosystem service in Common International Classification of Ecosystem Goods and Services (CICES) V5.1 and indicator metadata.

Ecosystem service		Organic matter decomposition
CICES V5.1	Section	Regulation and Maintenance
	Division	Regulation of physical, chemical, biological conditions
	Group(s)	Regulation of soil quality
	Class(es)	Decomposition and fixing processes and their effect on soil quality
	Class code(s)	2.2.4.2
Indicandum		**Ecosystem capacity to decompose organic matter**
Indicator		**Ratio of organic carbon to total nitrogen in soil**
Acronym		**C/N**
Short description		Calculation of the ratio of organic carbon (C) to nitrogen (N) in soil based on reference field data and laboratory analysis; shows the degree of decomposition of organic matter in soil and - indirectly - soil biological activity (inverse relationship)
Direct/indirect		Indirect
Simple/compound/complex		Compound
Estimated/calculated		Calculated
Scale		Ratio
Value range		9–29
Unit		–
Spatial unit		Ecosystem
Value interpretation		<10.0 → very high potential; 10.0–15.0 → high; 15.1–20.0 → medium; 20.1–25.0 → low; >25.0 very low
Source data		Field and laboratory studies at selected points representing particular types of terrestrial ecosystem supplemented by literature data

diameter to remove the skeleton fractions. Further analyzes were performed in earthy parts. However, the organic samples were precrushed and then ground in a knife mill made of tungsten carbide. The organic carbon content in mineral samples was analyzed using the Tiurin method and in organic samples using the Alten method. Total nitrogen content was determined using the Kjeldahl method by means of VELP UDK 127 distillation machine. The obtained results have been referred to literature data and calibrated.

Indicator values

Peat grassland, fens, and wet grassland on mineral soil characterized by high mineralization were assigned the highest potential to decompose organic matter (C/N ratio = 9 to 11) (Table 6.43, Fig. 6.24), which is consistent with natural content of chemical components in the studied soil types, resulting from their trophy. The lowest potential (C/N ratio > 25) was obtained for soils with the lowest biological activity, i.e., dry grassland, as well as young coniferous and mixed forest (C/N ratio between 20 and 25). It should be emphasized that these data relate to organic and humus horizons in which the bark of trees, which is characterized by a C/N ratio approximately 100/1, undergoes decomposition. In deeper soil levels, the ratio is gradually decreasing.

As C/N indicator values follow both soil differentiation and forest stand age, the spatial pattern of ecosystem potential represent a complicated mosaic-like type with patches of very different sizes (Fig. 6.24). The largest patch covers 7% of the study area. Patches of medium and high potential dominate (covering over 53% of the entire area), but they are not evenly distributed. Patches of low and medium potential dominate on afforested areas in the east and south, perforated by small patches representing other levels of potential. Big patches of high potential prevail in the north and west. Small patches of the lowest level of potential are quite evenly distributed, but the interspersion of patches of different level of potential is the second highest compared with other ES potentials. Large patches are generally more irregular in shape compared with small patches.

6.1.2.9 Oxygen emission
Ecosystem service description

This service belongs to *Regulation and Maintenance* section and *Regulation of physical, chemical, biological conditions* division in CICES V5.1

Table 6.43 Ecosystem potential to decompose organic matter (1 - very low; 5 - very high) indicated by ratio of organic carbon to total nitrogen in soil.

Ecosystem type	Ecosystem acronym	C/N indicator value	Ecosystem potential
Settlement area	SETTLE	Not considered	
Arable field on poor sandy soil	ARB1	17	3
Arable field on fertile mesic soil	ARB2	13	4
Arable field on fertile moist soil	ARB3	10	4
Dry grassland	GRAS1	29	1
Wet grassland on mineral soil	GRAS2	11	4
Peat grassland	GRAS3	9	5
Alder forest <40 years	ALD1	18	3
Alder forest 40−60 years	ALD2	15	4
Alder forest 60−80 years	ALD3	15	4
Alder forest 80−120 years	ALD4	14	4
Alder forest >120 years	ALD5	12	4
Riparian forest <40 years	RIP1	14	4
Riparian forest 40−60 years	RIP2	15	4
Riparian forest 60−80 years	RIP3	14	4
Riparian forest 80−120 years	RIP4	16	3
Riparian forest >120 years	RIP5	17	3
Oak-hornbeam forest <40 years	OAK1	16	3
Oak-hornbeam forest 40−60 years	OAK2	15	4
Oak-hornbeam forest 60−80 years	OAK3	14	4
Oak-hornbeam forest 80−120 years	OAK4	15	4
Oak-hornbeam forest >120 years	OAK5	11	4
Coniferous and mixed forest <40 years	CON1	28	1
Coniferous and mixed forest 40−60 years	CON2	22	2
Coniferous and mixed forest 60−80 years	CON3	21	2
Coniferous and mixed forest 80−120 years	CON4	18	3

Continued

Table 6.43 Ecosystem potential to decompose organic matter (1 - very low; 5 - very high) indicated by ratio of organic carbon to total nitrogen in soil. (cont'd)

Ecosystem type	Ecosystem acronym	C/N indicator value	Ecosystem potential
Coniferous and mixed forest >120 years	CON5	16	3
Swamp coniferous forest <40 years	SWP1	20	3
Swamp coniferous forest 40 −60 years	SWP2	25	2
Swamp coniferous forest 60 −80 years	SWP3	20	3
Swamp coniferous forest 80 −120 years	SWP4	18	3
Swamp coniferous forest >120 years	SWP5	18	3
Wetlands - reed beds and sedge swamps	REED1	11	4
Wetlands - fens	FEN	9	5
Wetlands - bogs	BOG	16	3
Lakes (all types)	LAKE	Not considered	

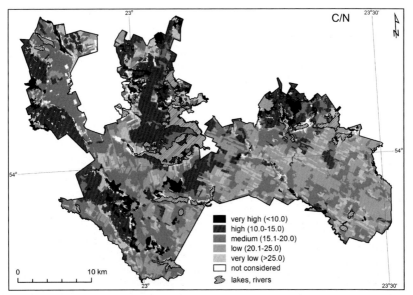

Figure 6.24 Ecosystem potential to decompose organic matter indicated by ratio of organic carbon to total nitrogen in soil (C/N indicator).

(Table 6.44). It can be further described as the regulation of the concentration of oxygen in the atmosphere that impact on global climate. The annual net production of oxygen has been proposed to indicate the capacity of ecosystems to provide it.

Theoretical framework[5]

The concept of ES in its current understanding started to be intensively developed since the end of the 20th century, triggered by publications of de Groot (1992), Costanza et al. (1997), and others. Many authors of older ecological works on the connections between nature and humans described in fact ES, although not knowing that concept yet (among others Odum, 1953).

In Poland, one of those authors was Krzymowska-Kostrowicka who in the book *Geoecology of tourism and recreation* published in 1997 analyzed the links between natural environment and tourism and recreation from the perspective of benefits to people. She characterized plant communities on different dimensions in terms of their importance for tourism and recreation. The classification of vegetation types was based on the book of Matuszkiewicz (2001), and the assessment of their recreational suitability was made on the basis of rich literature and own research in Poland and abroad. About 80 syntaxons of various ranks were distinguished, covering more than 200 plant associations.

Next, these classes were combined into 20 groups, adopting a functional criterion according to similarity in forming the recreational bioclimate, biotherapeutic, and psychoregulatory properties as well as esthetic values. Each group was described in a unified way in terms of general vegetation characteristics (structure, habitat, occurrence in Poland, and ecological, geographic, and anthropogenic diversity) and bioclimate recreational potential (e.g., oxygen production and emission of volatile substances by plants - phytoaerosols).

Forests are the terrestrial ecosystems that play the largest role in oxygen emission. According to various popular science sources (e.g., Śniegocki, 2018), the forests of the Earth meet a total of half of the oxygen demand of all people and animals, producing about 26 billion tons per year. A typical tree absorbs an average of 1 ton of carbon dioxide per cubic meter of growth and produces 727 kg of life-giving oxygen. According to many studies on the intensity of the photosynthesis process, from 0.5 to more

---

[5] The theoretical framework presented here applies to the two ES discussed in this book: oxygen emission and plant aerosol emission.

Table 6.44 Ecosystem service in Common International Classification of Ecosystem Services (CICES) V5.1 and indicator metadata.

Ecosystem service		Oxygen emission
CICES V5.1	Section	Regulation and Maintenance
	Division	Regulation of physical, chemical, biological conditions
	Group(s)	Atmospheric composition and conditions
	Class(es)	Regulation of chemical composition of atmosphere and oceans
	Class code(s)	2.2.6.1
Indicandum		**Ecosystem capacity to emit oxygen**
Indicator		**Annual net production of oxygen**
Acronym		**OXYGEN**
Short description		Estimation of annual net production of oxygen by plants (photosynthesis minus respiration) based on literature data
Direct/indirect		Direct
Simple/compound/ complex		Simple
Estimated/calculated		Estimated
Scale		Ratio
Value range		0–40
Unit		$t\ ha^{-1}\ year^{-1}$
Spatial unit		Ecosystem
Value interpretation		<5.0 → very low ecosystem potential; 5.0–10.0 → low; 10.1–15.0 → medium; 15.1–20.0 → high; >20 → very high
Source data		Krzymowska-Kostrowicka (1997) and other literature data

than 1 kg of pure oxygen (O_2) enters the atmosphere through 1 m^2 of leaf surface during the growing season.

The trees producing the largest amounts of oxygen include beech, maple and robinia (1.1 kg), willow and oak (0.8 kg), and lime and ash (0.7 kg). Similar amounts of oxygen are emitted by coniferous trees. One 60-year-old pine tree produces daily enough oxygen for three persons, and one tree produces as much oxygen throughout the year as the human being consumes within 2 years (one person uses $14-18$ m^3 of air per day).

Trees that grow in large complexes create a specific microclimate that is beneficial to humans. The forest air is clean and saturated with oxygen (necessary for the nourishment and regeneration of all body cells). According to the researchers' calculations, 1 ha of deciduous forest produces about 700 kg of pure oxygen within 24 h. This amount of gas satisfies the daily oxygen demand of 2500 people. Forests are often referred to as "large oxygen factories" or "green lungs".

Assessment method

The assessment of oxygen production by ecosystems was performed with the use of data provided by Krzymowska-Kostrowicka (1997). The OXYGEN indicator values are given in tons per hectare per year and show vegetation net production (photosynthesis minus respiration), without taking account of oxygen consumption by other organisms in the ecosystem.

Apart from forest ecosystems, also grassland, arable fields, and wetlands were taken into account. Aquatic ecosystems, which are still poorly researched in terms of net oxygen production, are not included, as available data are not comparable with data for terrestrial ecosystems.

Indicator values

Mature alder forests (from 60 to over 120 years) and oak—hornbeam forests to 60 years have the highest potential to produce oxygen (>20 t ha^{-1} year^{-1}). Other age classes of alder and oak-hornbeam forests, as well as mature coniferous and mixed forests (from 80 to over 120 years), have slightly lower potential ranging from 15 to 20 t ha^{-1} year^{-1}. Most forest ecosystems riparian forests (all age classes), the youngest alder forests, coniferous and mixed forests up to 80 years old, and both dry and wet grassland on mineral soil produce moderate amounts of oxygen ($10-15$ t ha^{-1} year^{-1}). In turn, all swamp coniferous forests, wetlands, and arable fields have low oxygen emission efficiency ($5-10$ t ha^{-1} year^{-1}). Peat grassland, fens, and bogs have the lowest potential for oxygen production - below 5 t ha^{-1} year^{-1} (Table 6.45).

Spatial pattern of OXYGEN indicator values is mosaic type with patches of very different sizes (Fig. 6.25). The largest patch covers over 12% of the study area. Areas of medium ecosystem potential dominate, covering over 40% of the total area, but patches of low and high potential subdominate, covering together almost 42%. Larger patches are generally more irregular compared with small patches. Small patches of the same level of potential are quite evenly distributed, and the interspersion of patches of different level of potential is of average level compared with other ES potentials.

Table 6.45 Ecosystem potential to emit oxygen (1 - very low; 5 - very high) indicated by annual net production of oxygen (t ha^{-1} year^{-1}).

Ecosystem type	Ecosystem acronym	OXYGEN indicator value	Ecosystem potential
Settlement area	SETTLE	Not considered	
Arable field on poor sandy soil	ARB1	5.0—10.0	2
Arable field on fertile mesic soil	ARB2	5.0—10.0	2
Arable field on fertile moist soil	ARB3	5.0—10.0	2
Dry grassland	GRAS1	10.1—15.0	3
Wet grassland on mineral soil	GRAS2	10.1—15.0	3
Peat grassland	GRAS3	<5.0	1
Alder forest <40 years	ALD1	10.1—15.0	3
Alder forest 40—60 years	ALD2	15.1—20.0	4
Alder forest 60—80 years	ALD3	>20.0	5
Alder forest 80—120 years	ALD4	>20.0	5
Alder forest >120 years	ALD5	>20.0	5
Riparian forest <40 years	RIP1	10.1—15.0	3
Riparian forest 40—60 years	RIP2	10.1—15.0	3
Riparian forest 60—80 years	RIP3	10.1—15.0	3
Riparian forest 80—120 years	RIP4	10.1—15.0	3
Riparian forest >120 years	RIP5	10.1—15.0	3
Oak-hornbeam forest <40 years	OAK1	>20.0	5
Oak-hornbeam forest 40—60 years	OAK2	>20.0	5
Oak-hornbeam forest 60—80 years	OAK3	15.1—20.0	4
Oak-hornbeam forest 80—120 years	OAK4	15.1—20.0	4
Oak-hornbeam forest >120 years	OAK5	15.1—20.0	4
Coniferous and mixed forest <40 years	CON1	10.1—15.0	3
Coniferous and mixed forest 40—60 years	CON2	10.1—15.0	3

Table 6.45 Ecosystem potential to emit oxygen (1 - very low; 5 - very high) indicated by annual net production of oxygen (t ha^{-1} year^{-1}). (cont'd)

Ecosystem type	Ecosystem acronym	OXYGEN indicator value	Ecosystem potential
Coniferous and mixed forest 60—80 years	CON3	10.1—15.0	3
Coniferous and mixed forest 80—120 years	CON4	15.1—20.0	4
Coniferous and mixed forest >120 years	CON5	15.1—20.0	4
Swamp coniferous forest <40 years	SWP1	5.0—10.0	2
Swamp coniferous forest 40—60 years	SWP2	5.0—10.0	2
Swamp coniferous forest 60—80 years	SWP3	5.0—10.0	2
Swamp coniferous forest 80—120 years	SWP4	5.0—10.0	2
Swamp coniferous forest >120 years	SWP5	5.0—10.0	2
Wetlands - reed beds and sedge swamps	REED1	5.0—10.0	2
Wetlands - fens	FEN	<5.0	1
Wetlands - bogs	BOG	<5.0	1
Lakes (all types)	LAKE	Not considered	

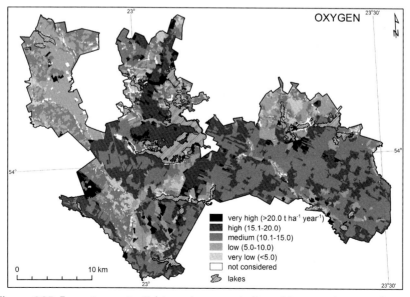

Figure 6.25 Ecosystem potential to emit oxygen indicated by annual net production of oxygen (OXYGEN indicator; t ha^{-1} year^{-1}).

6.1.2.10 Carbon sequestration
Ecosystem service description

This service belongs to *Regulation and Maintenance* section and *Regulation of physical, chemical, biological conditions* division in CICES V5.1 (Table 6.46). It can be further described as the regulation of the concentration of carbon dioxide in the atmosphere that impacts on global climate. The total ecosystem organic carbon stock (in soil, herb layer, and trees) has been proposed to indicate the capacity of ecosystems to provide it.

Theoretical framework

Organic carbon sequestration by ecosystems is one of the most important functions affecting all elements of the natural environment (Lal, 2005, 2004; Wall et al., 2012). The value of carbon sequestration as an ES is significant because it is not only limited to the direct ecosystem-recipient relationship but also affects the generation of other final ES. An important reservoir of organic carbon is soil cover and plant biomass (Fig. 6.26).

The primary way carbon is stored in soil is soil organic matter (SOM). The functioning of soils and shaping of their physical, chemical, and biological properties are closely related to the content of SOM, which improves soil structure and reduces erosion, leading to improved water quality in groundwater and surface waters. SOM is the basic reservoir of organic carbon in soil. The amount of total organic carbon (TOC) in soil cover represents a significant proportion of the carbon found in terrestrial ecosystems (Lal, 2008). Carbon input and losses are dependent on the environment, including geomorphological processes, climatic conditions, soil cover, and vegetation. Its decrease can lead to deterioration of soil conditions that have a direct negative impact on biomass. As a result, the functioning of ecosystems can be disturbed and thus the loss or limitation of the ability to provide ES. Soil carbon losses and increase in atmospheric CO_2 are closely related to human activity, mainly land-use changes including intensive forest management and agriculture (Lal, 2004). It should be noted that carbon cycle in nature is a continuous phenomenon and occurs with the participation of all elements of the natural environment, and for this reason, degradation of even one of them has huge impact on the functioning of the entire system. In this study, the authors focused on organic carbon content in soil and plant biomass (herb layer and trees).

Table 6.46 Ecosystem service in Common International Classification of Ecosystem Services (CICES) V5.1 and indicator metadata.

Ecosystem service		Carbon sequestration
CICES V5.1	Section	Regulation and Maintenance
	Division	Regulation of physical, chemical, biological conditions
	Group(s)	Atmospheric composition and conditions
	Class(es)	Regulation of chemical composition of atmosphere and oceans
	Class code(s)	2.2.6.1
Indicandum		**Ecosystem capacity to sequestrate carbon**
Indicator		**Organic carbon stock in soil, herb layer, and trees**
Acronym		**CARBON**
Short description		Calculation of total carbon stock in ecosystem by adding organic carbon stock in soil (to the depth of 50 cm), herb layer, and trees; organic carbon content in soil and herb layer calculated based on reference field data and laboratory analysis; organic carbon content in trees calculated using forest inventory data (timber volume, tree stand composition, and density) and average carbon content in single trees by species
Direct/indirect		Direct
Simple/compound/complex		Compound
Estimated/calculated		Calculated
Scale		Ratio
Value range		51.9–554.3
Unit		$t\ ha^{-1}$
Spatial unit		Ecosystem
Value interpretation		$<100 \rightarrow$ very low ecosystem potential; $100-150 \rightarrow$ low; $151-200 \rightarrow$ medium; $201-400 \rightarrow$ high; $>400 \rightarrow$ very high
Source data		— Field and laboratory studies at selected points representing particular types of terrestrial ecosystem — State Forests Information System (SILP); Forest Digital Map (LMN) of the four forest districts — Protection Plan of the Wigry National Park — Literature: Jagodziński (2011); Jelonek and Tomczak (2011); Skorupski et al. (2011)

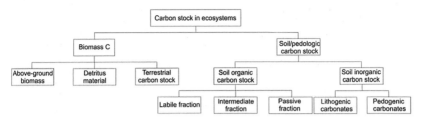

Figure 6.26 Scheme of components of the terrestrial carbon stock. *(After Lal, R., 2005. Forest soils and carbon sequestration. Forest Ecology and Management 220 (1—3), 242—258.)*

Assessment method

The value of carbon stock in ecosystems is the sum of TOC stored in soil, herb layer, and trees of particular ecosystem types. To estimate TOC in soil, 49 disturbed soil samples were collected from organic and humus horizons from 18 points located in different ecosystem types. Direct field studies were conducted in ecosystems for which there were no available literature data. Other values were taken from Skorupski et al. (2011). The content of organic carbon (TOC) in mineral and mineral—organic samples was determined using the Tiurin method, while in organic samples Alten method was used. The stock of organic carbon in soil up to a depth of 50 cm was calculated using formula:

$$TOCs = \frac{h\,D\,TOC}{10}(1 - q),$$

where *TOCs* is the organic carbon stock, *h* is the soil horizon thickness (m), *D* is the bulk density, *TOC* is the organic carbon content in soil horizon, and *q* is $\frac{skeletal\ fraction\ (\%)}{100}$

The organic carbon stock in herb layer was calculated on the basis of direct data from field studies. At selected points located in different types of ecosystems, herb layer samples were collected, in which the content of organic carbon was determined using Alten method.

The calculation of carbon content in trees was based on direct data on species structure of the studied stands (Jagodziński, 2011). Determination of the percentage share of dominant tree species in the distinguished types of forest ecosystem, data on timber volume, and data on the density of wood of individual species allowed calculating biomass in kilograms (Jelonek and Tomczak, 2011). The final result was obtained by summing up carbon content in dominant tree species of particular ecosystems, being the product of the following elements: biomass data (kg), average organic carbon content for habitats, and age classes of tree stands.

Indicator values

The organic carbon content in soils of the studied terrestrial ecosystems closely corresponds to the grain size of surface horizons. Highest content of TOC (up to 548 t ha^{-1}) was observed in organic substrate (peat, muck) found in ecosystems classified as wetlands of various types and peat grassland, while the lowest in loose and weakly sands present in all age categories of coniferous and mixed forests (minimum value 31 t ha^{-1}). The low content of organic carbon (below 60 t ha^{-1}) is also recorded in dry grassland and arable fields on poor sandy soil (Table 6.47).

Values of TOC content in herb layer ranged from 0.1 t ha^{-1} in the ecosystem of young (0–40 years) oak-hornbeam forests to 6.5 t ha^{-1} on fens (Table 6.47). Among the forest ecosystems, the richest in organic carbon was the herb layer of riparian forests at the age of 60–80 years - of 3.8 t ha^{-1}. Relatively high values were also recorded in grassland ecosystems (on average around 2.0 t ha^{-1}).

The calculated TOC content in trees varies from 14 t ha^{-1} in the category of riparian forests over 120 years old to 93 t ha^{-1} in the ecosystem of 80–120 year old coniferous and mixed forests (Table 6.47). In most types of forest ecosystems, the increase of TOC content with the tree stand age was noted. The maximum values for particular habitat types were characteristic for tree stands in the 80–120 age category, while the minimum values are for the youngest stands (0–40 years). In the oldest forest ecosystems, the TOC content decreases.

The overall content of TOC (CARBON indicator) in each type of ecosystem closely refers to the soil texture and humidity conditions (Table 6.47). An additional important factor is land use. Reed beds and sedge swamps stand out in terms of organic carbon content among the studied ecosystems. In these ecosystems, values up to 554 t ha^{-1} were recorded. Dry grasslands (52 t ha^{-1}) are the least rich in organic carbon. The significant role of forests in carbon sequestration and storage is also indisputable, bringing together as much as 74% of the TOC content in the study area. An increase in TOC content in older forest ecosystems was observed. The results show that among studied forest ecosystems, all age categories of swamp coniferous forests are the most abundant in TOC

Table 6.47 Ecosystem potential to sequestrate carbon (1 - very low; 5 - very high) indicated by organic carbon stock in soil, herb layer, and trees (t ha^{-1}).

Ecosystem type	Acronym	Total organic carbon (TOC) content in soil	TOC content in herb layer	TOC content in trees	CARBON indicator value	Ecosystem potential
Settlement area	SETTLE	Not considered				
Arable field on poor sandy soil	ARB1	58.7	0.0	0.0	58.7	1
Arable field on fertile mesic soil	ARB2	120.7	0.0	0.0	120.7	2
Arable field on fertile moist soil	ARB3	191.6	0.0	0.0	191.6	3
Dry grassland	GRAS1	49.9	2.0	0.0	51.9	1
Wet grassland on mineral soil	GRAS2	51.2	2.0	0.0	53.2	1
Peat grassland	GRAS3	289.1	2.1	0.0	291.2	4
Alder forest <40 years	ALD1	218.4	1.4	29.5	249.3	4
Alder forest 40–60 years	ALD2	340.0	2.2	46.0	388.2	4
Alder forest 60–80 years	ALD3	328.2	0.8	52.4	381.4	4
Alder forest 80–120 years	ALD4	414.0	1.0	66.1	481.1	5
Alder forest >120 years	ALD5	303.0	0.7	34.8	338.5	4
Riparian forest <40 years	RIP1	244.2	1.8	29.7	275.7	4
Riparian forest 40–60 years	RIP2	249.2	1.9	30.3	281.4	4
Riparian forest 60–80 years	RIP3	310.0	3.8	60.5	374.3	4

Continued

Riparian forest 80–120 years	RIP4	304.4	3.7	59.4	367.5	4
Riparian forest >120 years	RIP5	213.5	1.6	13.6	228.7	4
Oak-hornbeam forest <40 years	OAK1	63.0	0.1	22.3	85.4	1
Oak-hornbeam forest 40–60 years	OAK2	71.9	0.4	61.8	134.1	2
Oak-hornbeam forest 60–80 years	OAK3	92.1	0.5	79.1	171.7	3
Oak-hornbeam forest 80–120 years	OAK4	104.3	0.6	89.6	194.5	3
Oak-hornbeam forest >120 years	OAK5	89.2	0.5	76.6	166.3	3
Coniferous and mixed forest <40 years	CON1	31.3	0.9	26.9	59.1	1
Coniferous and mixed forest 40–60 years	CON2	43.1	2.1	64.7	109.9	2
Coniferous and mixed forest 60–80 years	CON3	52.2	1.5	80.2	133.9	2
Coniferous and mixed forest 80–120 years	CON4	64.0	1.7	92.6	158.3	3
Coniferous and mixed forest >120 years	CON5	82.9	3.2	89.5	175.6	3
Swamp coniferous forest <40 years	SWP1	235.0	0.9	26.0	261.9	4

Table 6.47 Ecosystem potential to sequestrate carbon (1 - very low; 5 - very high) indicated by organic carbon stock in soil, herb layer, and trees (t ha^{-1}). (cont'd)

Ecosystem type	Acronym	Total organic carbon (TOC) content in soil	TOC content in herb layer	TOC content in trees	CARBON indicator value	Ecosystem potential
Swamp coniferous forest 40–60 years	SWP2	354.0	1.9	51.4	407.3	5
Swamp coniferous forest 60–80 years	SWP3	380.8	2.4	50.4	433.6	5
Swamp coniferous forest 80–120 years	SWP4	430.2	2.7	57.0	489.9	5
Swamp coniferous forest >120 years	SWP5	468.3	2.9	62.0	533.2	5
Wetlands – reed beds and sedge swamps	REED1	547.8	6.5	0.0	554.3	5
Wetlands – fens	FEN	525.0	6.5	0.0	531.5	5
Wetlands – bogs	BOG	481.4	1.7	0.0	483.1	5
Lakes (all types)	LAKE	Not considered				

($>$400 t ha^{-1} in most categories) and also alder and riparian forests with TOC content of 200−400 t ha^{-1}.

Spatial pattern of CARBON indicator values is mosaic type with patches of various sizes (Fig. 6.27). The largest patch covers almost 10% of the study area. Areas of very low, low, and medium ecosystem potential codominate, covering together over 80% of the total area. Large patches are generally more irregular in shape compared with small patches. Small patches of the same level of potential are quite evenly distributed, and the interspersion of patches of different level of potential is of average level compared with other ES potentials.

6.1.2.11 Plant aerosol emission
Ecosystem service description
This service belongs to *Regulation and Maintenance* section and *Regulation of physical, chemical, biological conditions* division in CICES V5.1 (Table 6.48). It can be further described as the regulation of the concentration of plant aerosols in the atmosphere that impacts on micro- and mesoscale climates and improves living conditions for people. The

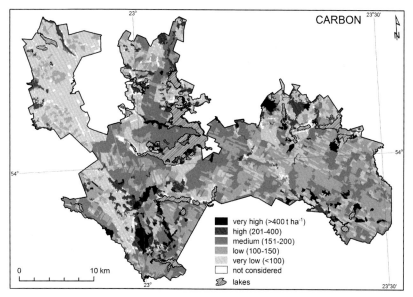

Figure 6.27 Ecosystem potential to sequestrate carbon indicated by organic carbon stock in soil, herb layer, and trees (CARBON indicator; t ha^{-1}).

efficiency of plant aerosol emission has been proposed to indicate the capacity of ecosystems to provide it.

Theoretical framework[6]

Various types of chemicals floating in the air of biological origin known as organic aerosols have significant impact on the quality of the broadly understood bioclimate. About 70%—80% of volatile substances participate in the creation of organic aerosol. These are both simple substances (oxygen, hydrogen, water, nitrogen, simple hydrocarbons, and their derivatives) and so-called secondary metabolites (alkaloids, amino acids, carbohydrates, and others).

The latter plays a special role, for they include essential oils and other aromatic compounds having basic biotherapeutic and psychoregulatory importance, also characterized by strong germicidal and bacteriostatic effects. The concentration of organic aerosols in the air depends not only on external factors, mainly weather conditions, but also on internal factors directly related to their producers, i.e., plants.

Paduch et al. (2007) describe plant aerosols (mainly terpenes) as substances useful in human health care. Terpenes are naturally occurring substances produced by a wide variety of plants. The authors show a broad range of biological properties of terpenoids, including cancer chemopreventive effects, antimicrobial, antifungal, antiviral, antihyperglycemic, antiinflammatory, and antiparasitic activities. Paduch et al. (2007) suggest that large-scale use of terpenoids in modern medicine should be taken into consideration. Marchese et al. (2017) summarize available scientific data, as reported by the most recent studies describing monoterpene found in over 100 plant species used for medicine and food purposes. It shows a range of biological activity including antioxidant, antiinflammatory, antinociceptive, anxiolytic, anticancer, and antimicrobial effects.

Not only the condition of plants has significant impact on the quality and quantity of the emitted substances but also soil fertility - both excessive and too small supply of nutrients lead to a reduction in the amount of substances released. For instance, pine trees growing on fertile soil produce about 1/3 less amount of essential oils compared with those growing on poor sandy soil. Similarly, in extremely poor conditions (e.g., in dry coniferous forests or in bogs), the amount of emitted oils is

---

[6] The general theoretical framework for this service is presented jointly with oxygen emission in Section 6.1.2.9.

Table 6.48 Ecosystem service in Common International Classification of Ecosystem Goods and Services (CICES) V5.1 and indicator metadata.

Ecosystem service		Plant aerosol emission
CICES V5.1	Section	Regulation and Maintenance
	Division	Regulation of physical, chemical, biological conditions
	Group(s)	Atmospheric composition and conditions
	Class(es)	Regulation of chemical composition of atmosphere and oceans
	Class code(s)	2.2.6.1
Indicandum		**Ecosystem capacity to emit plant aerosols**
Indicator		**Efficiency of plant aerosol emission**
Acronym		**AEROS**
Short description		Estimation of emission efficiency of plant aerosols supporting human health based on literature data
Direct/indirect		Direct
Simple/compound/ complex		Simple
Estimated/calculated		Estimated
Scale		Ordinal
Value range		1—5
Unit		—
Spatial unit		Ecosystem
Value interpretation		1 → very low potential; 2 → low; 3 → medium; 4 → high; 5 → very high
Source data		Krzymowska-Kostrowicka (1997) and other literature data

about 25%—50% lower than in the typical pine forest. Weather conditions also influence it - during warm and sunny days, pine trees emit five times more volatile substances than during cold days. A seemingly small amount of volatile substances emitted into the air is often sufficient to cause therapeutic effects in the human body (Krzymowska-Kostrowicka, 1997).

Biological properties of plant aerosols and their impact on human health formed the basis for the evaluation of ecosystem potential. Nonetheless, plant aerosols impact on human well-being also in other ways. Plant aerosol studies show on the example of boreal forests that vegetation can emit particles directly into the atmosphere. Primary biological aerosol particles include spores, fungi, and leaf matter. Aerosols directly scatter and absorb radiation. The scattering of radiation causes atmospheric cooling, whereas absorption can cause atmospheric warming. Aerosols modify the properties of clouds through a subset of the aerosol population called cloud condensation nuclei (CCN) (Spracklen et al., 2008).

Assessment method

The assessment of plant aerosol emission by ecosystems was performed with the use of data provided by Krzymowska-Kostrowicka (1997). The AEROS indicator values are on a 1—5 rank scale and were assigned to ecosystems based on expert knowledge.

Indicator values

Coniferous and mixed forests are ecosystems that stand out from others by very high intensity of emission of plant aerosols (rank 5). Slightly lower emission (though also high - rank 4) was noted in oak-hornbeam forests, swamp coniferous forests, as well as fens and bogs. This group also comprises dry grassland. Alder and riparian forests were assigned medium potential to emit plant aerosols (rank 3). Both peat grassland and wet grassland on mineral soil received the same rank. In the case of reed beds and sedge swamps, the aerosol emission efficiency was estimated as low (rank 2) and in the case of agroecosystems as very low (rank 1) (Table 6.49).

Because of the domination of agroecosystems in the western and northern parts of the study area, and forest prevailence in the remaining part, spatial pattern of AEROS indicator values is twofold (Fig. 6.28). Central and eastern part is dominated by uniform matrix of very high ecosystem potential with small number of patches of high potential randomly scattered on its background. In the northern and western parts, the pattern changes into more mosaic-like with high share of large patches of very low potential. The largest patch covers over 18% of the study area. Areas of very high potential dominate, covering over 42% of the total area. Large patches are generally more irregular in shape compared with small patches, which are characterized by MSI below 2. Small patches of the same level of potential are quite randomly scattered, and the interspersion of patches of different level of potential is medium.

Table 6.49 Ecosystem potential to emit plant aerosols indicated by the efficiency of plant aerosol emission (1 - very low; 5 - very high).

Ecosystem type	Ecosystem acronym	AEROS indicator value/ ecosystem potential
Settlement area	SETTLE	Not considered
Arable field on poor sandy soil	ARB1	1
Arable field on fertile mesic soil	ARB2	1
Arable field on fertile moist soil	ARB3	1
Dry grassland	GRAS1	4
Wet grassland on mineral soil	GRAS2	3
Peat grassland	GRAS3	3
Alder forest <40 years	ALD1	3
Alder forest 40–60 years	ALD2	3
Alder forest 60–80 years	ALD3	3
Alder forest 80–120 years	ALD4	3
Alder forest >120 years	ALD5	3
Riparian forest <40 years	RIP1	3
Riparian forest 40–60 years	RIP2	3
Riparian forest 60–80 years	RIP3	3
Riparian forest 80–120 years	RIP4	3
Riparian forest >120 years	RIP5	3
Oak-hornbeam forest <40 years	OAK1	4
Oak-hornbeam forest 40–60 years	OAK2	4
Oak-hornbeam forest 60–80 years	OAK3	4
Oak-hornbeam forest 80–120 years	OAK4	4
Oak-hornbeam forest >120 years	OAK5	4
Coniferous and mixed forest <40 years	CON1	5
Coniferous and mixed forest 40 –60 years	CON2	5
Coniferous and mixed forest 60 –80 years	CON3	5
Coniferous and mixed forest 80 –120 years	CON4	5
Coniferous and mixed forest >120 years	CON5	5
Swamp coniferous forest <40 years	SWP1	4
Swamp coniferous forest 40–60 years	SWP2	4
Swamp coniferous forest 60–80 years	SWP3	4
Swamp coniferous forest 80–120 years	SWP4	4

Continued

Table 6.49 Ecosystem potential to emit plant aerosols indicated by the efficiency of plant aerosol emission (1 - very low; 5 - very high). (cont'd)

Ecosystem type	Ecosystem acronym	AEROS indicator value/ ecosystem potential
Swamp coniferous forest >120 years	SWP5	4
Wetlands - reed beds and sedge swamps	REED1	2
Wetlands - fens	FEN	4
Wetlands - bogs	BOG	4
Lakes (all types)	LAKE	Not considered

6.1.2.12 Atmospheric heavy metal accumulation
Ecosystem service description

This service belongs to *Regulation and Maintenance* section and *Regulation of physical, chemical, biological conditions* division in CICES V5.1 (Table 6.50). It can be further described as the regulation of the concentration of heavy metals (HM) in the atmosphere that impacts on micro- and mesoscale

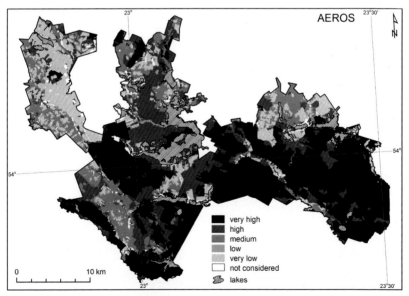

Figure 6.28 Ecosystem potential to emit plant aerosols indicated by the efficiency of plant aerosol emission (AEROS indicator).

climates and improves living conditions for people. The concentration of HM in soil has been proposed to indicate the capacity of ecosystems to provide it.

Table 6.50 Ecosystem service in Common International Classification of Ecosystem Services (CICES) V5.1 and indicator metadata.

Ecosystem service		Atmospheric heavy metal accumulation
CICES V5.1	Section	Regulation and Maintenance
	Division	Regulation of physical, chemical, biological conditions
	Group(s)	Atmospheric composition and conditions
	Class(es)	Regulation of chemical composition of atmosphere and oceans
	Class code(s)	2.2.6.1
Indicandum		**Ecosystem capacity to accumulate atmospheric heavy metals**
Indicator		**Concentration of heavy metals in soil**
Acronym		**METAL**
Short description		Calculation of total content of Cu, Zn, and Ni in organic and mineral soil horizons based on reference field data and laboratory analysis
Direct/indirect		Direct
Simple/compound/ complex		Compound
Estimated/calculated		Calculated
Scale		Ratio
Value range		49−235
Unit		$mg\ kg^{-1}$
Spatial unit		Ecosystem
Value interpretation		<70.0 → very low potential; 70.0−100.0 → low; 100.1−150.0 → medium; 150.1−200.0 → high; >200.0 → very high
Source data		Field and laboratory studies at selected points representing particular types of terrestrial ecosystem

Theoretical framework

The regulation of chemical composition of atmosphere belongs to one of the very important regulating ES, deciding not only about the sanitary state of the environment but also about the well-being of human life. The efficiency of the human body, its well-being, and health condition depend on the sanitary state of the air to a large extent.

Air pollution can be caused by an increased content of dust, HM, or nitrogen and sulfur compounds. Among the group of HM treated as air pollutants, the content of copper (Cu), nickel (Ni), and zinc (Zn) is very important, followed by mercury (Hg), arsenic (As), and cadmium (Cd). An important factor affecting the solubility of chemical compounds of HM is the pH of the solution (atmospheric precipitation) in which the compound is located. Lowering the pH value promotes solubility but depends both on the metal itself and on the compound in which the metal is present.

Soils are the major sink for HM released into the environment by anthropogenic activities (mostly by emissions into atmosphere, but also by disposal of high metal wastes, land application of fertilizers, animal manures, sewage sludge, pesticides, wastewater irrigation) (Wuana and Okieimen, 2011). HM and rainfall are transferred to the soil, where they accumulate, mainly in the form of sorption, displacing alkaline cations from the sorption complex. Therefore, the content of HM in soils can be treated as an indicator of air quality, additionally considering the low dynamics of HM content changes in the soil sorption complex compared with their content in atmospheric air.

Assessment method

The METAL indicator was calculated as the sum of three HM content (Cu, Ni, Zn) in the humus and organic soil horizons. The content of HM was determined in soil samples taken during field studies by atomic emission spectrometry with microwave excited plasma (Agilent 4100 MP-AES spectrometer) after mineralization of incinerated samples in *aqua regia*. Then, the values obtained for individual HM were summed up for each of the tested soil horizon in particular ecosystem types and presented as the average total value for two genetic horizons. The obtained results were compared with the geochemical background of the studied soil types in a specific region of Poland and referred to the standards for HM content in soils regarding Polish regulation.

Indicator values

The studied ecosystems are characterized by a very low total content of HM, which indicates a very high quality of the environment from the point of view of its sanitary condition. The best sorption and buffer properties are found in the rich moist soils containing the largest amount of organic matter. Ecosystems with the obtained rich soil can retain the largest amount of HM. These include old alder and riparian forests growing on peat, high peat bogs with fiber peat soils, and fens formed on fibrous sap peat soils. These soils are also characterized by the largest natural content of HM (Table 6.51).

The highest contents of the investigated HM were found in the eco-systems of the oldest alder forests (over 80 years old) and riparian forests growing on peat fens (60–120 years old), and peat grassland, which is consistent with the natural content of these chemical components in the soil types studied, resulting from their trophy. High HM content in soils was obtained in riparian forest (up to 60 years old and over 120 years old). The lowest values were characteristic for some arable fields, dry grassland, very young alder forest (<40 years old), coniferous and mixed forest (60–120 years old), swamp coniferous forest (over 80 years old), and bogs.

Spatial pattern of METAL indicator values is mosaic type with patches of very different sizes (Fig. 6.29). The largest patch covers almost 12% of the study area. Areas of low and very low potential codominate, covering together over 82% of the total area. Large patches are generally more irregular in shape compared with small patches. Small patches of the same level of potential are

Table 6.51 Ecosystem potential to accumulate atmospheric heavy metals (1 - very low; 5 - very high) indicated by heavy metal concentration in soil (mg kg^{-1}).

Ecosystem type	Ecosystem acronym	METAL indicator value	Ecosystem potential
Settlement area	SETTLE	Not considered	
Arable field on poor sandy soil	ARB1	49	1
Arable field on fertile mesic soil	ARB2	99	2
Arable field on fertile moist soil	ARB3	51	1
Dry grassland	GRAS1	59	1
Wet grassland on mineral soil	GRAS2	89	2
Peat grassland	GRAS3	222	5
Alder forest <40 years	ALD1	65	1
Alder forest 40–60 years	ALD2	82	2
Alder forest 60–80 years	ALD3	100	2

Continued

Table 6.51 Ecosystem potential to accumulate atmospheric heavy metals (1 - very low; 5 - very high) indicated by heavy metal concentration in soil (mg kg^{-1}). (cont'd)

Ecosystem type	Ecosystem acronym	METAL indicator value	Ecosystem potential
Alder forest 80—120 years	ALD4	230	5
Alder forest >120 years	ALD5	230	5
Riparian forest <40 years	RIP1	160	4
Riparian forest 40—60 years	RIP2	194	4
Riparian forest 60—80 years	RIP3	235	5
Riparian forest 80—120 years	RIP4	200	4
Riparian forest >120 years	RIP5	167	4
Oak-hornbeam forest <40 years	OAK1	86	2
Oak-hornbeam forest 40—60 years	OAK2	85	2
Oak-hornbeam forest 60—80 years	OAK3	102	3
Oak-hornbeam forest 80—120 years	OAK4	102	3
Oak-hornbeam forest >120 years	OAK5	90	2
Coniferous and mixed forest <40 years	CON1	82	2
Coniferous and mixed forest 40—60 years	CON2	78	2
Coniferous and mixed forest 60—80 years	CON3	52	1
Coniferous and mixed forest 80—120 years	CON4	67	1
Coniferous and mixed forest >120 years	CON5	86	2
Swamp coniferous forest <40 years	SWP1	68	1
Swamp coniferous forest 40—60 years	SWP2	83	2
Swamp coniferous forest 60—80 years	SWP3	88	2
Swamp coniferous forest 80—120 years	SWP4	68	1
Swamp coniferous forest >120 years	SWP5	62	1
Wetlands - reed beds and sedge swamps	REED1	86	2
Wetlands - fens	FEN	72	2
Wetlands - bogs	BOG	66	1
Lakes (all types)	LAKE	Not considered	

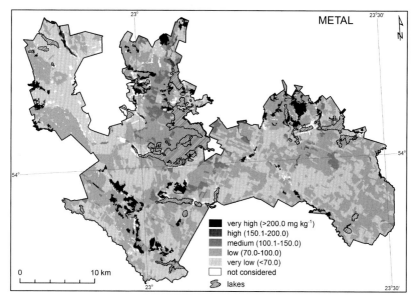

Figure 6.29 Ecosystem potentials to accumulate atmospheric heavy metals indicated by heavy metal concentration in soil (METAL indicator; mg kg^{-1}).

quite evenly distributed, and the interspersion of patches of different level of potential is the lowest compared with other ES potentials.

6.2 Potentials to provide ecosystem services - social assessment

6.2.1 Theoretical framework

The potential of ecosystems to provide services can be estimated in a variety of ways. In addition to the often difficult-to-calculate direct measures and subjective - although supported by scientific knowledge - expert assessment, one can also use the opinion of direct ES beneficiaries.

Although subjective, the opinions of direct ES users are based on long-term experience with the use of the goods and services local ecosystems provide (García-Nieto et al., 2015). And, while the assessment of a single person does not tell us much about actual ecosystem potential (being based on individual experience stemming from personal characteristics and history), the mean value from a representative, adequate sample of direct beneficiaries can provide an irreplaceable source of information of even greater reliability than other valuations. This regularity concerns all types of ES: provisioning, regulating, and cultural, as research on social awareness of ES and their identification by direct users shows how people easily recognize the vast majority of goods and services nature offers, even where these are regulating services seemingly hard to grasp (Affek and Kowalska, 2014). Respondents forming a representative sample of ES beneficiaries listed collectively all services included in popular classifications, as well as those ES that should probably also be there, but have not been recognized by experts so far.

One of the most common methods used to identify public opinion is questionnaire research (Scholte et al., 2015). In this study, public opinion was used to assess the potential of ecosystems to provide selected provisioning, regulating, and cultural services (Table 6.52) in seven types of ecosystems. The considered provisioning and cultural services cover the majority of divisions and classes distinguished in CICES V5.1. Users rated, among others, the ecosystem capacity to provide biomass for nutritional, construction, and medicinal purposes and ecosystem usefulness for sport and recreation and creative work. In contrast, out of regulating services, much more difficult to assess by nonspecialists, only water-related were included (retention and purification together as one service), which we assumed might be ranked variously as regards ecosystem potentials.

A detailed presentation of the applied approach, along with a broad literature review and thorough discussion of the results, can be found in Affek and Kowalska (2017).

6.2.2 Assessment method

The anonymous questionnaire survey was carried out over two seasons (summer 2014 and spring 2015) among residents living - and tourists staying - in the vicinity of WNP. Our target was to reach representatives of the entire population of ecosystem users, including dwellers and part-time visitors (tourists), men and women, young people and the elderly, and rich and poor. This accounted for the use of a time-consuming door-to-door method of collecting data. Every door of first and second houses in selected localities was thus knocked on, at different times of the day during working days and holidays, with those found to be in asked to participate in the study. To ensure full representativeness of the sample, we divided villages by dominant function, i.e.,

Table 6.52 Ecosystem service in Common International Classification of Ecosystem Services (CICES) V5.1 and indicator metadata.

Ecosystem service CICES V5.1	Section	Biomass for							Characteristics of living systems that enable/support			
		Nutritional purposes	Medicinal purposes	Construction purposes	The production of fertilizer and fodder	Ornamental purposes	Energy production	Water regulation	Sport and recreation	Education and research	Creative work	Spiritual experience
	Section	Provisioning						Regulation and Maintenance	Cultural			
	Class code(s)	1.1.1.1; 1.1.2.1; 1.1.3.1; 1.1.4.1; 1.1.5.1; 1.1.6.1	1.1.1.2; 1.1.2.2; 1.1.3.2; 1.1.4.2; 1.1.5.2; 1.1.6.2	1.1.1.2; 1.1.2.2; 1.1.3.2; 1.1.4.2;	1.1.1.2; 1.1.2.2; 1.1.3.2; 1.1.4.2; 1.1.5.2; 1.1.6.2	1.1.1.2; 1.1.2.2; 1.1.3.2; 1.1.4.2; 1.1.5.2; 1.1.6.2	1.1.1.3; 1.1.2.3; 1.1.3.3; 1.1.4.3; 1.1.5.3; 1.1.6.3	2.1.1.2; 2.2.1.3	3.1.1.1; 3.1.1.2	3.1.2.1; 3.1.2.2	3.1.2.4	3.2.1.2

Continued

Table 6.52 Ecosystem service in Common International Classification of Ecosystem Services (CICES) V5.1 and indicator metadata. (cont'd)

Ecosystem service	Biomass for						Characteristics of living systems that enable/support				
	Nutritional purposes	Medicinal purposes	Construction purposes	The production of fertilizer and fodder	Ornamental purposes	Energy production	Water regulation	Sport and recreation	Education and research	Creative work	Spiritual experience
Indicandum	Ecosystem capacity to provide biomass for		1.1.5.2; 1.1.6.2				Ecosystem capacity to regulate water	Ecosystem usefulness for			
	Nutritional purposes	Medicinal purposes	Construction purposes	The production of fertilizer and fodder	Ornamental purposes	Energy production		Sport and recreation	Education and research	Creative work	Spiritual experience
Indicator	Opinion of direct users on ecosystem (capacity to provide/capacity to regulate/usefulness for)										
Acronym	NUTRITION	MEDICINE	CONSTRUCT	FODDER	ORNAMENT	ENERGY	H2OREGUL	RECREATION	EDUCATION	CREATION	SPIRIT
Short description	Estimation of ecosystem potentials by means of participatory method; final values obtained through statistical analysis of 251 questionnaires filled by residents and tourists in the study area										
Direct/indirect	Direct										
Simple/compound/complex	Simple										
Estimated/calculated	Estimated										
Scale	Interval										
Value range	3.64–13.03										
Unit	–										
Spatial unit	Ecosystem (7 classes)										
Value interpretation	<5.00 – very low potential; 5.01–7.00 – low; 7.01–10.00 – medium, 10.01–12.00 – high; >12.00 – very high										
Source data	251 filled questionnaires distributed door-to-door in the study area										

in relation to farming, forestry, or tourism. Those to be surveyed were selected at random, independently in the case of each group. We left questionnaires with those expressing a willingness to participate, and returned after a day or two to collect them. More than one revisit was often made necessary by people's tendency to forget to complete the survey. We assumed that inhabitants' presences or absences were random. About half of the people encountered refused to take part in the study, while a further large fraction accepted the questionnaire, but never actually filled it in.

In the third part of the survey, the respondents were asked to indicate the capacity of ecosystems to provide each of the 11 listed services (six provisioning, four cultural, and one regulating) (Fig. 6.30). The scale of the evaluation (1−11) has been extended in such a way that respondents have the possibility to fully diversify the capacity of ecosystems to provide 11 ES in question. It was also possible to enter the answer NA (not applicable), if according to the respondents, the given type of ecosystem is not able to provide a given service. The scientific term "ecosystem services" was absent from the questionnaire, with its place being taken by the more colloquial and intelligible old Polish phrase "dobrodziejstwa przyrody", hence the approximate translation here of "gifts of nature".

Because of the need to keep the questionnaire in a reasonable volume and the predicted difficulties in distinguishing several dozen classes of ecosystems, the pool of ecosystem types has been reduced from 42 to 7. This simplified division corresponds approximately to MAES level 2 typology of ecosystems used for mapping and assessment of ecosystems on a European scale (Maes et al., 2013). Only forests (in line with the high proportion of the study area accounted for) were further divided into the three subtypes of deciduous and coniferous forest (both on mineral soil) and swamp coniferous forest.

Demographic data (including age, gender, education), for which the respondents were asked in the last part of the questionnaire, were used to analyze the representativeness of the research sample and for intergroup comparisons.

Data from on-paper questionnaires were digitized and uploaded to the statistical program (SPSS Statistics). We assumed that the intervals between numbers (from 1 to 11) assigned to services by the respondents are equal (interval scale), with this permitting the use of parametric analysis (e.g., ANOVA, t-tests, r-Pearson correlation).

To facilitate perception of the results, the responses from the questionnaire were recoded in such a way that the entry NA (not applicable) corresponds to the value 0, while the 1−11 scale is reversed. To further

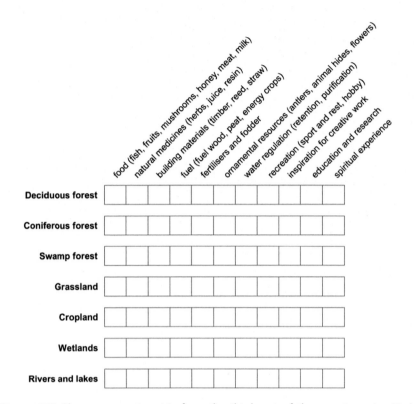

Figure 6.30 The assessment matrix from the third part of the questionnaire. Task: Listed above are seven different types of land cover (ecosystem). Please refer to your own intuition and indicate the capacity of each ecosystem to provide gifts of nature by typing in the number 1—11 (1 - highest capacity, 11 - lowest capacity). If you feel a benefit is not delivered by a given ecosystem at all, please write "NA" - for not applicable. *(Based on Affek, A., Kowalska, A., 2017. Ecosystem potentials to provide services in the view of direct users. Ecosystem Services 26, 183—196.)*

separate the NA answer from the lowest capacity, we shifted the scale by 4 points. As a result, the value of 15 corresponds to the highest ecosystem potential to provide a service, 5 to the lowest potential, and the 0 value to lack of potential (Table 6.53). For visualization purposes, the obtained mean values were reclassified to a five-point ordinal scale (0.00—5.00: extremely low, 5.01—7.00: low, 7.01—10.00: medium, 10.01—12.00: high, 12.01—15.00: very high).

Analysis of variance (ANOVA) was used to test whether differences among ecosystem potentials determined by reference to the entire sample were statistically significant. Because the Levene's test (Levene, 1960)

Table 6.53 Recoding of the questionnaire responses.

Ecosystem potential to provide a service	Highest—lowest											
Original rank	1	2	3	4	5	6	7	8	9	10	11	NA
Rank after recoding	15	14	13	12	11	10	9	8	7	6	5	0

showed that variances across ecosystems were significantly different for almost every service, Tamhane's T2 (Tamhane, 1977) was used for post hoc pairwise comparisons.

Spatial patterns of ecosystem potentials estimated by beneficiaries to provide 11 groups of services were derived from the detailed map of ecosystems generalized to 7 classes (see Section 5.2).

6.2.3 Indicator values

From among the 251 respondents surveyed, 190 filled in the third section of the questionnaire. Of these, 69% were female and 31% male. The majority (73%) were between 30 and 60 years old, 12% were under 30, and about 15% were above 60. Most respondents (45% and 43%, respectively) declared secondary or higher education. Mental work (31%), farming (27%), or a pension (22%) was these people's most frequently cited source of income (multiple answers allowed). About 14% of respondents work in tourism services. Almost 69% were permanent rural residents, while 31% came as tourists from towns and cities.

One-way ANOVA shows significant differences among ecosystems as regards their perceived potentials to provide each of the service considered (F values from 5.1 in the case of education and research to 113.9 in the case of fuel, $P = .000$) (Tables 6.54 and 6.55). Tamhane's post hoc tests between pairs of ecosystems demonstrate that most differences of more than 1.6 points achieve statistical significance. Such results show that ES beneficiaries (local residents and tourists) are aware of the diversification of ecosystems in terms of their potential to provide different services.

Greatest cohesion among respondents was achieved as the potential of forests to provide fuel was rated (SD = 3.0), while this was most limited for the assessed capacities of forests and wetlands to offer water retention and purification services (SD = 5.3—5.5) and of wetlands to act as a source of fuel (SD = 5.6).

Table 6.54 Indicator values (interval ascending scale from 0 to 15) showing the average opinion of direct users on ecosystem potential to provide the 11 ecosystem services (ES).

Ecosystem type	Ecosystem service (ES)										
	Provisioning ES						Regulating ES	Cultural ES			
	Biomass for							Characteristics of living systems that enable/support			
	Nutritional purposes	Medicinal purposes	Construction purposes	The production of fertilizer and fodder	Ornamental purposes	Energy production	Water regulation	Sport and recreation	Education and research	Creative work	Spiritual experience
Deciduous forest	10.39	10.91	12.39	4.73	8.82	13.03	6.88	11.38	10.54	9.22	8.67
Pine forest	9.75	10.81	12.66	4.26	10.11	13.02	6.12	11.31	10.51	9.16	8.72
Swamp forest	7.96	10.17	9.18	4.62	7.65	10.32	6.76	8.04	9.69	8.55	7.55
Grassland	11.19	12.14	5.57	11.94	7.72	5.38	5.76	9.83	10.16	9.47	7.91
Cropland	12.88	9.43	5.89	11.12	5.15	5.28	5.18	5.95	8.61	7.30	6.12
Wetland	6.26	9.31	5.83	5.34	6.47	9.26	7.20	6.42	10.32	7.90	7.50
Rivers and lakes	12.87	5.20	6.52	3.68	8.06	3.64	12.04	12.55	10.98	10.56	9.64

Table 6.55 Ecosystem potentials to provide services (1 - very low; 5 - very high) indicated by the opinions of direct users.

	Ecosystem service (ES)										
	Provisioning ES						Regulating ES	Cultural ES			
	Biomass for							Characteristics of living systems that enable/support			
Ecosystem type	Nutritional purposes	Medicinal purposes	Construction purposes	The production of fertilizer and fodder	Ornamental purposes	Energy production	Water regulation	Sport and recreation	Education and research	Creative work	Spiritual experience
Deciduous forest	4	4	5	1	3	5	2	4	4	3	3
Pine forest	3	4	5	1	4	5	2	4	4	3	3
Swamp forest	3	4	3	1	3	4	2	3	3	3	3
Grassland	4	5	2	4	3	2	2	3	4	3	3
Cropland	5	3	2	4	2	2	2	2	3	3	2
Wetland	2	3	2	2	2	3	3	2	4	3	3
Rivers and lakes	5	2	2	1	3	1	5	5	4	4	3

Participants in our study above all perceived coniferous and deciduous forests as sources of fuel and building materials. Nonetheless, their recreational potential was also acknowledged. The high ranks attributed to both tangible and intangible outputs may suggest that respondents expect forests to provide the often conflicting provisioning and cultural services simultaneously.

Swamp forests were perceived as having a distribution of potentials considerably different from the two other forest types. Above all, it is thought to supply biomass for natural medicines and energy and beyond that also possibilities for the pursuit of education and research. In this regard, this type of forest bore a stronger resemblance to wetlands, where support for education and research was identified as the most important service. This may stem from the well-established, and scientifically sustained, belief (Meli et al., 2014) that these are true hotspots for biodiversity.

In turn, the perceived potential of grassland and cropland is determined mostly by their productive function. Although recent agricultural land abandonment has limited this function in some areas (Pereira et al., 2005), agricultural ecosystems are inherently managed primarily to ensure the cultivation of the edible plants used in human and animal nutrition (Swinton et al., 2007). Surprisingly, the potential to support education and research also emerged as well-recognized. Hitherto, ecosystems of this type have hardly been perceived as related to educational services (García-Nieto et al., 2015; Plieninger et al., 2013). Natural medicines together with natural fertilizers and fodder were the key grassland outputs, while the suitability of cropland for spiritual and physical activities received the lowest rating.

Services have been differentiated clearly in terms of their universality, i.e., the feasibility of their being provided by the various ecosystems. For instance, all ecosystems were assessed similarly (quite favorably) as regards their suitability for education and research. In turn, the potential of ecosystems to deliver biomass for fertilizer and fodder was assigned a wide range of values, from extremely low in the case of forests and waters to high for cropland and grassland. Lakes and rivers are among the ecosystems for which the perceived feasibility of services being provided proved most diverse. Thus, a high potential to provide edible biomass and cultural services contrasted with virtually no capacity to provide biomass for non-nutritional purposes (materials and energy). Their attributed high potential to supply food is clear, as fish represents a significant ingredient of regional cuisine. Aquatic ecosystems stand out particularly in their exceptionally high perceived potentials to support physical activity (sport and recreation). This result is in line with the observations of Plieninger et al. (2013). Most respondents also recognized their well-developed capacity to retain and treat water. In the other ecosystems (e.g., wetlands and forests), the perceived potentials as regards this service are more diversified.

In general, the perceived high potential of all ecosystems to provide nonmaterial, cultural services can be explained in part by the specificity of the selected study area. The complex landscape pattern and high share of protected land apparently promote these services. Protected areas are considered to have a higher capacity to deliver regulating and cultural services than nonprotected areas (García-Nieto et al., 2015).

Spatial distribution of the perceived ES potential is strongly associated with land use. Agricultural lands (cropland and grassland) that cover approximately 32% of the area have the greatest potential to provide biomass for nutritional (Fig. 6.31) and medicinal purposes (Fig. 6.32) and for production of fertilizers and fodder (Fig. 6.35). They are thought to provide medium and low potential toward most cultural services (Figs. 6.38—6.41).

Forests that constitute 60% of the area show very high potential to provide biomass for construction purposes (Fig. 6.33) and energy production (Fig. 6.36) as well as high potential to provide medicines (Fig. 6.32).

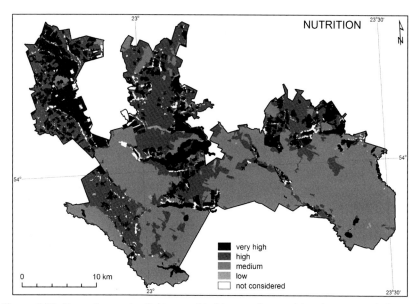

Figure 6.31 Ecosystem potential to provide biomass for nutritional purposes indicated by the opinions of direct users (NUTRITION indicator).

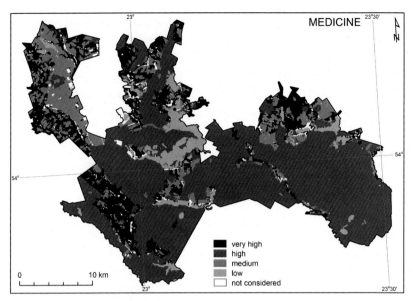

Figure 6.32 Ecosystem potential to provide biomass for medicinal purposes indicated by the opinions of direct users (MEDICINE indicator).

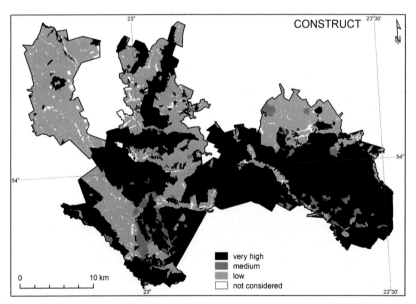

Figure 6.33 Ecosystem potential to provide biomass for construction purposes indicated by the opinions of direct users (CONSTRUCT indicator).

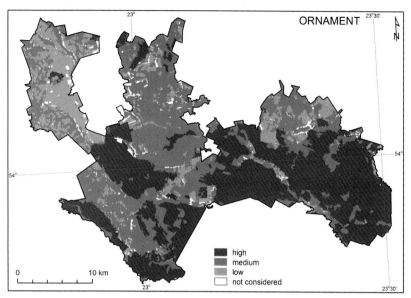

Figure 6.34 Ecosystem potential to provide biomass for ornamental purposes indicated by the opinions of direct users (ORNAMENT indicator).

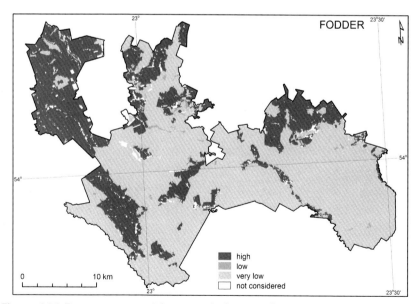

Figure 6.35 Ecosystem potential to provide biomass for the production of fertilizer and fodder indicated by the opinions of direct users (FODDER indicator).

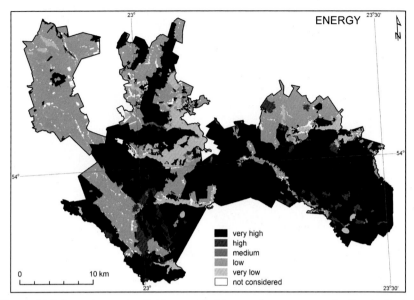

Figure 6.36 Ecosystem potential to provide biomass for energy production indicated by the opinions of direct users (ENERGY indicator).

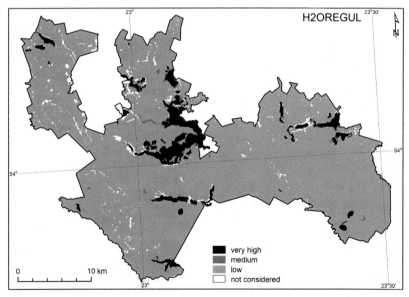

Figure 6.37 Ecosystem potential to regulate water indicated by the opinions of direct users (H2OREGUL indicator).

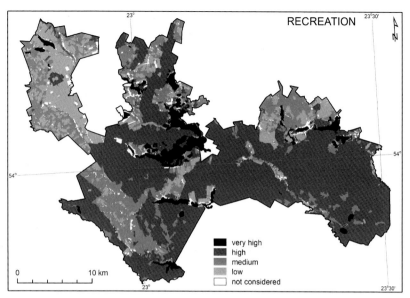

Figure 6.38 Ecosystem usefulness for sport and recreation indicated by the opinions of direct users (RECREATION indicator).

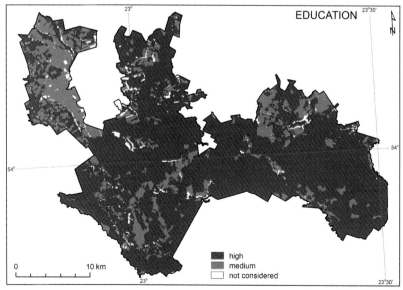

Figure 6.39 Ecosystem usefulness for scientific investigation and education indicated by the opinions of direct users (EDUCATION indicator).

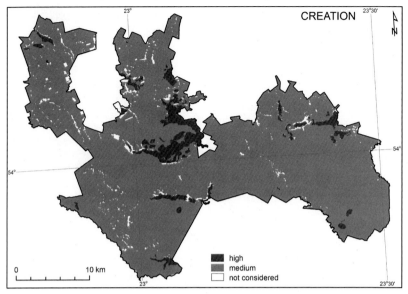

Figure 6.40 Ecosystem usefulness for creative work indicated by the opinions of direct users (CREATION indicator).

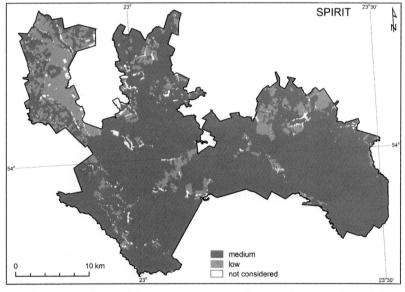

Figure 6.41 Ecosystem usefulness for spiritual experience indicated by the opinions of direct users (SPIRIT indicator).

Coniferous and mixed forests, dominating in the study area, are also thought to supply generously ornamental materials (Fig. 6.34). Both coniferous and deciduous forests are considered to have high recreational potential (Fig. 6.38).

Rivers and lakes that cover almost 6% of the study area are also thought to have high potential to provide biomass for nutritional purposes (Fig. 6.31) and to support sport and recreation (Fig. 6.38). Moreover, in case of creative work, aquatic ecosystems seem to be a big inspiration for their users compared with other ecosystems (Fig. 6.40). They, together with wetlands, play the biggest role in water regulation (Fig. 6.37)

Almost the entire study area, according to the respondents, has high potential for supporting education and research (Fig. 6.39). Only arable fields show medium potential for such activities. A similar pattern, although one level lower, is observed in terms of spiritual experience (Fig. 6.41).

CHAPTER 7

Ecosystem potentials to provide ecosystem services - synthetic approach

Contents

In this chapter, we synthetized the results presented in the previous analytical chapter. The investigation was conducted only on 29 ecosystem services (ES) indicators for which the ecosystem (detailed or Mapping and Assessment of Ecosystems and their Services [MAES]—derived) was the spatial reference unit. Therefore, the six remaining indicators developed for landscape or hunting units were not taken into account. It was dictated by the fact that it was not possible to reliably convert (downscale) values from the heterogeneous reference units to the ecosystem level.

To start with, synthetic assessment matrices were generated, both for original and ranked indicator values. These matrices served as a basis for the majority of further analyses. In the first step, rank values obtained through expert and social assessment were compared, where possible. In the next step, the aggregated potential for each ecosystem type was calculated. Additionally, ecosystems that potentially could be ES hotspots were identified. Both aggregated ES potentials and ES hotspots were mapped, separately for each ES section and in total. Then, the interactions among services and similarities among ecosystem types were investigated. Finally, spatial patterns of ES potentials, described in short for each indicator separately in the analytical chapter, were compared to each other and further investigated by means of landscape metrics.

Ecosystem Service Potentials and Their Indicators in Postglacial Landscapes
ISBN 978-0-12-816134-0
https://doi.org/10.1016/B978-0-12-816134-0.00007-9
Copyright © 2020 Elsevier Inc.
All rights reserved.

7.1 Assessment matrix of ecosystem potentials

Here, we present in the synthetic way the rank values of 29 ES indicators for 42 ecosystem types (Table 7.1). The 1—5 rank scale was applied, as in the previous analytical chapter. In some cases, also 0 or N was assigned, meaning no relevant ES potential and service not considered, respectively. The assessment of beneficiaries, although conducted only on seven generalized ecosystem types, was also included in the assessment matrix. It was possible, as the general ecosystem classification is the hierarchical aggregation of the detailed classification.

The assessments of ecosystem potentials using the expert method (based on direct and indirect measurements) and the social method (based on the respondents' opinions) differ significantly. The expert method includes selected, quite narrowly understood ES and refers to the detailed authors' typology of ecosystems. In turn, the social method assumes to cover all ES provided in the study area. However, here, the ES groups rather than the individual ES were evaluated and only for very general ecosystem types.

Such a characteristic of the applied assessment methods, associated with the specificity of the examined reality (respondents with limited-time and discriminatory abilities vs. databases concerning only selected aspects of the environment), resulted in the fact that only a few indicators and only for some ecosystems can be reasonably compared with each other.

Although the ES of these two assessment groups are not identical, several of them, mainly in the provisioning section, are quite similar and therefore suitable for comparison. For instance, both beneficiaries and direct measurements indicate that swamp forests are characterized by lower potential for providing services related to the use of wood (TIMBER, ORNAMENT, ENERGY) than coniferous and oak-hornbeam forests.

In turn, the comparison of assessments for the potential of forest ecosystems for food production (HONEY, BERRY, NUTRITION) shows that the opinions of beneficiaries differ from expert assessment. Beneficiaries notice higher potential of deciduous forests compared with coniferous and swamp forests, while expert assessment indicates that it is exactly the opposite. Such a difference may result from the fact that in the expert assessment, no mushrooms and wild game were considered, which in turn were probably taken into account by the respondents when assessing the overall forest potential for food production.

Other indicators that can be compared are SOILH2O and H2ORE-GUL, belonging to the regulation and maintenance section. Both refer

Table 7.1 Assessment matrix showing the potential of ecosystems to provide services according to expert and social assessment. Potential on a 0–5 scale, where 0 denotes no potential, 1 very low potential, 5 very high potential, and N not considered.

Ecosystem types assessed by experts	YIELD	TIMBER	HONEY	MAXLSU	BERRY	FISH	EROSION	SOILH2O	POLLIN	NATURA	RICHNESS	INVAS	SATURATION	C/N	OXYGEN	CARBON	AEROS	METAL	NUTRITION	MEDICINE	CONSTRUCT	FODDER	ORNAMENT	ENERGY	H2OREGUL	RECREATION	EDUCATION	CREATION	SPIRIT	Ecosystem types assessed by beneficiaries
	Provisioning						Regulation & Maintenance												Provisioning						R&M	Cultural				
SETTLE	N	N	3	N	0	N	N	N	4	1	N	N	N	N	N	N	N	N	N	N	N	N	N	N	N	N	N	N	N	–
ARB1	2	0	4	N	0	N	1	1	2	1	3	N	5	3	2	1	1	1	5	3	2	4	2	2	2	2	3	3	2	Cropland
ARB2	4	0	5	N	0	N	1	1	2	1	3	N	5	4	2	2	1	2	5	3	2	4	2	2	2	2	3	3	2	
ARB3	3	0	4	N	0	N	1	1	1	1	3	N	2	4	2	3	1	1	5	3	2	4	2	2	2	2	3	3	2	
GRAS1	N	0	5	4	3	N	3	1	5	4	3	5	2	1	3	1	4	1	4	5	2	4	3	2	2	3	4	3	3	Grassland
GRAS2	N	0	3	5	1	N	5	1	3	3	2	3	2	4	3	1	3	2	4	5	2	4	3	2	2	3	4	3	3	
GRAS3	N	0	3	4	0	N	5	4	2	5	3	3	4	5	1	4	3	5	4	5	2	4	3	2	2	3	4	3	3	
ALD1	N	2	1	N	3	N	4	2	1	1	3	1	2	3	4	3	1	4	4	4	5	1	3	5	2	4	4	3	3	Deciduous forest
ALD2	N	3	1	N	3	N	4	4	1	1	3	2	2	4	4	4	3	2	4	4	5	1	3	5	2	4	4	3	3	
ALD3	N	3	1	N	2	N	4	3	1	2	5	2	2	4	5	4	3	2	4	4	5	1	3	5	2	4	4	3	3	
ALD4	N	3	1	N	2	N	4	2	1	3	5	2	4	5	5	3	5	4	4	4	5	1	3	5	2	4	4	3	3	
ALD5	N	4	1	N	2	N	4	2	1	3	5	2	2	4	5	4	3	5	4	4	5	1	3	5	2	4	4	3	3	
RIP1	N	2	2	N	3	N	4	4	2	3	3	1	2	4	3	4	3	4	4	4	5	1	3	5	2	4	4	3	3	
RIP2	N	3	1	N	3	N	4	4	1	3	4	1	4	4	3	4	3	4	4	4	5	1	3	5	2	4	4	3	3	
RIP3	N	3	1	N	3	N	4	5	1	4	4	1	5	4	3	4	3	5	4	4	5	1	3	5	2	4	4	3	3	
RIP4	N	4	1	N	3	N	4	5	1	5	5	1	5	3	3	4	3	4	4	4	5	1	3	5	2	4	4	3	3	
RIP5	N	4	2	N	3	N	4	2	1	5	5	5	3	3	4	3	4	4	4	4	5	1	3	5	2	4	4	3	3	
OAK1	N	1	2	N	5	N	4	2	3	3	5	2	2	3	5	1	4	2	4	4	5	1	3	5	2	4	4	3	3	
OAK2	N	4	2	N	3	N	4	2	2	3	5	2	2	4	5	2	4	2	4	4	5	1	3	5	2	4	4	3	3	
OAK3	N	5	2	N	3	N	4	1	2	4	4	2	2	4	4	3	4	3	4	4	5	1	3	5	2	4	4	3	3	
OAK4	N	5	2	N	3	N	4	2	1	5	4	2	2	4	3	4	3	4	4	4	5	1	3	5	2	4	4	3	3	
OAK5	N	4	2	N	3	N	4	1	2	5	4	2	2	4	3	4	2	4	4	4	5	1	3	5	2	4	4	3	3	
CON1	N	2	4	N	3	N	2	1	5	1	3	4	1	1	3	1	5	2	3	4	5	1	4	5	2	4	4	3	3	Coniferous forest
CON2	N	3	3	N	3	N	3	1	4	1	4	3	1	2	5	2	3	4	3	4	5	1	4	5	2	4	4	3	3	
CON3	N	4	3	N	4	N	3	2	4	2	3	4	1	2	3	2	5	1	3	4	5	1	4	5	2	4	4	3	3	
CON4	N	5	3	N	5	N	3	1	4	3	4	5	1	3	4	3	5	1	3	4	5	1	4	5	2	4	4	3	3	
CON5	N	5	3	N	5	N	3	1	4	3	4	4	1	3	4	3	5	2	3	4	5	1	4	5	2	4	4	3	3	
SWP1	N	2	5	N	5	N	4	3	4	3	2	4	2	3	2	4	4	1	3	4	3	1	3	4	2	3	3	3	3	Swamp forest
SWP2	N	2	4	N	5	N	4	5	3	3	2	4	2	2	2	5	4	2	3	4	3	1	3	4	2	3	3	3	3	
SWP3	N	3	4	N	5	N	4	5	3	4	2	3	3	3	2	5	4	2	3	4	3	1	3	4	2	3	3	3	3	
SWP4	N	3	4	N	5	N	4	5	2	5	5	4	3	3	2	5	4	1	3	4	3	1	3	4	2	3	3	3	3	
SWP5	N	3	4	N	5	N	4	5	3	5	5	5	3	2	5	4	1		3	4	3	1	3	4	2	3	3	3	3	
REED1	N	0	1	N	0	N	5	3	1	1	1	2	3	4	2	5	2	2	2	3	2	2	2	3	3	2	4	3	3	Wetlands
FEN	N	0	1	N	1	N	5	5	1	4	3	5	5	5	1	5	4	2	2	3	2	2	2	3	3	2	4	3	3	
BOG	N	0	2	N	4	N	5	5	1	5	1	4	2	3	1	5	4	1	2	3	2	2	2	3	3	2	4	3	3	
LAKE1	N	0	0	N	0	3	N	N	0	4	1	N	N	N	N	N	N	N	5	2	2	1	3	1	5	5	4	4	3	Rivers and lakes
LAKE2	N	0	0	N	0	4	N	N	0	4	1	N	N	N	N	N	N	N	5	2	2	1	3	1	5	5	4	4	3	
LAKE3	N	0	0	N	0	4	N	N	0	4	1	N	N	N	N	N	N	N	5	2	2	1	3	1	5	5	4	4	3	
LAKE4	N	0	0	N	0	5	N	N	0	4	1	N	N	N	N	N	N	N	5	2	2	1	3	1	5	5	4	4	3	
LAKE5	N	0	0	N	0	2	N	N	0	4	1	N	N	N	N	N	N	N	5	2	2	1	3	1	5	5	4	4	3	
LAKE6	N	0	0	N	0	1	N	N	0	5	1	N	N	N	N	N	N	N	5	2	2	1	3	1	5	5	4	4	3	
LAKE7	N	0	0	N	0	4	N	N	0	1	1	N	N	N	N	N	N	N	5	2	2	1	3	1	5	5	4	4	3	

more or less exclusively to the potential of ecosystems for water retention. In this case, expert and social assessment is more consistent. The only apparent difference (rank 5 in the expert assessment against rank 2 in the social assessment) concerns the potential of swamp forests. Although a similar difference was obtained in the case of riparian forests, it can be assumed that the respondents thinking about deciduous forests referred rather to oak-hornbeam forests, for which high rank correspondence was noted. Comparing expert and beneficiary assessments for other services and ecosystems due to the nature of the indicators does not seem to be justified.

There are many similarities but at the same time also several differences between our assessment matrix and the one presented in the publications of Burkhard et al. (2014, 2012, 2009). Several dissimilarities were noted in the evaluation of wetlands potential to provide mainly regulating and cultural services. In our study, their regulating potential was higher in case of erosion control (EROSION), water control (SOILH2O), and nutrient regulation (C/N) and lower in case of recreation (RECREATION). The observed differences concerned both expert and social assessments. In the works mentioned, the matrix values are based on experience from different case studies in different European regions but were derived mainly by the authors as hypotheses linking different land cover types (CORINE land cover classes) with ES supply capacities. In our case, matrix values are derived from public databases, field measurements, ecological modeling, and surveying direct ES users. The differences may also result from the adopted land cover/ecosystem classifications. The detailed classification of ecosystem types used in our study by taking account of habitat conditions and forest stand age contributes inevitably to greater differentiation in the obtained levels of ecosystem potentials. Nevertheless, the differences can be recorded as well in the case of more general ecosystem classification used in social assessment (e.g., in the potential of different forest types).

7.2 Aggregated potential of ecosystems

The aggregated ecosystem potential to provide biomass for nutrition comprises six indicators (YIELD, HONEY, MAXLSU, BERRY, FISH, and NUTRITION) (Table 7.2). The obtained values range from 1.33 for fens (FEN) to 5.00 for medium-sized eutrophic lakes (LAKE4). Other lakes, swamp forests, and arable fields also received high values.

The potential to provide biomass for materials and energy is the average from other six indicators (TIMBER, MEDICINE, CONSTRUCT, FODDER, ORNAMENT, ENERGY) evaluated mostly by direct bene-ficiaries. The obtained values range from 1.80 for all types of lake to 4.00 for older coniferous and mixed forests (CON4 and CON5).

Table 7.2 Aggregated potentials of ecosystems to provide services.

Ecosystem	Provisioning potential (mean from nutrition and materials)	Biomass for nutrition (mean from 6 ES)	Biomass for materials and energy (mean from 6 ES)	Regulating potential (mean from 12 ES)	Cultural potential (mean from 4 ES)	Overall potential (mean from 3 ES sections)
ARB1	3.13	3.67	2.60	1.91	2.50	2.51
ARB2	3.63	4.67	2.60	2.18	2.50	2.77
ARB3	3.30	4.00	2.60	1.82	2.50	2.54
GRAS1	3.60	4.00	3.20	2.63	3.25	3.16
GRAS2	3.23	3.25	3.20	2.63	3.25	3.03
GRAS3	3.43	3.67	3.20	3.50	3.25	3.39
ALD1	3.00	2.67	3.33	2.33	3.50	2.94
ALD2	3.08	2.67	3.50	2.83	3.50	3.14
ALD3	2.92	2.33	3.50	2.96	3.50	3.13
ALD4	2.92	2.33	3.50	3.25	3.50	3.22
ALD5	3.00	2.33	3.67	3.17	3.50	3.22
RIP1	3.17	3.00	3.33	3.00	3.50	3.22
RIP2	3.08	2.67	3.50	3.13	3.50	3.24
RIP3	3.08	2.67	3.50	3.42	3.50	3.33
RIP4	3.17	2.67	3.67	3.33	3.50	3.33
RIP5	3.33	3.00	3.67	3.17	3.50	3.33
OAK1	3.42	3.67	3.17	2.83	3.50	3.25
OAK2	3.33	3.00	3.67	2.92	3.50	3.25
OAK3	3.42	3.00	3.83	2.92	3.50	3.28
OAK4	3.42	3.00	3.83	2.96	3.50	3.29
OAK5	3.33	3.00	3.67	2.88	3.50	3.24

Continued

Table 7.2 Aggregated potentials of ecosystems to provide services. (cont'd)

Ecosystem	Provisioning potential (mean from nutrition and materials)	Biomass for nutrition (mean from 6 ES)	Biomass for materials and energy (mean from 6 ES)	Regulating potential (mean from 12 ES)	Cultural potential (mean from 4 ES)	Overall potential (mean from 3 ES sections)
CON1	3.42	3.33	3.50	2.42	3.50	3.11
CON2	3.42	3.00	3.83	2.54	3.50	3.15
CON3	3.58	3.33	3.83	2.63	3.50	3.24
CON4	3.83	3.67	4.00	2.96	3.50	3.43
CON5	3.83	3.67	4.00	2.96	3.50	3.43
SWP1	3.58	4.33	2.83	2.96	3.00	3.18
SWP2	3.42	4.00	2.83	3.13	3.00	3.18
SWP3	3.50	4.00	3.00	3.25	3.00	3.25
SWP4	3.33	3.67	3.00	3.33	3.00	3.22
SWP5	3.50	4.00	3.00	3.67	3.00	3.39
REED1	1.95	1.50	2.40	2.75	3.00	2.57
FEN	1.87	1.33	2.40	3.71	3.00	2.86
BOG	2.53	2.67	2.40	3.08	3.00	2.87
LAKE1	2.90	4.00	1.80	3.75	4.00	3.55
LAKE2	3.15	4.50	1.80	3.75	4.00	3.63
LAKE3	3.15	4.50	1.80	3.75	4.00	3.63
LAKE4	3.40	5.00	1.80	3.75	4.00	3.72
LAKE5	2.65	3.50	1.80	3.75	4.00	3.47
LAKE6	2.40	3.00	1.80	4.00	4.00	3.47
LAKE7	3.15	4.50	1.80	3.00	4.00	3.38

ES, ecosystem services.

The aggregated potential to provide provision services is the mean from the above two partial means (Table 7.2, Figs. 7.1 and 7.2). The obtained values range from 1.87 for fens (FEN) to 3.83 for older coniferous and mixed forests (CON4 and CON5). Although the range is only slightly narrowed compared with the two partial potentials covered, and the same ecosystems are on the extremes, the potential to provide nutrition is significantly negatively correlated with the potential to provide materials and energy ($r = -0.40$, $P < .05$).

The aggregated ecosystem potential to provide regulating services comprises 13 indicators (EROSION, SOILH2O, POLLIN, INVAS, SATURATION, C/N, OXYGEN, CARBON, AEROS, METAL, H2OREGUL, and previously averaged NATURA and RICHNESS) (Table 7.2, Figs. 7.1 and 7.2). The obtained values range from 1.82 for moist arable fields (ARB3) to 4.00 for small dystrophic lakes (LAKE6). Other lakes also ranked high, while other arable fields and young coniferous and alder forest quite low. The regulating potential appeared not to be correlated with the overall provisioning potential, although we noticed a weak but significant negative correlation with the potential to provide materials and energy ($r = -0.39$, $P < .05$).

The aggregated cultural potential comprises four indicators (RECREATION, EDUCATION, CREATION, SPIRIT), all of which were evaluated only through social assessment. Therefore, the assigned values are equal for those detailed ecosystems that fall within the same ecosystem

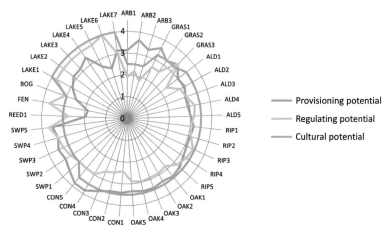

Figure 7.1 Aggregated provisioning, regulating, and cultural potential to provide ecosystem services.

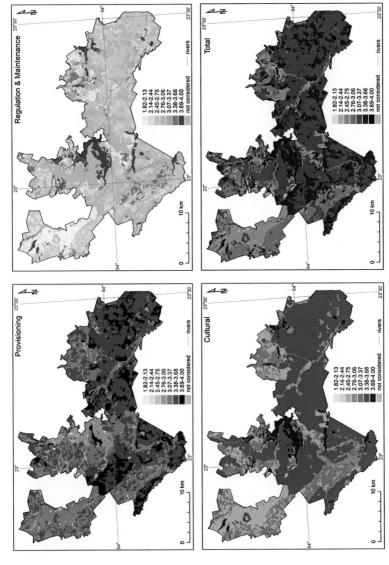

Figure 7.2 Spatial distribution of aggregated ecosystem potential to provide ecosystem services (ES) (out of each ES section and in total). The scale range on all maps was unified by adjusting to the extreme values obtained, i.e., 1.82 and 4.00. Values were classified into seven equal intervals.

category used in the survey. The obtained values range from 2.50 for arable fields to 4.00 for lakes. All forests except swamp forests were evaluated equally as having the potential slightly higher than medium (3.50). Cultural potential appeared to be positively correlated with regulating potential (r = 0.58, $P < .05$).

The overall aggregated potential being the arithmetic mean of provisioning, regulating, and cultural potential ranges from 2.51 for poor sandy arable fields (ARB1) to 3.72 for medium-sized eutrophic lakes (LAKE4). The relatively narrow range is the natural effect of averaging three weakly correlated partial potentials. 83% of the ecosystem types received scores above medium, with all types of lake and older forest taking the lead. In turn, all types of arable field and wetlands were ranked below medium.

Despite differences in the assessment of ecosystem potential to provide particular services, the aggregated ES potentials in our study and in those of Burkhard et al. (2014, 2012, 2009) show some similarities. In both studies, the highest aggregated potential to provide biomass for nutrition is observed for lakes and arable fields and to provide biomass for materials and energy for forests. In the case of regulating aggregated potential, all forest types take leading positon in Burkhard et al. (2014, 2012, 2009), while in our study, lakes have the higher aggregated potential. The cultural aggregated potential is the highest for lakes and forests in both evaluations.

7.2.1 Hotspot analysis

Areas (ecosystems) with the highest potential to produce and deliver a given ES are often recognized as ES hotspots. The "highest potential" is defined differently in the literature. For instance, Anderson et al. (2009) recognized as hotspots the highest rated 10%, 20%, or 30% of the evaluated area, while Holt et al. (2015) used two thresholds, the top 10% and 25% of the area with the highest ES supply. In our research, we set the rank scale of ES potential not in relation to the obtained spatial distribution of indicator values in the study area (a posteriori) but to the a priori-defined thresholds taken from literature. In this way, we aimed to assess the potential of postglacial ecosystems in the wider perspective, reaching beyond the selected study area.

From a management point of view, it is important not only to determine the individual and aggregated (averaged) ES potential of particular ecosystem types and hence hotspot ecosystems but also to find out which ecosystems are multifunctional (i.e., have high or very high potential to provide multiple services) and what is their coverage and spatial distribution. In this section, we focus on multiservice hotspots, i.e., ecosystem types that show high or very high (rank 4 or 5) potential to provide multiple services. We assumed that an ecosystem may be regarded as multiservice hotspot when it is capable to efficiently provide at least half of the analyzed services.

The conducted analysis showed that older coniferous and mixed forests are major provisioning ES hotspots (the potential to provide six out of nine provisioning ES is high or very high) followed by younger coniferous forests, oak-hornbeam forests, and older riparian and alder forests (five out of nine ES) (Table 7.3). This proportion translates to 67% and 56% of provisioning ES analyzed, which indicates that these ecosystems are true multiservice hotspots. They cover 60% of the study area (Fig. 7.3). The capacity to efficiently provide several regulating services was demonstrated by selected forest and nonforest communities associated with wet and swampy habitats, primarily older swamp and alder forests, mid-age riparian forests, fens, and wet grassland (high or very high potential to provide 6 or 7 out of 12 regulating ES; 50% and 58% of the analyzed regulating ES). They cover 7% of the study area. Cultural multiservice hotspots are those ecosystems that have the potential to efficiently provide at least two out of four cultural ES analyzed. These include all types of forest ecosystem (alder, riparian, oak-hornbeam, coniferous, and mixed), with the exception of swamp forests. Cultural multiservice hotspots cover 56% of the study area.

When all services from all three ES sections are taken together, only 3 out of 34 terrestrial ecosystem types may be regarded as true multiservice hotspots: the oldest alder forests and riparian forests in the age of 60–80 and 80–120. They have the capacity to efficiently provide 52% of the analyzed services (13 out of 25 ES) and cover less than 0.1% of the study area. However, many other ecosystems having high coverage in the study area (e.g., older coniferous and mixed forests, oak-hornbeam forests, and peat grassland) can efficiently provide only one service less (12 out of 25 ES; 48%). They cover together almost one-third of the study area. At this point, it is worth noting that lakes, which cover 6% of the study area, were excluded from the hotspot analysis due to the incomparably small number of services assessed. However, without any doubt they constitute hotspots for several services not addressed in this book.

Spatial distribution of multiservice hotspots in the study area does not have the same pattern for all ES sections (Fig. 7.3). Cultural hotspots show highly clustered (clumped) distribution, while provisioning and regulating

Table 7.3 Number and percent of ecosystem services (ES) showing high or very high (rank 4 or 5) potential to provide a given service.

	Provisioning (9 ES)		Regulating (12 ES)		Cultural (4 ES)		Total (25 ES)	
	N	%	N	%	N	%	N	%
ARB1	3	33.3	1	8.3	0	0.0	4	16.0
ARB2	4	44.4	2	16.7	0	0.0	6	24.0
ARB3	3	33.3	1	8.3	0	0.0	4	16.0
GRAS1	5	55.6	3	25.0	1	25.0	9	36.0
GRAS2	4	44.4	2	16.7	1	25.0	7	28.0
GRAS3	4	44.4	7	58.3	1	25.0	12	48.0
ALD1	4	44.4	2	16.7	2	50.0	8	32.0
ALD2	4	44.4	5	41.7	2	50.0	11	44.0
ALD3	4	44.4	4	33.3	2	50.0	10	40.0
ALD4	4	44.4	6	50.0	2	50.0	12	48.0
ALD5	5	55.6	6	50.0	2	50.0	13	52.0
RIP1	4	44.4	5	41.7	2	50.0	11	44.0
RIP2	4	44.4	6	50.0	2	50.0	12	48.0
RIP3	4	44.4	7	58.3	2	50.0	13	52.0
RIP4	5	55.6	6	50.0	2	50.0	13	52.0
RIP5	5	55.6	5	41.7	2	50.0	12	48.0
OAK1	5	55.6	4	33.3	2	50.0	11	44.0
OAK2	5	55.6	5	41.7	2	50.0	12	48.0
OAK3	5	55.6	5	41.7	2	50.0	12	48.0
OAK4	5	55.6	5	41.7	2	50.0	12	48.0
OAK5	5	55.6	5	41.7	2	50.0	12	48.0

Continued

Table 7.3 Number and percent of ecosystem services (ES) showing high or very high (rank 4 or 5) potential to provide a given service. (cont'd)

	Provisioning (9 ES)		Regulating (12 ES)		Cultural (4 ES)		Total (25 ES)	
	N	%	N	%	N	%	N	%
CON1	5	55.6	3	25.0	2	50.0	10	40.0
CON2	5	55.6	2	16.7	2	50.0	9	36.0
CON3	6	66.7	3	25.0	2	50.0	11	44.0
CON4	6	66.7	4	33.3	2	50.0	12	48.0
CON5	6	66.7	4	33.3	2	50.0	12	48.0
SWP1	4	44.4	5	41.7	0	0.0	9	36.0
SWP2	4	44.4	5	41.7	0	0.0	9	36.0
SWP3	4	44.4	4	33.3	0	0.0	8	32.0
SWP4	3	33.3	6	50.0	0	0.0	9	36.0
SWP5	4	44.4	7	58.3	0	0.0	11	44.0
REED1	0	0.0	3	25.0	1	25.0	4	16.0
FEN	0	0.0	7	58.3	1	25.0	8	32.0
BOG	1	11.1	5	41.7	1	25.0	7	28.0

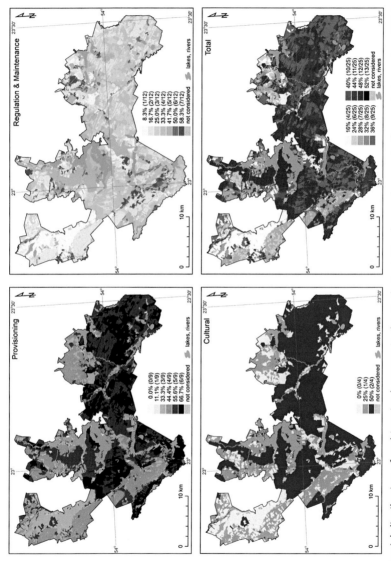

Figure 7.3 Spatial distribution of multiservice hotspots in the study area. Number and percent of ecosystem services (ES) (out of each ES section and in total) that can be efficiently provided by a given patch (ecosystem) is shown.

hotspots a more random distribution. This is undoubtedly related to the high diversity of the number of hotspot patches and their coverage in the study area.

7.3 Bundles of ecosystem services, synergies, and trade-offs

In line with the approach adopted in this study, the potential of ecosystems to provide services is evaluated on the gradual rather than on a binary scale. This causes that the coexistence of specific ES potentials changes in the more or less continuous way and therefore any possible ES bundles are inherently fuzzy. Various grouping methods implemented to define clusters, including fuzzy K-Means and hierarchical clustering, confirm this observation. Depending on the parameters adopted and the grouping method applied, different ES bundles can be separated. However, the cores of those bundles seem to maintain consistency and repeat over consecutive grouping solutions. Moreover, in each solution, there is a relatively stable number of ES that show similar probability (approximately 0.5) of belonging to two different clusters. In the case of normal and fuzzy K-Means grouping methods, the most unambiguous results are obtained when distinguishing 2 groups (and when also including FISH, MAXLSU, and YIELD indicators – three groups). However, the grouping seems not to be based on high intragroup consistency but is rather determined by strong negative intergroup correlations. This may suggest that ES interactions are predominantly trade-offs rather than synergies.

Considering the significant simplifications resulting from the grouping algorithms and their limited consistency, we decided to define ES bundles based on direct visual interpretation of ES correlation matrix (Table 7.4). The adopted solution assumes the existence of eight overlapping ES groups, including four single-service groups and four multiservice bundles (Fig. 7.4).

The first ES bundle comprises six services: biomass for the production of fertilizer and fodder (FODDER), honey (HONEY), pollination (POLLIN), edible wild berries (BERRY), invasive plant species control (INVAS), and plant aerosol emission (AEROS). It is characterized by relatively weak internal associations and numerous significant negative correlations with the majority of other services. A distinguishing feature of this group is that each service is provided by a separate, narrowly defined ecosystem service provider (ESP), usually a narrow set of species (plants or

Table 7.4 Ecosystem services (ES) pairwise correlation matrix. Pearson's correlations shown when significant with $P < .05$ (with $P < .01$ in bold) divided into positive (green) and negative (red) correlations. Number of ecosystem types analyzed within each ES pair is shown under the diagonal.

	NUTRITION	FODDER	HONEY	INVAS	BERRY	POLLIN	AEROS	TIMBER	C/N	SATURATION	CARBON	SOILH2O	NATURA	METAL	RECREATION	SPIRIT	CREATION	EDUCATION	EROSION	H2OREGUL	OXYGEN	ORNAMENT	ENERGY	CONSTRUCT	RICHNESS	MEDICINE	MAXLSU	YIELD	FISH
NUTRITION																													
FODDER	0.49																												
HONEY	34	34					-0.51	-0.37																					
INVAS	31	31	31		0.68																								
BERRY	31	31	31	31		0.54	0.51																						
POLLIN	34	34	35	31	34		0.57	0.57																					
AEROS	34	34	34	31	31	34		0.62	0.53	0.57																			
TIMBER	25	25	25	25	25	25	25																						
C/N	34	34	34	31	31	34	34	34		-0.48	-0.39	-0.40		-0.76	-0.50	0.41													
SATURATION	34	34	34	31	31	34	34	34	34		-0.67	-0.72	0.52																
CARBON	34	34	34	31	31	34	34	34	34	34		-0.68	-0.63	-0.37	0.47	0.45	0.83												
SOILH2O	34	34	34	31	31	34	34	34	34	34	34		-0.63																
NATURA	41	41	34	31	35	34	34	34	34	34	34	34		0.44															
METAL	34	34	34	31	31	34	34	34	34	34	34	34	34		0.35	0.46		-0.44	-0.53	-0.44									
RECREATION	7	34	34	31	31	34	34	34	34	34	34	34	41	34		0.42		-0.52	-0.58			-0.36							
SPIRIT	7	34	34	31	31	34	34	34	34	34	34	34	41	34	7		0.37	-0.65	-0.60										
CREATION	7	34	34	31	31	34	34	34	34	34	34	34	41	34	7	7		-0.43	-0.60		0.40	0.93	0.93						
EDUCATION	7	34	34	31	31	34	34	34	34	34	34	34	41	34	7	7	7		-0.74	-0.44	-0.59	0.35	0.79	0.94	0.81				
EROSION	34	34	34	31	31	34	34	34	34	34	34	34	41	34	34	34	34	34		-0.60	0.46		0.62	0.54	0.53	0.48	0.49	0.56	0.35
H2OREGUL	34	34	34	31	31	34	34	34	34	34	34	34	41	34	34	34	34	34	34		-0.40		0.77	0.73	0.58	0.57	0.60	0.53	0.45
OXYGEN	34	34	34	31	31	34	34	34	34	34	34	34	34	34	34	34	34	34	34	34		-0.41	-0.39	-0.41	-0.35		0.58	0.52	0.45
ORNAMENT	7	34	34	31	31	34	34	34	34	34	34	34	41	34	7	7	7	7	34	34	34		-0.52				0.42		
ENERGY	7	34	34	31	31	34	34	34	34	34	34	34	41	34	7	7	7	7	34	34	34	7		-0.42	-0.61	-0.49		0.39	0.72
CONSTRUCT	7	34	34	31	31	34	34	34	34	34	34	34	41	34	7	7	7	7	34	34	34	7	7		-0.48	-0.51	0.43	0.50	0.52
RICHNESS	41	34	34	31	31	34	34	34	34	34	34	34	41	34	41	41	41	41	34	34	34	34	34	34		-0.41	0.36	0.56	0.39
MEDICINE	7	34	34	31	31	34	34	34	34	34	34	34	41	34	7	7	7	7	34	34	34	7	7	7	41		0.36	0.68	0.71
MAXLSU	3	34	34	3	3	3	3	0	3	3	3	3	3	3	3	3	3	3	3	3	3	3	3	3	41	7		0.37	0.78
YIELD	3	34	34	3	3	3	3	0	3	3	3	3	3	3	3	3	3	3	3	3	3	3	3	3	3	3	3		-1.00
FISH	7	7	34	0	0	0	0	0	0	0	0	0	7	0	7	7	7	7	0	0	0	7	7	7	7	7	3	3	

Additional negative correlations read in the upper region: RICHNESS–OXYGEN 0.45; RICHNESS–ENERGY 0.77; RICHNESS–CONSTRUCT 0.73; RICHNESS–MEDICINE 0.68; RICHNESS–ORNAMENT 0.87; MEDICINE–MAXLSU -0.88; MEDICINE–YIELD -0.57; OXYGEN–CARBON -0.42; OXYGEN–SOILH2O -0.39.

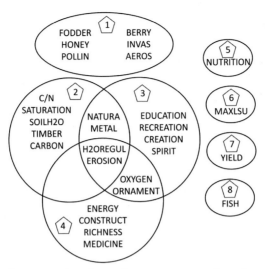

Figure 7.4 Graphic representation of ecosystem services bundles and their interrelations.

animals) occurring with varying densities across forest and nonforest ecosystems. In all other ES groups, the ESP is either the entire ecosystem (together with the abiotic part) or a widely understood functional group (e.g., forest stand) with diverse and variable species composition.

The core of the second ES bundle is formed by five services: organic matter decomposition (C/N), soil formation (SATURATION), water retention (SOILH2O), tree biomass for nonnutritional purposes (TIMBER), and carbon sequestration (CARBON). They are significantly positively correlated with each other and show inverse relationships with many ES from the first bundle. A characteristic feature of ES belonging to this bundle is their close dependance on soil properties, and moisture in particular (C/N, SATURATION, SOILH2O, CARBON), and through these properties also with tree stand (TIMBER).

The core of the third ES bundle is formed by all four cultural services analyzed in this study: characteristics of living systems that enable/support scientific investigation and education (EDUCATION), sport and recreation (RECREATION), creative work (CREATION), and spiritual experience (SPIRIT). All the above ES are closely interrelated, and the mean coefficient of pairwise correlations is 0.62. They also show very strong negative correlations with the majority of the ES from the first group. A characteristic feature of ES belonging to this bundle is that they depend on the level of development, naturalness, and maturity of the entire ecological systems, while much less on the presence of specific species.

The core of the fourth ES bundle is formed by four other services: biomass for energy production (ENERGY), biomass for construction purposes (CONSTRUCT), nursery habitat maintenance (RICHNESS), and biomass for medicinal purposes (MEDICINE). They are quite strongly interrelated with each other and show negative correlations with the majority of the ES from the first group and some ES from the second and third group. It seems that the common feature of ES belonging to this bundle is that when they are provided by forest ecosystems, they are determined primarily by the type of forest, regardless of its age and maturity.

ES belonging to bundles 2, 3, and 4 are out of all considered terrestrial ecosystems most efficiently provided by forests and thus are not clearly separated from each other. There is a large group of ES whose affiliation is fuzzy. They belong with similar probability to at least two different bundles.

Nursery habitat maintenance indicated by NATURA indicator and atmospheric heavy metal accumulation (METAL) are the two services belonging both to the second and to the third bundle. They show numerous similar correlations with different ES from these bundles. Their intermediate location is conditioned both by the dependence on soil properties and on the degree of ecosystem maturity.

The next two services, oxygen emission (OXYGEN) and biomass for ornamental purposes (ORNAMENT), form a common part of the third and the forth bundle. OXYGEN appears to be more closely linked with the fourth group, while ORNAMENT with the third group. Their intermediate location results from the dependence on both the type and maturity of forest ecosystem.

A special place is occupied by the other two services: erosion control (EROSION) and water regulation (H2OREGUL). They cannot be assigned unambiguously to any of the already defined ES bundles. Although they are most strongly correlated with the ES from the third group, their links with ES from the other two bundles are only slightly weaker. These services do not seem to be associated with any particular type of terrestrial ecosystem.

The four remaining groups are single-service groups. Biomass for nutritional purposes (NUTRITION), a broadly defined service constituting ES group number 5 alone, turned out not to be positively correlated with any other service. It shows only inverse relationships with several ES from bundle 1, 2, and 3.

The other three single-service groups are formed by very specific services provided only by selected ecosystem types. The service livestock and their outputs (MAXLSU) is calculated only for grassland and is correlated (negatively) only with one ES - nursery habitat maintenance indicated by RICHNESS indicator. The last two services, edible biomass of cultivated plants (YIELD) calculated only for arable fields and edible biomass of wild animals (FISH) calculated only for lakes, are completely independent from all other ES analyzed.

The observed interactions among ES are very complex and ambiguous. Most studies reported trade-offs among regulating and provisioning services (Raudsepp-Hearne et al., 2010; Turner et al., 2014). However, in our study there are many positive relationships between services of these two ES sections, for instance, between HONEY/BERRY and INVAS or between TIMBER and C/N. They show that the increase in the provisioning ES potential does not always result in the reduction of regulating potential. On the contrary, it may lead to high environmental benefits (Yang et al., 2015). For instance, Boreux et al. (2013) demonstrated that pollination is in synergetic relationship with food provision. Moreover, synergies between TIMBER and regulating services forming the common bundle indicate the high potential of forest ecosystems to deliver them all, provided that they will not be used too intensively. The strong positive associations between different cultural services have been observed in many studies (Daniel et al., 2012; Yang et al., 2015). In most cases, cultural services are also positively correlated with regulating services (Turner et al., 2014), what was revealed in our study as well (positive relationship of cultural ES with EROSION and H2OREGUL). In turn, neutral relationships were often found between cultural and provisioning services (Lee and Lautenbach, 2016). The negative relationships revealed in our study between some cultural ES and HONEY and BERRY are rather an exception from the general rule.

As one of the practical applications, the identified ES bundles and complex interrelations among ES might encourage decision-makers to perceive multifunctional ecosystems more holistically and raise awareness of possible side effects of intensive exploitation or land-use change. The described relationships can be also used to predict potentials of some ES based on the known potentials of other ES in the situation of data scarcity.

7.4 Similarities among ecosystem types

The analysis of ES bundles is complemented by similarity analysis conducted on ecosystems. Apart from distinguishing ecosystem clusters, we also aimed to determine the main dimensions ordering ecosystems according to the demonstrated potentials to provide ES.

Dendrogram of hierarchical cluster analysis shows that ecosystems group on several levels according to their potential to provide services (Fig. 7.5). Grouping of ecosystems at 30% of maximum distance into six types (arable fields, grassland, coniferous forests, deciduous forests, swamp forests, and wetlands) strictly corresponds to the aggregated MAES-derived ecosystem division used in the social assessment. On the one hand, this may be due to the real and natural similarity of those detailed ecosystems that fall within the same MAES-derived categories. On the other hand, the obtained close similarity certainly also result from the fact that 11 out of 29 ES indicators were originally assessed for generalized MAES-derived ecosystems, and their values downscaled to detailed ecosystem typology. Nevertheless, ecosystem clustering at a higher and lower level of similarity is already independent of this effect.

At the smallest distances, forests of the same ecological type group according to their maturity, younger with the youngest and older with the oldest. The largest distance was reported for alder forest classes. Two younger classes group firstly with riparian forests and then with older alder forests. This clustering pattern reflects the impact of forest succession on ES potential and confirms the desirability of including forest age in ES assessments. At the more general level, wetlands group with swamp forests

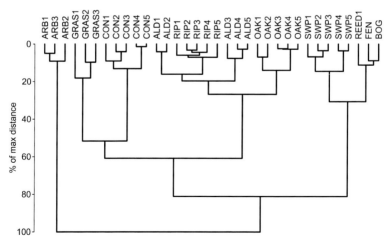

Figure 7.5 Hierarchical clustering of ecosystems in relation to their potential to provide services (Euclidean distance; Ward grouping method, standardized original ecosystem services indicator values as input data).

and grassland with coniferous and mixed forests. Arable fields constitute the most separate type of land use in terms of ES potential.

To determine the main dimensions ordering ecosystems according to the demonstrated potentials to provide ES, the principal component analysis (PCA) was used. The first four components accounted for 83% of the total variation in ES potentials (Table 7.5). One more PCA component reached eigenvalue >1 (1.34), but its informative value (4.78% of variance) was considered irrelevant.

The first principal component explains 38% of variance and orders ecosystems on the gradient of structural complexity including layering and the amount of aboveground biomass (Fig. 7.6). Biologically poor arable fields are on the opposite extreme in relation to multilayered and rich in biomass mature forests, both coniferous and deciduous. These ecosystem features seem to determine primarily the cultural ES potential, as of all considered ES, cultural services are most closely related to this component (Table 7.6). Apparently, cultural attractiveness of ecosystems increases with increasing ecosystem maturity and amount of aboveground biomass. The next two groups of ES, also correlated with this component although with opposite signs, are those related directly to trees (TIMBER, ENERGY, AEROS, OXYGEN, CONSTRUCT) and annually harvested crops (YIELD, HONEY, FODDER). This indicates that the first PCA dimension does not refer to the amount of biomass extracted annually from the ecosystem but to the overall aboveground biomass in stock, which reflects exactly the potential approach adopted in this book.

The dominance of productive function in case of arable lands has been highlighted frequently. Swinton et al. (2007) pointed out that agricultural ecosystems are inherently managed primarily to ensure the cultivation of the edible plants used in human and animal nutrition. Their cultural potential has been hardly ever recognized (Plieninger et al., 2013; Stallman,

Table 7.5 Four major principal component analysis (PCA) components with eigenvalues and percent of variance explained.

PCA component	Eigenvalue	Cumulative eigenvalue	% of variance explained	Cumulative % of variance explained
1	10.63	10.63	37.98	37.98
2	5.05	15.68	18.03	56.00
3	4.44	20.12	15.84	71.84
4	3.11	23.22	11.09	82.94

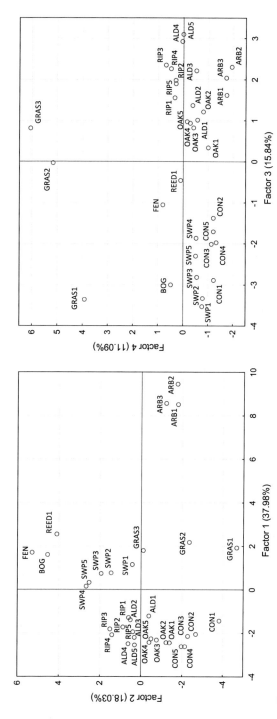

Figure 7.6 Four main dimensions (factors) extracted through principal component analysis that differentiate ecosystems due to their potential to provide services (standardized original ecosystem service indicator values as input data).

Table 7.6 R-Pearson correlations between ecosystem service indicators and four major principal component analysis components (factors). Correlations weaker than ±0.5 omitted, positive values underlined.

	Factor 1	Factor 2	Factor 3	Factor 4
YIELD	0.82			
TIMBER	−0.74			
HONEY	0.63			
MAXLSU				0.90
BERRY			−0.66	
EROSION	−0.65			
SOILH2O		0.86		
POLLIN		−0.64	−0.69	
NATURA				
RICHNESS	−0.55			
INVAS			−0.76	
SATURATION	0.51	0.56		
C/N			0.64	
OXYGEN	−0.50			
CARBON		0.91		
AEROS	−0.70		−0.64	
METAL			0.61	
NUTRITION	−0.71	0.61		
MEDICINE	−0.52	−0.53		0.62
CONSTRUCT	−0.84			
FODDER	0.77			0.51
ORNAMENT	−0.91			
ENERGY	−0.89			
H2OREGUL	−0.66	0.67		
RECREATION	−0.86			
EDUCATION	−0.91			
CREATION	−0.83			
SPIRIT	−0.96			

2011; van Brekel and Verburg, 2014). In contrast, forest ecosystems that are on the opposite extreme in PCA diagram seem to have high potential to provide the often conflicting provisioning and cultural services simultaneously, and this is a common social expectation (Agbenyega et al., 2009; Grilli et al., 2016; Kikulski, 2011).

The second principal component explaining 18% of variance differentiates terrestrial ecosystems in terms of water content, both in soil and in the entire ecosystem. Ecosystems are ordered from the very wet wetlands and swamp forests to dry coniferous forests and dry grassland. This component is loaded primarily by services whose potential is dependent directly

(SOILH2O, H2OREGUL) or indirectly (SATURATION, CARBON) on water content and by services that are closely connected to dry and grassy ecosystems (POLLIN and MEDICINE).

The high potentials of wetland ecosystems and riparian forests to retain and treat water are well known. They function as buffers reducing pollution from agricultural land to streams (Hefting et al., 2005) and mitigate effects of hydrological stress (Okruszko et al., 2011). They are also thought to store large quantities of carbon because of their relatively high rates of productivity and saturated conditions that can favor its belowground storage (Giese et al., 2003). In turn, dry grasslands are highly regarded for their floral resources rich in honey plant species (Affek, 2018) and herbs (Willner et al., 2019).

The third principal component explaining 16% of variance can be interpreted as the gradient of habitat acidity/fertility or in the more general terms as the gradient of ecosystem productivity. Ecosystems are ordered from the alkaline and fertile deciduous forests to acid and poor swamp coniferous forests. Also, the distribution of grassland ecosystems (by increasing productivity) and location of wetlands (close to other ecosystems with acid soils) supports this interpretation. This component is loaded by services whose high potential is related on the one hand to ecosystems with acid and sandy soils (BERRY, AEROS, INVAS, POLLIN) and on the other hand to ecosystems with rich and biologically active soil (C/N, METAL).

Eutrophic soil conditions characteristic for deciduous forests provide good decomposition rate (Wanic et al., 2011), but on the other hand, they promote invasive alien species, especially in case of ecosystem degradation (Hood and Naiman, 2000; Richardson et al., 2000a,b). Plant species composition of coniferous forests and dry grasslands makes these ecosystems a good habitat for pollinators (Kremen et al., 2007; Westrich, 1996; Winfree, 2010) as well as a valuable source of berries and phytoncides of prohealth nature (Marchese et al., 2017; Paduch et al., 2007). Grassland ES potential in relation to soil fertility and biodiversity was identified by Lamarque et al. (2011), while the therapeutic role of forests was presented by Falencka-Jabłońska (2012).

The fourth principal component explaining 11% of variance separates one specific type of land use (grassland) from all other ecosystems and land uses. It is loaded by services provided most efficiently by grassland (MAXLSU, MEDICINE, FODDER) and even in the case of MAXLSU solely by grassland. Livestock grazing in forests, together with raking forest litter as a replacement for straw in husbandry, both still popular in the post-World War II period (Broda, 1965), has now been left behind.

To sum up the results of PCA analysis, we can conclude that the potential of ecosystems to provide services is primarily determined by the following ecosystem properties (listed by importance):

- stock of aboveground biomass;
- human pressure and the resulting ecosystem maturity;
- water content;
- soil fertility/acidity and the resulting overall productivity;
- specific human use of anthropogenic ecosystems.

7.5 Spatial patterns of ecosystem potentials

In this book, 35 ES indicators were analyzed in total, of which the landscape was the spatial unit of assessment for 2 indicators, hunting unit for 4 indicators, MAES-derived ecosystems for 11 indicators, and ecosystems in a narrow sense for 18 indicators. Out of the last 18 indicators, 3 were dedicated only to a single land-use type, hence only the small part of the study area was evaluated. As the results of the spatial structure analysis depend on the degree of generalization of reference spatial units, as well as on a map grain size, which was shown for landscape metrics (Solon, 2004a) and for autocorrelation (Grêt-Regamey et al., 2014), we decided to conduct the comparative spatial analysis on 15 indicators, for which values were assigned to all ecosystem types in a narrow sense only. As the spatial analysis was conducted only on ES indicators evaluated in ecosystem units, the spatial pattern of ES indicator values was determined primarily by the spatial distribution of ecosystems. However, when different ecosystem types were assigned the same values of ES potential, the resulting output was less complex and often formed completely new spatial pattern. Although the spatial pattern of each ES potential is unique, they could be grouped into more or less distinct types and characterized with the help of different spatial metrics (Table 7.7).

According to the PCA of the 15 landscape metrics, spatial patterns of ecosystems and TIMBER and SOILH2O indicators occupy isolated positions, and their similarity to the other patterns is relatively low (Fig. 7.7). The considerable difference between ecosystem spatial pattern and other patterns is obvious, as the former includes 42 types of ecosystem, while indicator patterns consist of up to 5 unique values (1−5 rank scale of ES potentials). This inherent dissimilarity translates to the specific distribution of landscape metric values. Of the 15 metrics, 9 take extreme values compared with all other spatial patterns of ES potentials, with NumP, SSPI, and MNN being the highest, while MPS, PSCoV, HOLE, MSI, AWMSI, and MPI being the

Table 7.7 Values of selected landscape metrics characterizing spatial patterns of ecosystem service indicators compared with reference spatial distribution of ecosystem types.

Ecosystems	NumP	MPS	PSCoV	LPI	i60	HOLE	SEI	MSI	AWMSI	FDIM	SSPI	CNCV	IJI	MPI	MNN
TIMBER	3143	26.45	384.16	4.40	239	118	0.78	1.78	3.17	1.32	0.66	0.79	68.39	152.28	476.90
SOILH2O	746	106.17	556.68	3.67	670	139	0.79	2.29	4.41	1.35	0.58	0.72	81.32	502.36	283.40
METAL	1133	69.90	1418.00	40.46	6	522	0.62	1.93	10.32	1.31	0.63	0.79	68.93	5675.10	249.40
RICHNESS	1244	63.67	685.44	11.85	22	420	0.69	2.03	6.57	1.36	0.62	0.80	65.85	2231.10	215.60
C/N	1235	64.13	544.84	8.26	22	430	0.78	2.01	5.88	1.51	0.63	0.79	71.74	1463.30	239.80
NATURA	1674	47.31	539.62	7.31	52	342	0.92	2.11	4.64	1.37	0.61	0.78	89.40	819.65	195.60
CARBON	1399	56.74	399.40	5.19	65	399	0.92	2.12	4.52	1.37	0.60	0.74	88.76	955.39	163.20
SATURATION	1506	52.59	519.89	9.89	65	444	0.88	2.03	4.88	1.46	0.62	0.78	83.15	863.26	207.20
AEROS	1469	53.91	869.83	18.31	21	304	0.87	2.03	4.21	1.52	0.61	0.79	95.08	975.63	207.30
INVAS	1329	59.59	837.45	18.31	15	332	0.82	1.99	4.42	1.53	0.62	0.80	83.86	1149.80	190.30
POLLIN	1415	55.97	423.40	4.46	139	373	0.91	2.17	4.39	1.36	0.60	0.75	84.11	734.80	206.30
BERRY	1561	50.85	776.18	6.59	25	379	0.85	2.12	5.48	1.54	0.60	0.77	83.82	1191.70	199.20
EROSION	1066	74.30	493.10	6.41	106	399	0.83	2.12	5.83	1.40	0.60	0.76	69.72	2416.80	233.20
OXYGEN	1379	57.43	785.31	22.36	23	405	0.79	2.05	5.84	1.53	0.61	0.79	80.21	1034.40	213.90
HONEY	1489	53.19	640.91	12.01	39	368	0.84	2.07	5.60	1.49	0.61	0.79	81.90	1273.90	210.70
	1250	63.51	945.91	18.19	17	382	0.77	2.24	7.01	1.50	0.59	0.75	76.73	2955.40	200.30

For the description of the used metrics and their abbreviations see Section 3.3.

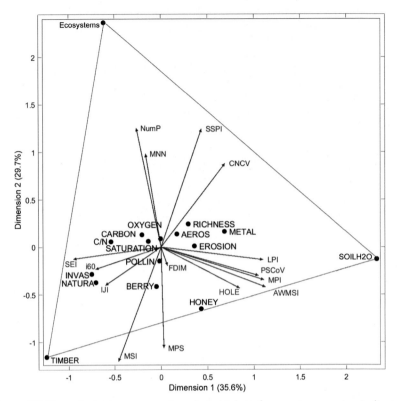

Figure 7.7 The principal component analysis biplot of ecosystem service indicators (and ecosystem typology) and landscape metrics (variables standardized).

lowest. Such values indicate a fine grain mosaic, built of the large number of patches, relatively small on average, with a low variance of patch size, with ecosystems of the same type well interspersed in space, relatively regular in shape, and with a low level of perforation. For TIMBER indicator, six metrics take extreme values. Maximal are MPS and MSI, while NumP, LPI, SSPI, and CNCV take minimal values. Such value distribution indicates a mosaic-like pattern of relatively big patches, mostly irregular in shape and with an average level of interspersion. For SOILH2O indicator, eight metrics take extreme values. Maximal are PSCoV, LPI, HOLE, AWMSI, and MPI, while minimal are i60, SEI, and FDIM. Such value distribution indicates a pattern composed of several very large and heavily perforated patches, forming a matrix, in which embedded are patches of different shape and size, but generally rather small, forming clusters. All other patterns occupy intermediate positions within the triangle determined by the above-described patterns of ecosystems, TIMBER, and SOILH2O.

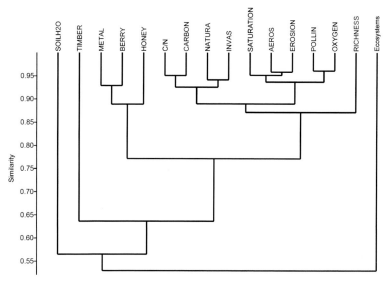

Figure 7.8 Hierarchical clustering of ecosystem service indicators (and ecosystem typology) in relation to their spatial distribution in the study area (Bray–Curtis similarity index, unweighted pair group method, variables standardized). Variables (landscape metrics) describing spatial distribution are shown in Table 7.7.

According to hierarchical clustering, the remaining spatial patterns can be divided into two main groups (Fig. 7.8). The first comprises METAL, HONEY, and BERRY indicators, and the second RICHNESS (being on partly intermediate position) and the other nine indicators. The main differentiating metrics are AWMSI and MPI. In the first group, their values are significantly higher compared with the second group (for AWMSI 6.47 and 4.89, and for MPI 2534 and 1000, respectively). It means that the spatial patterns of METAL, HONEY, and BERRY show lower isolation of patches of the same type, and patches have more complex shape compared with spatial patterns from the second group. The first group is not further divided into subgroups, while the second group is divided into three subgroups. The first of them (NATURA, CARBON, INVAS, C/N), with the most pronounced specificity, differs from the other subgroups by the lower values of MPI, LPI, and PSCoV, as well as by the higher values of i60 and SEI (the latter is the highest among all the patterns studied). This value distribution indicates patterns with lower differentiation in patch size, higher isolation of patches of the same type, and more even abundance of patches from different classes compared with spatial patterns from the other subgroups. Differences between the two remaining subgroups are smaller but still well visible. POLLIN and OXYGEN differ from SATURATION, EROSION, and AEROS by lower values of PSCoV and LPI as well as by higher values of i60 and MPI, which means lower differentiation in patch size, smaller area covered by the largest patches, and lower isolation of patches of the same type. Specific position is occupied by RICHNESS. This pattern is located between the first and the second group, but with a little higher similarity to the second group.

The above analysis, although giving valuable information about spatial patterns of ES potentials, does not inform in detail about the dispersion type and spatial arrangement of patches, so additional analyses are necessary. According to the rough evaluation of spatial distribution randomness, based on the ratio of the variance to the mean of the nearest neighbor distance (VMR) (Cox and Lewis, 1966), all value classes of all indicators are globally overdispersed (VMR > 1), which means that the same values are clumped in short distances.

More detailed information on the repeatability is given by autocorrelation analysis with Moran's I indicator. In general, Moran's I values range from −1 to 1, where 1 describes a highly clustered (correlated) pattern. With increasing distance, this relationship weakens, and autocorrelation becomes insignificant as Moran's I approaches zero. Increasingly negative values of I indicate greater dispersion, with −1 value indicating perfect, regular dispersion.

The obtained autocorrelation curves of 15 ES indicators take positive values at distances up to 5 km, with nonlinear decrease with distance (Fig. 7.9). Based on the Euclidean distances between autocorrelation curves, three main types of curves describing 10 indicators can be distinguished. The curves of the other five indicators are of an intermediate nature. The first type is represented only by RICHNESS. It is characterized by the very steep drop in Moran's I to 0.08 at the distance of 500 m and 0.037 at the distance of 1000 m, and then showing a much more gentle decline to 0.005 at 3000 m. The second type is of opposite character and is represented solely by TIMBER. In this case, we may observe rather smooth and regular decline of Moran's I, from 0.62 at the distance of 500 m and 0.47 at the distance of 1000 m to 0.21 at 3000 m. The third type is represented by eight indicators: OXYGEN, CARBON, METAL, C/N, SOILH2O, AEROS, SATURATION, and EROSION. It is interesting that at

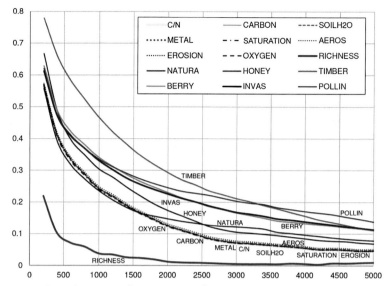

Figure 7.9 Spatial autocorrelation curves of ecosystem service indicators. Horizontal axis - upper lag distance in meters, vertical axis - Moran's I index.

distances from 1000 to 5000 m autocorrelation values for this type are related to the values of the second type according to the equation:

$$[\text{Third type value}]_{\text{lag}} = [\text{Second type value}]_{\text{lag}} - (0.2445 - 0.0000359*\text{lag})$$

where lag denotes upper lag limit. The values of other indicators change individually, but three of them (INVAS, BERRY, POLLIN) are very similar to each other at distances of up to 1000 m.

The analysis of autocorrelation curves' variability suggests that different spatial properties of the patches influence the level of aggregation of indicator values in different ways and that these relationships change as the spatial scale changes. This issue was analyzed for two different distance ranges: 0—1000 m and 1000—2000 m (Fig. 7.10). It turned out that in both cases, the most important factor differentiating autocorrelation curves is the area of large patches. At short distances in particular, the area occupied by the patches from the 5th to the 15th by size separates the TIMBER and RICHNESS indicators from all the others. Other differentiating metrics are the share of the largest patch (LPI) and the area occupied by the 10 largest patches (but only for short distances). At short distances (0—1000 m), the shape of patches can also significantly influence autocorrelation, while at longer distances (1000—2000 m) spatial distribution of patches with the same ES potential (measured by the level of proximity MPI and nearest neighbor distance MNN).

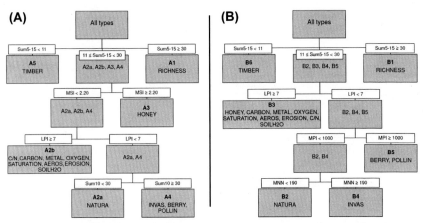

Figure 7.10 Decision trees separating autocorrelation curves (A - at the distance 0—1000 m; B - at the distance 1000—2000 m) of ecosystem service indicators. Types of curves distinguished on the basis of Euclidean distances between them. Sum10 describes the joint percentage area of the first 10 largest patches, while Sum5—15 describes the joint percentage area of patches from the largest 5th to 15th. Other metrics are as in Table 7.7.

The analysis of landscape metrics concerns the distribution and characteristics of patches with specific values, while the analysis of autocorrelation is focused only on the distribution of values. These are complementary methods, like the two sides of the same coin, and together they provide more complete information on the relationship between landscape structure and ES potential.

Many authors emphasize the relationship between different aspects of landscape pattern and potential of many services (Mitchell et al., 2015; Syrbe and Walz, 2012; Verhagen et al., 2016). Generally, it is assumed that combination and configuration of patches (measured by the landscape metrics) can modify (decrease or increase) the ES potential of a given area, so the effective potential on the landscape level is not the simple sum of potentials of particular ecosystem patches (Mitchell et al., 2015). From the landscape ecological point of view, the influence of landscape structure on ES potentials may be separated into three different aspects: (a) location of a given ecosystem in topography, (b) neighborhood and adjacency to other ecosystems, and (c) spatial characteristics of ecosystem patches related to size, shape, and distribution. Only the third aspect will be further discussed.

Out of the three provisioning ES analyzed in this chapter (tree biomass for nonnutritional purposes measured by TIMBER indicator, honey measured by HONEY indicator, and edible wild berries measured by

BERRY indicator), only in one case (BERRY) we may suspect that the shape of ecosystem patches (measured by such metrics as MSI, AWMSI, FDIM, SSPI, CNCV) can modify (enhance) the ES potential, as some of the berry-producing plant species prefer forest edges.

Different situation is observed in the case of potential to provide regulating ES. According to data and discussions in Mitchell et al. (2015), Leitão and Ahern (2002), Schindler et al. (2013), Su et al. (2012), Verhagen et al. (2016), Vos et al. (2014), and Zhang and Gao (2016), influence of landscape spatial pattern on ES potential may be more complex. For instance, the potential to control invasive plant species (measured by INVAS indicator) depends not only on the intrinsic characteristics of an ecosystem but is also modified by fragmentation level (measured by MPS metrics), proximity (MPI, MNN metrics), and patch shape (MSI and similar metrics). As a result, effective ecosystem potential is smaller when patches are small with more complex shape, mean nearest neighbor distance is low, and when proximity values are high. In turn, potential to provide pollination (measured by POLLIN indicator) is positively correlated with proximity (MPI metric), according to logistic model on distances 1—2 km (Vos et al., 2014). However, the most complex and multifaceted influence of landscape structure on ES potential concerns the maintenance of nursery habitats (measured on the ecosystem level by NATURA and RICHNESS indicators). In general, the smaller and more complex the patch (measured by MPS, MSI, and similar metrics), the lower the number of core species and the higher number of satellite and multiecosystem species. In most cases, this kind of species composition influences negatively the potential to maintain nursery habitats. On the other hand, species richness is modified by the neighborhood structure (MPI, MNN, IJI metrics), with positive correlation to proximity according to logistic model on distances 1—2 km (Vos et al., 2014) and probably negative correlation to IJI. Moreover, the total landscape effect depends on the taxon examined and can be different in different regions. For other indicators analyzed in this chapter, either there is no significant relation to landscape pattern or dependencies are not known (see Verhagen et al., 2016).

Based on the above analysis, we can conclude that the landscape structure influences mainly the potential of these services, which are provided by narrowly understood groups of species (ESPs), modifying either their habitats or connectivity with other ecosystems. It is also worth to underline that the same landscape characteristics (e.g., fragmentation level) may directly exert positive effect on one ES potential (e.g., pollination) and negative on the other (e.g., invasive plant species control), being neutral for the others (e.g., edible wild berries).

At the end it is worth to remind that the influence of landscape structure on ES is wider than only modifications of the ES potential. It can also modify usability of and accessibility to potentials as well as flows of services (Mitchell et al., 2015). These issues are, however, beyond the scope of this book.

CHAPTER 8

Use of ecosystem services and its impact on the perceived ecosystem potentials

Contents

8.1 Actual use of services

The actual use of local ecosystem services (ES) was quantified on the basis of written declarations of direct ecosystem users: residents and tourists (see Section 3.3). Of 251 respondents interviewed, 69% were female and 31% male. The majority of them (73%) were between 30 and 60 years old, 12% were under 30, and about 15% were above 60. Most respondents declared secondary (45%) or higher (43%) education. Mental work (31%), farming (27%), or a pension (22%) was the most frequent income source among surveyed people (multiple answers allowed). A significant group worked also in tourism services (14%). More than 69% of the respondents were permanent rural residents, whereas 31% came as tourists from towns and cities.

8.1.1 Ecosystem service use declared in open questions

In the first open-ended question, respondents were asked to list gifts of nature that they have used in the three preceding years. The answers were analyzed in detail and then classified into one of the 20 ES categories within one of three sections: provisioning, regulating, and cultural (Fig. 8.1). Few

Ecosystem Service Potentials and Their Indicators in Postglacial Landscapes
ISBN 978-0-12-816134-0
https://doi.org/10.1016/B978-0-12-816134-0.00008-0
Copyright © 2020 Elsevier Inc.
All rights reserved.

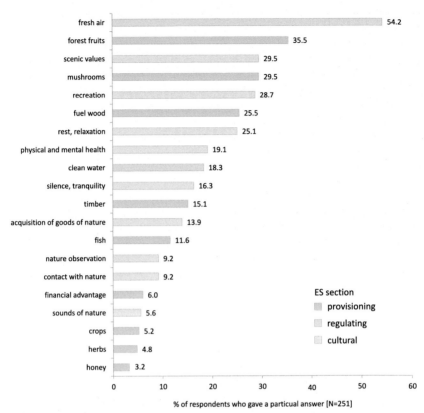

Figure 8.1 Gifts of nature recognized by respondents as received/experienced within 3 years preceding the survey.

single answers that do not fit in any category (e.g., syrup from pine buds, birch juice, gravel, few people, good influence on the condition of Polish nature in general) were excluded from further analysis. Fresh/clean air was the most frequently mentioned ecosystem service (54%) in respondents' answers. It was one of the two gifts of nature, along with clean water, which was classified as the regulating ES. The other services listed belonged to provisioning and cultural sections. The most frequently mentioned provisioning services included forest fruits (berries, raspberries), mushrooms, fish, and fuel wood and timber. In turn, the most popular cultural ES are landscape values (esthetic and natural values, e.g., relief diversity, beautiful views), recreation (e.g., sailing, canoeing, walking, cycling, swimming in the lake, etc.), and resting/relaxation. Many people also paid attention to the interaction with nature consisting in acquiring its goods (mushroom picking, fishing, and hunting) and on the health aspect of interaction (improvement of physical and mental health). Some of the answers were repeated so often that we decided not to ignore them but assign to a separate category, although according to the assumptions they are not strictly ES, but rather they are benefits (e.g., health) that are not always derived directly from ES (such as financial benefit: a workplace in tourism or in the wood industry).

On average, the respondents listed ES from four different categories (mean 3.7, SD 2.2) (Fig. 8.2). However, some of the applied ES categories were quite broad and comprised plenty of listed particular services (e.g., recreation category covered all sports, tourism, and strictly recreational activities). "Champions" reported services from up to 12 different categories.

The between-group comparison showed different approaches to ES among respondents, particularly when place of permanent residency was the grouping variable (Fig. 8.3). People who stayed in the study area only temporarily (mainly tourists) definitely more often mentioned cultural services as those from which they benefited in the preceding 3 years, than

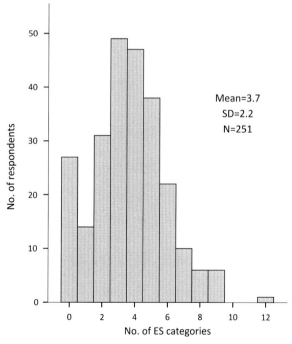

Figure 8.2 Number of respondents, from whose answers a given number of ecosystem service (ES) categories could be extracted.

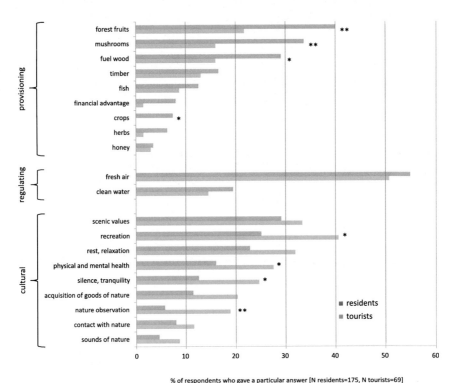

% of respondents who gave a particular answer [N residents=175, N tourists=69]

* difference sig. at the 0.05 level   ** difference sig. at the 0.01 level

Figure 8.3 Gifts of nature recognized by respondents as received/experienced within 3 years preceding the survey-comparison between residents and tourists.

the local residents. In particular, tourists more often spontaneously pointed out such services as nature observation ($\chi^2 = 10.0$, $P = .002$), recreation ($\chi^2 = 5.7$, $P = .017$), silence and tranquility ($\chi^2 = 5.4$, $P = .021$), and mental and physical health ($\chi^2 = 4.2$, $P = .040$). In turn, permanent residents paid much attention to provisioning services, such as forest fruits ($\chi^2 = 7.3$, $P = .007$), mushrooms ($\chi^2 = 7.6$, $P = .006$), agricultural crops ($\chi^2 = 5.4$, $P = .020$), and fuel wood ($\chi^2 = 4.6$, $P = .033$). However, both respondent groups indicated regulating services with similar frequency.

Other between-group comparisons did not show so distinct differences. Nevertheless, it is worth noting that better-educated people listed more often contact with nature ($\chi^2 = 7.3$, $P = .026$) and recreation ($\chi^2 = 6.1$, $P = .046$) as well as rest/relaxation ($\chi^2 = 8.1$, $P = .018$) as gifts of nature. Moreover, middle-aged women (30−60 years old) more often than other respondents mentioned fish ($\chi^2 = 9.7$, $P = .002$) as used ES. In turn, silence and tranquility were most often pointed out by men over 60 years old ($\chi^2 = 4.8$, $P = .028$). Women regardless of age more often than men spontaneously indicated landscape values ($\chi^2 = 4.2$, $P = .042$) as a service they used in the preceding 3 years.

8.1.2 Ecosystem service use declared in closed questions

According to the results of the second section of the questionnaire, obtained for the closed questions with 45 ES listed (Section 3.3), the vast majority of respondents declared the use of nutrition from local nature, mainly mushrooms (98%), vegetables (94%), fruits (92%), and eggs (91%) in the preceding 3 years (Fig. 8.4). Among the materials, the largest use was made of fuel wood. Over 82% of the respondents used it at least few times over the preceding 3 years. Among the cultural services, the most popular were rest/relaxation, swimming in the lake/sunbathing, walking/jogging, mushroom picking, and nature observation. Each of these services was regularly used by over 55% of the respondents. The least popular local gifts of nature included in the questionnaire were hunting, shellfish, and mollusks as well as other provisioning services-materials: wool and animal hides, reed, wicker, and energy: peat, biomass. About 90% of the respondents indicated that they did never use the above services in the preceding 3 years.

There were numerous differences in the declared use of ES between respondents' subgroups in the closed questions. A comparison of average ranks (Mann−Whitney U test) showed that men more often went fishing ($Z = -4.2$, $P = .000$) and used fuel wood ($Z = -2.7$; $P = .007$), whereas women more frequently used natural ornaments ($Z = -2.7$, $P = .008$). The results also indicate that better-educated people more often used 17 ES out of 45 included in the questionnaire. The difference in the frequency of use is particularly pronounced in the case of cultural services (13 out of 18), in physical (6 in 11), intellectual (5 in 5), and spiritual (2 in 2) interactions. The level of education correlates significantly with cultural services such as nature observation (Spearman's *rho* coefficient $= 0.35$, $P = .000$), prayers/ meditation close to nature (*rho* $= 0.24$, $P = .000$), sightseeing tours to enjoy nature (*rho* $= 0.29$, $P = .000$), or canoeing (*rho* $= 0.26$, $P = .000$) and sailing (*rho* $= 0.26$, $P = .000$). Better-educated people also consume significantly more selected local nutrition products, including fish (*rho* $= 0.24$, $P = .000$), game (*rho* $= 0.22$, $P = .001$), and honey (*rho* $= 0.27$, $P = .000$). Fodder necessary for breeding domestic animals is the only service, which is more frequently used by less-educated respondents (*rho* $= -0.25$, $P = .000$).

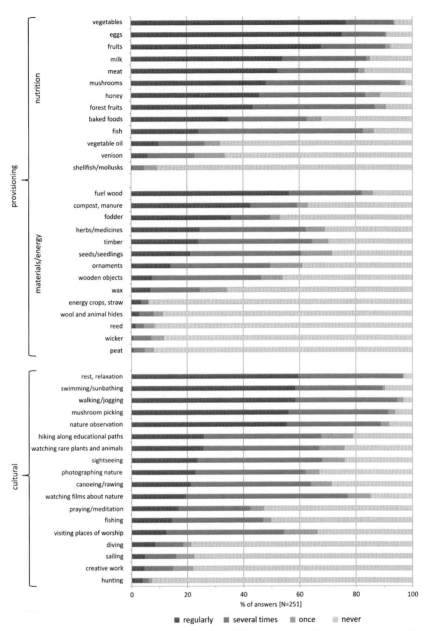

Figure 8.4 Frequency of ecosystem service (ES) use within 3 years preceding the survey, declared by the respondents.

The declared ES use by tourists and permanent local residents is slightly different. Tourists, like better-educated people, benefit more often from cultural services, e.g., they observe more often surrounding nature ($Z = -3.0$, $P = .002$) and rest close to nature ($Z = -3.1$, $P = .002$), whereas the permanent residents more frequently use natural raw materials from local ecosystems, such as timber ($Z = -2.9$, $P = .004$), fuel wood ($Z = -3.8$, $P = .000$), seeds/seedlings ($Z = -3.4$, $P = .001$), wax ($Z = -3.0$, $P = .003$), fodder ($Z = -6.4$, $P = .000$), and compost/manure ($Z = -5.0$, $P = .000$).

The frequency of ES use is also differentiated by respondents' age. Older people more often pick up ($rho = 0.19$, $P = .003$) and consume ($rho = 0.15$, $P = .024$) mushrooms from nearby forests, whereas younger ones more often swim in the lakes ($rho = -0.19$, $P = .003$), dive ($rho = -0.20$, $P = .002$), and make use of natural ornaments ($rho = -0.17$, $P = .010$).

8.1.3 Comparison of ecosystem service use declared in open and closed questions

Many of the ES spontaneously mentioned in the open questions had their equivalents in the second part of the questionnaire, where a closed list of 45 provisioning and cultural services was included. A comparison of the declared ES use obtained by the two methods (spontaneously listed in the open question and specifying frequency in the closed question) shows that, on average, respondents are able to recall spontaneously only some of the services used (Fig. 8.5).

Large differences, about 40–94 percentage points, occur in both provisioning and cultural services. Interestingly, many important differences in the ES use between local residents and tourists, calculated on the basis of an open question, have no confirmation in the closed questions. Most of the differences, apart from using fuel wood ($\chi^2 = 30.4$, $P = .000$), timber ($\chi^2 = 12.6$, $P = .000$), and herbs ($\chi^2 = 4.1$, $P = .042$), are statistically insignificant. Moreover, in the case of two provisioning and two cultural services, the difference has the opposite sign. A larger percentage of permanent residents than tourists declare nature observation and acquisition of nature goods in a closed question. Similarly, a greater percentage of tourists than local people report the use of forest fruits and fish. This result can be explained by the fact that tourists, recalling the ES used, focused on the interaction involving the acquisition of goods, whereas permanent residents

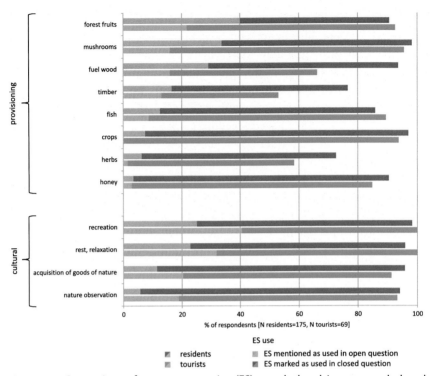

Figure 8.5 Comparison of ecosystem service (ES) use declared in open and closed questions.

focused on the already acquired goods; in reality, they both just as often collected and ate the fruits of nature.

8.2 The frequency of using ecosystem services and user characteristics as factors modifying the assessment of ecosystem potentials[1]

The assessment of the potential of 7 ecosystem types distinguished in the questionnaire to provide 11 groups of services, determined on the basis of the entire group of respondents, is presented in Section 6.2. At this point, the relationships between selected characteristics of the respondents and their assessment of the ecosystem potential are presented. Several analyses were computed to test the links between perceived ecosystem potentials

---

[1] A full description of the results obtained, along with their broad discussion, is included in the article of Affek and Kowalska (2017).

and selected sociodemographic variables (e.g., age, gender, education, place of residence), in line with the actual use of services declared in the second part of the questionnaire.

Between-group comparisons showed differences in the perception of ecosystem potentials among respondents with different sociodemographic characteristics (Table 8.1). In general, better-educated respondents ($t = -2.12$, $P = .04$) and urban residents ($t = 2.19$, $P = .03$) were inclined to indicate higher potentials as regards the delivery of cultural ES. For instance, respondents with higher education differed from those with secondary education in noting a greater potential for deciduous forests ($t = -2.24$, $P = .03$) and pine forests ($t = -2.22$, $P = .03$), as well as lakes and rivers ($t = 2.77$, $P = .01$), to supply spiritual experience. The better-educated people also pointed to the role of swamp forests in water regulation ($t = -2.27$, $P = .025$), while only offering a lower rating when it came to the capacity to provide some tangible services, i.e., building materials in the case of grassland ($t = 2.30$, $P = .023$) and fuel in the case of pine forests ($t = 2.09$, $P = .038$).

Urban residents, as visitors to the study area, perceived a higher potential of all ecosystem types to regulate water, compared with local inhabitants. All differences, except for those involving cropland and deciduous forest, achieved statistical significance. Also, the capacities of the three forest types to serve as sources of inspiration were the subject of higher ratings among tourists ($t = 2.27-2.31$, $P < .025$). The same applies to certain other cultural services (spiritual experience, education, and science) provided by wetlands ($t = 2.18-2.3$, $P < .031$). Only single provisioning services (natural medicines from cropland, ornamental resources from pine forests) received significantly higher ratings among local residents than outsiders ($t = -2.11-2.64$, $P < .04$).

When gender was taken into account, women were found to offer significantly more favorable evaluations of pine forests' suitability for education and science ($t = 2.33$, $P = .02$) and of arable lands' capacity to inspire creative work ($t = 2.03$, $P = .04$). The capacity of all kinds of forests ($t = 2.58-2.92$, $P < .01$) and cropland ($t = 2.97$, $P = .00$) to provide natural medicines was also higher rated by female than male respondents. The same regularity applies to the potential of wetlands to deliver fuel ($t = 2.46$, $P = .02$). In turn, men compared to women ascribed greater potential to deciduous and pine forests when it came to the retention and purification of water.

Table 8.1 Statistically significant ($P < .05$) differences in evaluations of ecosystem potentials between respondent subgroups broken down by gender, education, and place of residence (provisioning ES-green, regulating ES-blue, cultural ES-red).

	Ecosystem type	Ecosystem service	Mean 1	Mean 2	t	Significance (2-tailed)
			Women	Men		
Gender	Deciduous forest	Natural medicines (herbs, juice, resin)	11.45	9.59	2.92	0.004
		Water regulation (retention, purification/detoxification)	6.37	8.30	-2.02	0.047
	Pine forest	Natural medicines (herbs, juice, resin)	11.29	9.62	2.60	0.010
		Building materials (timber, reed, straw)	12.34	13.30	-1.98	0.050
		Water regulation (retention, purification/detoxification)	5.47	7.88	-2.49	0.014
		Education and studying (nature observation, research)	10.97	9.45	2.05	0.044
	Swamp forest	Natural medicines (herbs, juice, resin)	10.79	8.56	2.57	0.011
	Cropland	Natural medicines (herbs, juice, resin)	10.20	7.52	2.97	0.003
		Inspiration for creative work	7.82	6.07	2.03	0.044
	Wetland	Fuel (fuel wood, peat, energy crops)	9.92	7.53	2.46	0.015
			Secondary	Higher		
Education	Deciduous forest	Spiritual experience	8.24	9.81	-2.24	0.027
	Pine forest	Fuel (fuel wood, peat, energy crops)	13.57	12.57	2.09	0.038
		Inspiration for creative work	8.53	10.16	-2.42	0.017
		Spiritual experience	8.32	9.90	-2.22	0.028
	Swamp forest	Water regulation (retention, purification/detoxification)	6.03	8.11	-2.27	0.025
	Grassland	Building materials (timber, reed, straw)	6.40	4.39	2.30	0.023
	Rivers and lakes	Spiritual experience	9.00	11.01	-2.76	0.007
			Visitors	Locals		
Place of permanent residence	Deciduous forest	Inspiration for creative work	10.27	8.64	2.27	0.025
	Pine forest	Ornamental resources (antlers, animal hides, shells)	9.10	10.55	-2.11	0.036
		Water regulation (retention, purification/detoxification)	7.63	5.40	2.37	0.019
		Inspiration for creative work	10.22	8.57	2.31	0.023
	Swamp forest	Water regulation (retention, purification/detoxification)	8.71	5.70	3.27	0.001
		Inspiration for creative work	9.87	7.88	2.30	0.023
	Grassland	Water regulation (retention, purification/detoxification)	7.25	4.86	2.68	0.008
	Cropland	Natural medicines (herbs, juice, resin)	7.88	10.19	-2.64	0.009
	Wetland	Water regulation (retention, purification/detoxification)	8.52	6.40	2.22	0.028
		Education and studying (nature observation, research)	11.35	9.78	2.18	0.031
		Spiritual experience	8.73	6.83	2.26	0.028
	Rivers and lakes	Water regulation (retention, purification/detoxification)	12.80	11.58	2.01	0.047

Younger respondents highlighted the educational and scientific potential of both grasslands ($r = -0.22$, $P = .01$) and wetlands ($r = -0.17$, $P = .04$), while a capacity to inspire creative work was assigned to grasslands ($r = -0.2$, $P = .02$) and waters ($r = -0.2$, $P = .01$). Younger respondents were also more inclined than others to regard deciduous forests ($r = -0.26$, $P = .00$) and cropland ($r = -0.17$, $P = .04$) as sources of building materials, as well as pine forests as inspirations for science and education ($r = -0.16$, $P = .04$).

Above all, the analysis of the full correlation matrix shows that there is a considerable share of positive correlations between the intensity of use made of cultural services and the potential to provide them that people perceive, especially in the case of deciduous and pine forests (Table 8.2). For instance, the higher the declared frequency of occurrence of meditation/praying in nature, the higher the rating of this service rendered in relation

Table 8.2 Cumulative correlation matrix. Percent of significant ($P < .05$) positive and negative correlations (Spearman's *rho*) between perceived potentials to deliver services and actual usage intensity.

			Intensity of ES use							
			Provisioning ES				Cultural ES			
			Nutrition (13 ES)		Materials and energy (14 ES)		Education, inspiration, spiritual life (7 ES)		Sport and recreation (11 ES)	
			+	-	+	-	+	-	+	-
Perceived ecosystem potential	Deciduous forest	Provisioning	3	0	2	2	7	5	3	2
		Regulating	0	0	0	0	14	0	55	0
		Cultural	8	0	4	2	54	0	36	0
	Pine forest	Provisioning	5	0	11	5	10	0	8	2
		Regulating	0	0	0	0	43	0	64	0
		Cultural	10	0	0	0	61	0	27	0
	Swamp forest	Provisioning	3	5	4	8	10	0	5	0
		Regulating	0	8	0	14	29	0	45	0
		Cultural	4	2	0	0	11	0	18	2
	Grassland	Provisioning	3	4	10	2	10	5	11	0
		Regulating	0	8	0	0	57	0	55	0
		Cultural	8	2	2	0	25	0	16	2
	Cropland	Provisioning	4	4	10	0	10	0	5	2
		Regulating	0	0	0	0	14	0	36	0
		Cultural	10	2	4	0	4	0	7	2
	Wetland	Provisioning	6	1	4	1	12	0	14	0
		Regulating	0	0	0	0	57	0	55	0
		Cultural	4	2	0	2	4	0	7	5
	Rivers and lakes	Provisioning	0	5	2	6	5	5	12	2
		Regulating	0	0	0	0	0	14	0	0
		Cultural	4	0	5	0	18	0	18	0

to other services, in each ecosystem type (Spearman's *rho* from 0.18, $P = .031$ on grassland, up to 0.42, $P = .000$ in deciduous forests) (Fig. 8.6A).

The frequency with which people draw benefit from cultural services is also linked with evaluations of the potential of ecosystems to deliver ES other than the cultural. The more often respondents made use of cultural services (canoeing, swimming, hunting, sightseeing, photographing nature, or watching wildlife), the higher rank they were likely to assign to the only example of a regulating service included in the survey (i.e., water retention and purification) (Fig. 8.6B). This relationship achieved statistical significance in the cases of all the ecosystems considered (*rho* up to 0.37, $P = .000$), except for rivers and lakes. In general, the more those surveyed benefited from cultural ES, the higher they rated the potential of nature (and in particular forests) to provide cultural and regulating services (*rho* up to 0.37, $P = .000$).

Links between the use of cultural services and evaluations in regard to the potentials of ecosystems to render provisioning services proved to be relatively weak. However, one of the more interesting results is that the more often respondents benefited from cultural services (in particular wildlife observation), the higher the perceived capacity of wetlands to provide natural medicines (*rho* up to 0.27, $P = .001$).

The frequency of use made of tangible ecosystem outputs (nutrition, materials, and fuels) only affects the assessment of ecosystem potentials slightly. Only 6% (131 of 2079) of the relevant correlation coefficients proved to be significant statistically. This is only about one-third as many as in the case of intangible cultural services (234 of 1386). Nonetheless, some of the links involving the use of rendered services are very clear and highly significant. For instance, the more often the surveyed people obtained heat from local fuel wood, the higher their rating of the potential of deciduous and pine forests to provide fuel (*rho* up to 0.28, $P = .000$). There are nevertheless certain cases in which more intensive use of natural materials actually goes together with a tendency to assign lower ratings to the potentials for selected ecosystems to deliver some provisioning services. For instance, those who use timber from local forests more frequently perceived those forests as having a more limited potential to deliver food provisioning services (*rho* as low as -0.21, $P = .009$). Increased use of local wood also correlates with a lower perceived capacity of aquatic ecosystems to provide building materials (*rho* as low as -0.23, $P = .005$).

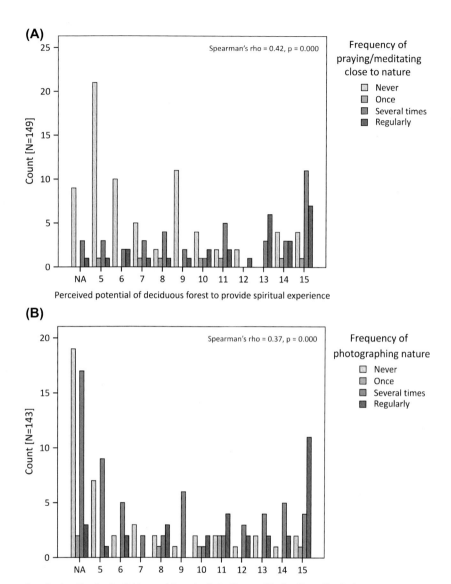

Figure 8.6 Relationships between perceived ecosystem potentials and the frequency of use of cultural services: (A) the perceived potential for deciduous forest to provide a spiritual experience in relation to the frequency with which praying/meditating close to nature takes place, (B) the perceived potential of wetlands to regulate water in relation to the frequency with which people photograph nature (5-lowest value, 15-highest value, NA-not applicable).

8.3 Conclusions

The presented analysis demonstrated how important are the individual characteristics of direct users in the ES perception and use. The results indicate that the intensity of ES use depends on the level of education. In general, better-educated people benefit more from nature. In turn, frequent use of the gifts of nature contributes to a better understanding of its value and appreciation of the services provided. It was also shown that the more often people interact with nature, the greater they perceive the potential of ecosystems to provide cultural and regulating services. However, this effect was not observed in the case of provisioning services. This might be because the use of goods from local nature (nutrition, materials) is often a necessity for the members of the studied community to ensure a decent life and is not the result of free choice, as is the case with nonessential cultural services. Knowledge and experience also seem to be those factors that account for differences in the results because of age. Older people prove to have a better understanding of what nature can give us and what its real value is. This clearly results from their greater experience and longer interaction with the environment they live in. In turn, younger people more often perceive and are able to name services that require specific knowledge not easily gained through experience but rather by way of ecological education. Moreover, people staying permanently and making their living in the area differ from tourists as regards their hierarchy of needs and expectations toward local nature. These differences influence their perception and use of ES. Local residents are more likely to use provisioning services, and some of them also value higher, whereas tourists more frequently take advantage of and give more importance to cultural services.

The applied survey-based method represents a convenient way to recognize the actual use of ES and to determine attitudes and dominant opinions regarding the potentials of ecosystems to provide services.

CHAPTER 9

Summing up and conclusions

Contents

9.1 Summary of the book content

The presented book is the first in the literature to report on a comprehensive assessment of the potential of nature to provide ecosystem services (ES) in postglacial landscape. It comprises nine main chapters, including this one. In the first, introductory part of Chapter 1 the conceptual framework of the book is outlined and the scope of the study and the research objectives are presented. These are:

(1) development of methods for estimating the potential of nature to provide ES;

(2) multifaceted assessment and mapping of ES potential in the postglacial study area;

(3) determination of ES similarities among ecosystems;

(4) determination of the impact of selected factors on the social assessment of ES potential.

The second part discusses the current developments in the ES concept, its origins, and evolution. Based on the broad review of scientific papers, books, and initiatives related to ES, the following issues have been addressed: ES terminology, classifications, indicators, spatiotemporal dimension and mapping, social perception, synergies and trade-offs, and links with nature potential and biodiversity.

Chapter 2 presents the characteristics of the study area against the background of European postglacial landscapes. It starts with the genesis, range, and general features of this kind of landscape and then describes in detail the selected study area (located in North-east Poland and covering

Ecosystem Service Potentials and Their Indicators in Postglacial Landscapes
ISBN 978-0-12-816134-0
https://doi.org/10.1016/B978-0-12-816134-0.00009-2
Copyright © 2020 Elsevier Inc.
All rights reserved.
337

792 km^2) including climate, morphogenesis, soils, waters, vegetation, fauna, and nature conservation forms.

Chapter 3 then features the general description of methods used to estimate potentials of postglacial ecosystems to provide services in expert and social assessments. Field works and laboratory analyses are described as well as methodological assumptions of the survey and the structure of the questionnaire distributed among direct ES users (residents and tourists) in the study area. Then, methods used to synthesize the analytical results are presented: creation of the assessment matrix, analysis of aggregated potential, multiservice hotspot analysis, analysis of ES interactions, and ecosystem similarities. Next, landscape metrics and autocorrelation analysis depicting spatial pattern of ES potentials are characterized. The chapter ends with the description of analysis of links between ES usage, user characteristics, and perceived ES potentials based on data collected in the survey.

Chapter 4 presents the basic assumptions of the most recent version of the Common International Classification of Ecosystem Services (CICES V5.1), selected as the framework of our research. In contrast to many other classification systems, it includes only three main sections (provisioning, regulation and maintenance, and cultural) and relates only to final ES, which are defined as the contributions that ecosystems make to human well-being.

Chapter 5 presents the spatial reference units applied in this study: ecosystem in the narrow sense, MAES (Mapping and Assessment of Ecosystems and their Services)-derived ecosystem, hunting unit, and landscape. When distinguishing ecosystems in the narrow sense, account was taken not only of land cover but also of habitat conditions (fertility and humidity) and forest succession stage. The extended typology of rural postglacial ecosystems, used in expert assessment, includes 35 terrestrial and 7 lake ecosystem types. If the type of data did not give a possibility to estimate the potential of nature for ecosystems, it was calculated for heterogeneous spatial units, covering a number of different ecosystems: 15 hunting units and 91 individual landscape units. In the questionnaire survey aimed at assessing ecosystem potentials from the beneficiary perspective, a simplified ecosystem division was introduced, based on the MAES level 2 typology. Only forests were further divided into the three subtypes.

Chapter 6 features the main analytical part of the book and forms the basis for the following synthetic chapters. It presents the entire path of indicator development for the evaluation of ES potentials in the postglacial landscape. Altogether, 29 ES were assessed by means of 35 indicators. Services are arranged first according to the type of assessment (expert/social) and then follow CICES V5.1 classification and coding. All subchapters devoted to particular services include ES description (a table with ES location in CICES V5.1, indicator metadata, and briefly mentioned data sources), theoretical framework, assessment method, and indicator values. Each subchapter ends with a map showing ES potential in the study area. Additionally, all services assessed for ecosystems in the narrow sense have their spatial pattern described by means of landscape metrics.

Chapter 7 is the synthesis of the analytical results presented in the previous chapter. However, the investigation was conducted only on 29 ES indicators for which the ecosystem (detailed or MAES-derived) was the spatial reference unit. To start with, synthetic assessment matrices were generated, which served as a basis for the majority of further analyses. Rank values obtained through expert and social assessment were compared, where possible. In the next step, the aggregated potential for each ecosystem type was calculated, and ES hotspots were identified and then both mapped, separately for each ES section and in total. Next, the interactions among services and similarities among ecosystem types were investigated. Finally, spatial patterns of ES potentials, described in short for each indicator separately in the analytical chapter, were compared with each other and further analyzed using landscape metrics and autocorrelation curves.

The last part of the results is presented in Chapter 8. Here, the impact of various factors on the social assessment of ES potentials is described. The factors comprise the frequency of ES use and sociodemographic characteristics of beneficiaries (e.g., age, gender, place of residence, level of education)–both obtained through questionnaire survey and proximity to selected ecosystem types (forests and lakes) derived from spatial analysis of respondents' localization and neighborhood.

9.2 Potential approach

The estimation of the value of nature in terms of benefits derived by humans is becoming increasingly popular in environmental assessments and spatial planning worldwide. In line with this approach, the value of any ecosystem can be determined on the basis of its potential (capacity) or the actual amount of goods and services delivered (flow). The idea to introduce ecosystem potentials (to provide services) into the ES concept arose from the need to distinguish the possibility of ES usage from their actual use. This research focused primarily on recognizing the potential (capacity) of ecosystems to provide services, while the use of local ES was determined only to estimate the impact of this factor on the social assessment of ecosystem potentials. We adopted the approach, in which the main emphasis is on final ES to be clearly separated from the resulting benefits for humans. Less clear distinction has been thus achieved between ecosystem function and ES, particularly in the case of regulating ES. In some cases, the function is basically the same as the service. When framing our research scope, we decided to focus only on biotic ecosystem outputs, and we restricted further ES to those that are provided by animals, plants, and fungi that were formed in the Holocene.

The study has further proceeded on the assumption that potential to provide provisioning services is based around the entire currently existing resources of an ecosystem, which are capable of being utilized (e.g., standing timber and fruits of the forest), with simultaneous independent treatment of different services supplied by these same ecosystems or the component parts thereof. This kind of approach is optimum from the methodological point of view, when it comes to the assessment of absolute potential and the comparison of ecosystem types. On the other hand, it is obviously not sufficient when it comes to planning of the sustainable use of space. In the latter case, there is a further need to define the available resources in another way, e.g., as a determined part of the potential or a level of growth of the resources over a given time, which may still be removed without any catastrophic shift taking place, with renewability thus being assured in an ongoing way, in line with sustainable development principles.

One of the objectives the authors of the study addressed was to formulate methodological solutions easy to transfer to other regions. This required the adoption of several assumptions, of which two seem to be the most important:

(a) maximum use made of public data collected using standardized methods (be these statistical or spatial data);
(b) assessment of as many services as possible in typological spatial units (ecosystem types) instead of regional spatial units (e.g., hunting units).

Next, it was necessary to propose reliable measurement tools by which to determine ES potentials. To this end, two types of assessment were proposed-expert assessment drawing on scientific knowledge and social (beneficiary) assessment obtained from the opinion of direct ES users (inhabitants and tourists) on the basis of questionnaire survey. In expert assessment conducted by a team of scientists specialized in various elements of the environment, only provisioning and regulating services were taken into account. Cultural services were evaluated only through social assessment due to their high subjectivity.

9.3 Expert assessment

The expert assessment of ecosystem potentials to provide all of the services used by people in the study area was not possible because of lack of data. For this reason, the values of several indicators have not been calculated directly, but only estimated based on literature data and expert knowledge. For some of the ES, even the very coarse estimates were unavailable; hence, they were excluded from the analysis. As a rule, one indicator corresponded to one service. However, two exceptions were introduced: one in the case of provisioning service edible biomass of wild animals (separate indicators related to fish and game animals) and another in the case of regulating service nursery habitat maintenance, where due to inherent service complexity, six different indicators were proposed. These are both direct and indirect indicators, simple and complex, and calculated and estimated. Many of them are new, original proposals. The characteristics of the services and availability of data also determined the choice of spatial reference unit. ES expert assessment was carried out in ecosystem types, landscapes, or hunting units. However, only ES indicator values assigned to ecosystems were type-specific. Landscapes and hunting units were assigned unique place-specific values.

The distinguishing feature of ecosystem-based expert assessment is that it took account of ecosystem characteristics other than land cover (such as forest successional stage and soil properties), a circumstance that is not common in the quantification and mapping of ES. Particular emphasis was placed on forest ecosystems, as these usually rank high where ES potentials are concerned. Through the application of a more-detailed ecosystem typology, significant differences in the potential supply of most of the services were found between areas falling within the same category in general land cover classifications (CORINE, MAES), which were used already in several ES assessments. In contrast to pan-European modeling, the present study offers a combination of specially developed indicators and a means of mapping ecosystem capacities in detail, which can be of value to particular stakeholders, and may act in support of decision-making processes regarding the use of ES. On the other hand, many ecosystems of early succession stages (such as clearings and areas of fallow land) are seen to be unstable and ephemeral, implying that any mapping carried out would soon become outdated.

9.4 Social assessment

Besides the biophysical, scientifically based assessment whose importance has been highlighted frequently, we utilized views expressed by direct beneficiaries of ES. Survey-based study, particularly when standardized and performed on a representative sample, represents a convenient way to determine attitudes and dominant opinions regarding potentials of ecosystems to provide services. Social valuations reflect the relative importance of ES to people and are crucial because ES flow is influenced not only by the given ecosystem's capacity but also by society's desired level of supply of given services (Affek and Kowalska, 2017).

In this study, we demonstrated that direct users of ecosystems (local community and tourists) draw distinctions between different ecosystems where their potentials to provide services are concerned. Their perception of the potential of nature to provide services largely corresponds to the scientific knowledge developed as part of the ES concept. This shows that people living in or visiting areas of high natural value are aware of the diversity of natural capital and the various potentials of particular ecosystems.

The way ecosystems are perceived by the individual ES user is shaped not only by its "objective" value but also by many other factors, as our study has shown. Furthermore, what drives decision-making processes is a value attributed subjectively, given that perception is a key process in the interaction between people and the natural environment. It is also important to realize how the perception of nature is linked with people's individual features and activities. We claim that the perceived usefulness of ecosystems influences actual ES usage and vice versa. On the one hand, something that is not perceived as attractive and worth the effort to acquire/interact with will not be utilized. On the other hand, as our study showed, experience in benefiting from a certain service results in a more adequate recognition of its true value. Furthermore, the use of services can itself modify the properties and potentials of ecosystems (Bastian et al., 2013). Nonetheless, the demonstrated relationship indicates how subjective and relative is the assessment of nature's potential to provide services, and this is particularly true of highly subjective cultural ES.

A great help in understanding the matters addressed in the questionnaire came as the term "ecosystem services" was substituted by a more accessible term from colloquial speech, i.e., "the gifts of nature". This seems to be a good keyword by which the ES concept can be promoted. After all, the survey also served an educational function, encouraging respondents to reflect on the relationships that connected them with surrounding nature and, as a result, raising their ecological awareness.

9.5 Study outcomes and possible applications

The main research outcomes include the determination of:

(a) nature potential to provide 29 ES on the basis of 35 ES indicators calculated for different spatial units;

(b) spatial patterns of ES potentials;

(c) aggregated ES potentials and interactions among services;

(d) similarities among ecosystem types in relation to their ES potentials;

(e) actual ES use and its impact on the social assessment of ES potentials.

The conducted research gives us the opportunity to verify the hypotheses formulated in the introduction (Section 1.1). We demonstrated unambiguously that ecosystems in postglacial landscape differ in terms of their potentials to provide services-not only in the case of particular ES potentials (as indicated by the similarity analysis) but also in terms of the overall aggregated potential and the level of multifunctionality. Only selected ecosystems could be considered multiservice hotspots. We also demonstrated that the results of ES assessment are highly dependent on the method used. The apparent inconsistencies were obtained between expert and social assessment. The third hypothesis was also confirmed through the comparison of the six indicators referring to the very broadly defined service: nursery habitat maintenance. The obtained high inconsistency among those six indicators shows how highly multifaceted this service is and how difficult it is to comprehensively assess it. The fourth hypothesis was duly verified by the principal component analysis and the conducted interpretation of the four main principal components (see Section 7.4). We demonstrated that the distribution of ES potentials is determined primarily by ecosystem features such as stock of aboveground biomass, human pressure and the resulting ecosystem maturity, water content, fertility/acidity, and specific human use of anthropogenic ecosystems. The last fifth hypothesis was confirmed through the analysis of links between the frequency of using ES and the social assessment of ecosystem potentials (Section 8.2). We demonstrated unambiguously that the social perception of ecosystem potential is determined by the way people interact with nature.

In the course of the research, we obtained other interesting results not anticipated at the stage of formulating hypotheses. An analysis of associations among ES potentials has revealed numerous interrelationships between the type of land use, the type and maturity of the ecosystem, the particular ES provider, habitat conditions, and the level of human pressure. However, it is worth noting that all specific results obtained in the synthetic part are strongly determined by the type and number of ES included in the analysis. It is highly probable that with a different set of ES, the result would also be somehow different.

A relatively full assessment of the potential of ecosystems can be achieved, where knowledge of the above relationships makes it possible to select a "minimum representative set" of indicators (especially where the potential of certain benefits can be assessed on the basis of several indicators). Our assessment also showed that different ecosystem types have the capacity to provide different bundles of services. On this basis, we can conclude that the diversity of landscape pattern has positive influence on the quality and range of services available at the local scale. However, it needs to be stressed that cause—effect relationships between ES indicators were not analyzed, making it impossible to determine unambiguously whether ES bundles identified are of general or merely local value.

Moreover, as indicator values were in majority assigned to types of ecosystem, maps presenting capacities to deliver services do not account for the spatial arrangement of ecosystem patches. If a region's potential to deliver services is to be revealed in full, assessments of services at the ecosystem level will need to be complemented by work at the landscape level (see the concept of landscape services, Bastian et al., 2014). However, indicator values would not then be type-specific but rather context-related, with each ecosystem patch characterized by the unique nontransferable value.

The calculation procedures applied in regard to the indicators have a universal character and may be transferred to other regions of the World, not only limited to postglacial landscapes (assuming spatial and tabular data achieving the similar level of detail). For this reason, each ES indicator developed is described in detail using a set of metadata that comprise indicandum (what is indicated), indicator name, acronym, short standardized description, type of logical link with the indicandum (direct/indirect), general characteristics of indicator construction (simple/compound/complex and estimated/calculated), scale of measurement (ratio/rank), unit of measurement, spatial assessment unit (ecosystem/hunting unit/landscape), the acquired value range in the study area, and value interpretation in terms of ES potential.

Similarly, obtained results mostly allow for extrapolation to other areas on either a local or regional scale, as well as even for generalization on a supraregional scale, if with certain limitations and reservations. Here, it would need to be stressed that, where most of the indicators are concerned, the calculations were done in real values on ratio scales, with results obtained allowing for direct comparisons to be made both spatially and temporally. In contrast, the final ranking is concerned with a particular area and does not take account of maxima or minima applying on the European scale. For this reason too, results processed in this way (potentials mapped by reference to 1–5 rank scale translating to the descriptive scale "very low-very high") are of limited relevance only, relating primarily to the study area, but partly also to the broader context of European postglacial landscapes.

The methodological solutions (definitions, indicators, methods of analysis, and data synthesis) developed within this research have the capacity to become a standard procedure to investigate ES potentials at the local and regional scale. The obtained ES potentials can be used in the economic valuation of natural capital. They can also be used as important information in the development of spatial planning documents, in particular those concerning spatial management at the commune level. The presented book may serve as a guide for researchers and practitioners dealing with ES and the assessment of the potential of nature in line with the benefits a person may derive from them.

References

Abildtrup, J., Garcia, S., Olsen, S.B., Stenger, S.A., 2013. Spatial preference heterogeneity in forest recreation. Ecological Economics 92, 67−77.

Affek, A., Kowalska, A., 2014. Benefits of nature. A pilot study on the perception of ecosystem services. Ekonomia i Środowisko 51, 154−160.

Affek, A., Kowalska, A., 2017. Ecosystem potentials to provide services in the view of direct users. Ecosystem Services 26 (A), 183−196.

Affek, A., 2018. Indicators of ecosystem potential for pollination and honey production. Ecological Indicators 94 (2), 33−45.

Agbenyega, O., Burgess, P.J., Cook, M., Morris, J., 2009. Application of an ecosystem function framework to perceptions of community woodlands. Land Use Policy 26 (3), 551−557.

Alam, M., Dupras, J., Messier, C., 2016. A framework towards a composite indicator for urban ecosystem services. Ecological Indicators 60, 38−44.

Albert, C., Bonn, A., Burkhard, B., Daube, S., Dietrich, K., Engels, B., Frommeri, J., Götzl, M., Grêt-Regamey, A., Job-Hoben, B., Koellner, T., Marzelli, S., Moning, C., Müller, F., Rabe, S.-E., Ring, I., Schwaiger, E., Schweppe-Kraft, B., Wüstemann, H., 2016. Towards a national set of ecosystem service indicators: insights from Germany. Ecological Indicators 61, 38−48.

Albizua, A., Williams, A., Hedlund, K., Pascual, U., 2015. Crop rotations including ley and manure can promote ecosystem services in conventional farming systems. Applied Soil Ecology 95, 54−61.

Allendorf, T.D., Yang, J., 2013. The role of ecosystem services in park-people relationships: the case of Gaoligongshan nature reserve in southwest China. Biological Conservation 167, 187−193.

Alvarez-Suarez, J.M., Tulipani, S., Romandini, S., Bertoli, E., Battino, M., 2010. Contribution of honey in nutrition and human health: a review. Mediterranean Journal of Nutrition and Metabolism 3, 15−23.

Amarasekare, P., 2003. Competitive coexistence in spatially structured environments: a synthesis. Ecology Letters 6 (12), 1109−1122.

Anderson, B.J., Armsworth, P.R., Eigenbrod, F., Thomas, C.D., Gillings, S., Heinemeyer, A., Roy, D.B., Gaston, K.J., 2009. Spatial covariance between biodiversity and other ecosystem service priorities. Journal of Applied Ecology 46 (4), 888−896.

Andersson, E., Nykvist, B., Malinga, R., Jaramillo, F., Lindborg, R., 2015. A social-ecological analysis of ecosystem services in two different farming systems. Ambio 44 (Suppl. 1), 102−112.

Anselin, L., Syabri, I., Kho, Y., 2006. GeoDa: an introduction to spatial data analysis. Geographical Analysis 38 (1), 5−22.

Badola, R., 1998. Attitudes of local people towards conservation and alternatives to forest resources: a case study from the lower Himalayas. Biodiversity & Conservation 7 (10), 1245−1259.

Balmford, A., Rodrigues, A.S.L., Walpole, M., ten Brink, P., Kettunen, M., Braat, L., de Groot, R., 2008. The Economics of Biodiversity and Ecosystems: Scoping the Science. European Commission, Cambridge, UK.

Balvanera, P., Kremen, C., Martinez-Ramos, M., 2005. Applying community structure analysis to ecosystem function: examples from pollination and carbon storage. Ecological Applications 15 (1), 360−375.

Balvanera, P., Siddique, I., Dee, L., Paquette, A., Isbell, F., Gonzalez, A., Byrnes, J., O'Connor, M.I., Hungate, B.A., Griffin, J.N., 2014. Linking biodiversity and ecosystem services: current uncertainties and the necessary next steps. BioScience 64 (1), 49—57.

Balvanera, P., Quijas, S., Martín-López, B., Barrios, E., Dee, L., Isbell, F., Durance, I., White, P., Blanchard, R., de Groot, R., 2016. The links between biodiversity and ecosystem services. In: Potchin, M., Haines-Young, R., Fish, R., Turner, K.R. (Eds.), Handbook of Ecosystems Services. Routledge, Taylor and Francis Group, pp. 45—61.

Banaszak, J., Cierzniak, T., 2000. Ocena stopnia zagrożeń i możliwości ochrony owadów w agroekosystemach. Wiadomości Entomologiczne 18 (2), 73—94 (in Polish).

Banaszak, J., Jaroszewicz, B., 2009. Bees of the Białowieża National Park and adjacent areas, NE Poland (Hymenoptera: Apoidea, Apiformes). Polish Journal of Entomology 78 (4), 281—313.

Banaszak, J., Krzysztofiak, A., 1992. Communities of bees in the forests of Poland. In: Banaszak, J. (Ed.), Natural Resources of Wild Bees in Poland. Pedagogical University, Bydgoszcz, pp. 33—40.

Banaszak, J., Krzysztofiak, A., 1996. The natural wild bee resources (Hymenoptera: Apoidea) of the Wigry National Park. Polish Journal of Entomology 65 (1—2), 33—50.

Banaszak, J., Szefer, P., 2013. Pszczoły (Hymenoptera: Apoidea) Równiny Sępopolskiej. Cz. I. Różnorodność gatunkowa. Wiadomości Entomologiczne 32 (3), 185—201 (in Polish).

Banaszak, J., 1980. Studies on methods of censusing the numbers of bees (Hymenoptera, Apoidea). Polish Ecological Studies 6, 355—366.

Banaszak, J., 1983. Ecology of bees (Apoidea) of agricultural landscape. Polish Ecological Studies 9, 421—505.

Banaszak, J., 1992. Strategy for conservation of wild bees in an agricultural landscape. Agriculture, Ecosystems & Environment 40 (1—4), 179—192.

Banaszak, J., 2009. Pollinating insects (Hymenoptera: Apoidea, Apiformes) as an example of changes in fauna. Fragmenta Faunistica 52 (2), 105—123.

Banaszak, J., 2010. Bees of the Masurian Landscape Park: diversity and ecology (Hymenoptera: Apoidea, Apiformes). Polish Journal of Entomology 79 (1), 25—53.

Banaszuk, H., 1985. Gleby. In: Stasiak, A. (Ed.), Województwo Suwalskie. Studia i Materiały, vol. 1. Ośrodek Badań Naukowych w Białymstoku, IGiPZ PAN w Warszawie, Białystok, pp. 59—80 (in Polish).

Barbier, E.B., 2007. Valuing ecosystem services as productive inputs. Economic Policy 22 (49), 177—229.

Barszcz, A., 2005. An Overview of the Socio-Economics of Non-Wood Forest Products in Poland. Proceedings of a Project Workshop in Krakow "Non-Wood Forest Products and Poverty Mitigation: Concepts, Overviews and Cases". Research Notes 166. University of Joensuu, Faculty of Forestry, pp. 1—20.

Barthlott, W., Kier, G., Mutke, J., 1999. Globale Artenvielfalt und ihre ungleiche Verteilung. Courier Forschungsinstitut Senckenberg 215, 7—22 (in German).

Bartkowski, T., 1977. Metody badań geografii fizycznej. Państwowe Wydawnictwo Naukowe, Warszawa—Poznań (in Polish).

Bastian, O., Syrbe, R.-U., Rosenberg, M., Rahe, D., Grunewald, K., 2013. The five pillar EPPS framework for quantifying, mapping and managing ecosystem services. Ecosystem Services 4, 15—24.

Bastian, O., Grunewald, K., Syrbe, R.-U., Walz, U., Wende, W., 2014. Landscape services: the concept and its practical relevance. Landscape Ecology 29 (9), 1463—1479.

Bastian, O., 1991. Biotische Komponenten in Landschaftsforschung und -planung. Probleme ihrer Erfassung und Bewertung. Institut für Geographie der Martin-Luther-Univ. Halle-Wittenberg, Dresden (in German).

Bastian, O., 2008. Landscape classification — between fact and fiction. Problemy Ekologii Krajobrazu 20, 13—20.

Beck, C.A., Campbell, D., Shrives, P.J., 2010. Content analysis in environmental reporting research: enrichment and rehearsal of the method in a British-German context. The British Accounting Review 42 (3), 207—222.

Becker, D., Verheul, J., Zickel, M., Willmes, C., 2015. LGM Paleoenvironment of Europe - Map. CRC806-Database. https://crc806db.uni-koeln.de/dataset/show/lgm-paleoenvironment-of-europe–map1449850675.

Bednarek, R., 1991. Wiek, geneza i stanowisko systematyczne gleb rdzawych w świetle badań paleopedologicznych w okolicach Osia (Bory Tucholskie). Rozprawy UMK, Wydawnictwo UMK, Toruń (in Polish).

Ber, A., 1972. Pojezierze Suwalskie. In: Galon, R. (Ed.), Geomorfologia Polski. Tom. 2. Niż Polski. Państwowe Wydawnictwo Naukowe, Warszawa, pp. 179—185 (in Polish).

Ber, A., 2000. Plejstocen Polski północno-wschodniej w nawiązaniu do głębszego podłoża i obszarów sąsiednich. Prace Państwowego Instytutu Geologicznego, vol. 170. Państwowy Instytut Geologiczny, Warszawa (in Polish).

Ber, A., 2009. Geneza jeziora Wigry w nawiązaniu do struktur głębokiego podłoża. Prace Geograficzne 41, 37—51 (in Polish).

Berbeć, A.K., 2014. Bioróżnorodność i usługi ekosystemowe w rolnictwie. Wieś Jutra 2 (179), 39—42 (in Polish).

Berry, P., Turkelboom, F., Verheyden, W., Martín-López, B., 2016. Ecosystem service bundles. In: Potschin, M.B., Jax, K. (Eds.), OpenNESS Ecosystem Services Reference Book. EC FP7 Grant Agreement No. 308428.

Bieniek, A., 2013. Gleby sandrów wewnętrznych Polski północno-wschodniej. Rozprawy i Monografie, 184. Wydawnictwo Uniwersytetu Warmińsko-Mazurskiego, Olsztyn (in Polish).

Biesiacki, A. (Ed.), 1982. Warunki przyrodnicze produkcji rolnej, woj. suwalskie. Instytut Uprawy, Nawożenia i Gleboznawstwa, Puławy (in Polish).

Billeter, R., Liira, J., Bailey, D., Bugter, R., Arens, P., Augenstein, I., Aviron, S., Baudry, J., Bukacek, R., Burel, F., Cerny, M., Blust, G.D., Cock, R.D., Diekötter, T., Dietz, H., Dirksen, J., Dormann, C., Durka, W., Frenzel, M., Hamersky, R., Hendrickx, F., Herzog, F., Klotz, S., Koolstra, B., Lausch, A., Coeur, D.L., Maelfait, J.P., Opdam, P., Roubalova, M., Schermann, A., Schermann, N., Schmidt, T., Schweiger, O., Smulders, M.J.M., Speelmans, M., Simova, P., Verboom, J., van Wingerden, W.K.R.E., Zobel, M., Edwards, P.J., 2008. Indicators for biodiversity in agricultural landscapes: a pan-European study. Journal of Applied Ecology 45 (1), 141—150.

Bingham, G., Bishop, R., Brody, M., Bromley, D., Clark, E., Cooper, W., Costanza, R., Hale, T., Hayden, G., Kellert, S., Norgaard, R., Norton, B., Payne, J., Russell, C., Suter, G., 1995. Issues in ecosystem valuation: improving information for decision making. Ecological Economics 14 (2), 73—90.

Błaszkiewicz, M., 2010. Kiedy zanikła wieloletnia zmarzlina na młodoglacjalnym obszarze Polski? In: Marks, L., Pochocka-Szwarc, K. (Eds.), Dynamika zaniku lądolodu podczas fazy pomorskiej w północno-wschodniej części Pojezierza Mazurskiego. XVII Konferencja "Stratygrafia Plejstocenu Polski", Państwowy Instytut Geologiczny — PIB, Warszawa, pp. 52—53 (in Polish).

Bobek, H., Schmithüsen, J., 1949. Die Landschaft im logischen System der Geographie. Erdkunde 3 (2), 112—120 (in German).

Boerema, A., Rebelo, A.J., Bodi, M.B., Esler, K.J., Meire, P., 2016. Are ecosystem services adequately quantified? Journal of Applied Ecology 54 (2), 358—370.

Bogacki, M., 1985. Budowa geologiczna i ukształtowanie powierzchni. In: Stasiak, A. (Ed.), Województwo suwalskie. Studia i materiały, vol. 1. Ośrodek Badań Naukowych w Białymstoku, IGiPZ PAN w Warszawie, Białystok, pp. 11—58 (in Polish).

Bohn, U., Gollub, G., Hettwer, C., Neuhäuslová, Z., Raus, T., Schlüter, H., Weber, H., 2000/2003. Map of the Natural Vegetation of Europe. Scale 1:2 500 000. Federal Agency for Nature Conservation, Bonn.

Bonczar, Z., 2004. Bonasa banasia (L. 1758) — jarząbek. In: Gromadzki, M. (Ed.), Ptaki (część I). Poradnik ochrony siedlisk i gatunków Natura 2000 — podręcznik metodyczny, vol. 7. Ministerstwo Środowiska, Warszawa, pp. 268—271 (in Polish).

Boreux, V., Kushalappa, C.G., Vaast, P., Ghazoul, J., 2013. Interactive effects among ecosystem services and management practices on crop production: pollination in coffee agroforestry systems. Proceedings of the National Academy of Sciences of the United States of America 110 (21), 8387—8392.

Boyd, J., Banzhaf, S., 2007. What are ecosystem services? The need for standardized environmental accounting units. Ecological Economics 63 (2—3), 616—626.

Boyd, J., 2007. Nonmarket benefits of nature: what should be counted in green GDP? Ecological Economics 61 (4), 716—723.

Boykin, K.G., Kepner, W.G., Bradford, D.F., Guy, R.K., Kopp, D.A., Leimer, A.K., Samson, E.A., East, N.F., Neale, A.C., Gergely, K.J., 2013. A national approach for mapping and quantifying habitat-based biodiversity metrics across multiple spatial scales. Ecological Indicators 33, 139—147.

Braat, L., ten Brink, P. (Eds.), 2008. The Cost of Policy Inaction: The Case of Not Meeting the 2010 Biodiversity Target. Wageningen, Brussels.

Braat, L.C., Boeraeve, F., Bouwma, I., van Daele, T., Dendoncker, N., Grêt-Regamy, A., Klok, C., Miguel-Ayala, L., Perez-Soba, M., Peterseil, J., Santos-Martin, F., Scholefield, P., Torre-Marin, A., Weibel, B., Weiss, M., 2013. Mapping of Ecosystems and their Services in the EU and its Member States (MESEU). Final Report. Part 1: Introduction, Summary & Conclusions. Alterra, Wageningen.

Braat, L.C., 2013. The value of the ecosystem services concept in economic and biodiversity policy (Chapter 10). In: Jacobs, S., Dendoncker, N., Keune, H. (Eds.), Ecosystem Services, Global Issues, Local Practices. Elsevier, Amsterdam.

Braat, L.C., 2014. Ecosystem services: the ecology and economics of current debates. Economics and Environment 4 (51), 20—35.

Braat, L.C., 2018. Five reasons why the science publication "assessing nature's contributions to people" (Diaz et al. 2018) would not have been accepted in ecosystem services. Ecosystem Services 30, A1—A2.

Brandt, P., Abson, D.J., DellaSala, D.A., Feller, R., von Wehrden, H., 2014. Multifunctionality and biodiversity: ecosystem services in temperate rainforests of the Pacific Northwest, USA. Biological Conservation 169, 362—371.

Braun-Blanquet, J., 1964. Pflanzensoziologie. Grundzüge der Vegetationskunde. Pflanzensoziologie Grundzüge der Vegetationskunde. Springer-Verlag, Wien (in German).

Britton, A.J., Fisher, J.M., 2007. Interactive effects of nitrogen deposition, fire and grazing on diversity and composition of low-alpine prostrate Calluna vulgaris heathland. Journal of Applied Ecology 44 (1), 125—135.

Broda, J. (Ed.), 1965. Dzieje lasów, leśnictwa i drzewnictwa w Polsce. Państwowe Wydawnictwo Rolnicze i Leśne, Warszawa (in Polish).

Brown, G., Fagerholm, N., 2015. Empirical PPGIS/PGIS mapping of ecosystem services: a review and evaluation. Ecosystem Services 13, 119—133.

Brown, C., Reyers, B., Ingwall-King, L., Mapendembe, A., Nel, J., O'Farrell, P., Dixon, M., Bowles-Newark, N.J., 2014. Measuring Ecosystem Services: Guidance on Developing Ecosystem Service Indicators. UNEP-WCMC, Cambridge, UK.

Brown, G., Hausner, V.H., Grodzińska-Jurczak, M., Pietrzyk-Kaszyńska, A., Olszańska, A., Peek, B., Rechciński, M., Lægreid, E., 2015a. Cross-cultural values and management preferences in protected areas of Norway and Poland. Journal for Nature Conservation 28, 89—104.

Brown, G., Hausner, V.H., Lægreid, E., 2015b. Physical landscape associations with mapped ecosystem values with implications for spatial value transfer: an empirical study from Norway. Ecosystem Services 15, 19—34.

Bryan, B.A., Barry, S., Marvanek, S., 2009. Agricultural commodity mapping for land use change assessment and environmental management: an application in the Murray-Darling Basin, Australia. Journal of Land Use Science 4 (3), 131—155.

Bukvareva, E., Zamolodchikov, D., Grunewald, K., 2019. National assessment of ecosystem services in Russia: methodology and main problems. Science of the Total Environment 655, 1181—1196.

Bunevič, A.N., Dackevič, V.A., 1985. Stacialnoe razmeščenie i pitanie enotovidnoj sobaki v Belovežskoj Pušče. Zapovedniki Belorussii 9, 114—120 (in Belarusian).

Burkhard, B., Maes, J. (Eds.), 2017. Mapping Ecosystem Services. Pensoft Publishers, Sofia.

Burkhard, B., Kroll, F., Müller, F., Windhorst, W., 2009. Landscapes' capacities to provide ecosystem services — a concept for land-cover based assessments. Landscape Online 15, 1—22.

Burkhard, B., Kroll, F., Nedkov, S., Müller, F., 2012. Mapping ecosystem service supply, demand and budgets. Ecological Indicators 21, 17—29.

Burkhard, B., Kandziora, M., Hou, Y., Müller, F., 2014. Ecosystem service potentials, flows and demands — concepts for spatial localisation, indication and quantification. Landscape Online 34, 1—32.

Calvet-Mir, L., Gómez-Baggethun, E., Reyes-Garcia, V., 2012. Beyond food production: ecosystem services provided by home gardens. A case study in Vall Fosca, Catalan Pyrenees, Northeastern Spain. Ecological Economics 74, 153—160.

Cardinale, B.J., 2011. Biodiversity improves water quality through niche partitioning. Nature 472, 86—91.

Carignan, V., Villard, M.-A., 2002. Selecting indicator species to monitor ecological integrity: a review. Environmental Monitoring and Assessment 78 (1), 5—61.

Carvalho-Ribeiro, S.M., Lovett, A., 2011. Is an attractive forest also considered well managed? Public preferences for forest cover and stand structure across a rural/urban gradient in northern Portugal. Forest Policy and Economics 13 (1), 46—54.

Casado-Arzuaga, I., Madariaga, I., Onaindia, M., 2013. Perception, demand and user contribution to ecosystem services in the Bilbao metropolitan greenbelt. Journal of Environmental Management 129, 33—43.

Castro, A.J., Martín-López, B., García-Llorente, M., Aguilera, P.A., López, E., Cabello, J., 2011. Social preferences regarding the delivery of ecosystem services in a semiarid Mediterranean region. Journal of Arid Environments 75 (11), 1201—1208.

Catt, J.A., 1988. Quaternary Geology for Scientists and Engineers. Ellis Horwood Limited, Chichester.

Chadwick, B.A., Bahr, H.M., Albrecht, S.L., 1984. Social Science Research Methods. Englewood Cliffs, Prentice-Hall, New Jersey.

Chan, K.M.A., Satterfield, T., Goldstein, J., 2012. Rethinking ecosystem services to better address and navigate cultural values. Ecological Economics 74, 8—18.

Chesson, P., 2000. Mechanisms of maintenance of species diversity. Annual Review of Ecology and Systematics 31, 343—366.

Chmielewski, S., 1988. Charakterystyka fizycznogeograficzna. In: Kostrowicki, A.S. (Ed.), Studium geoekologiczne rejonu jezior wigierskich, Prace Geograficzne, vol. 147. IGiPZ PAN, Warszawa, pp. 13—21 (in Polish).

Choiński, A., 2007. Limnologia fizyczna Polski. Wydawnictwo Naukowe UAM, Poznań (in Polish).

Chudecki, Z., Niedźwiecki, E., 1983. Nasilanie się erozji wodnej na obszarach słabo urzeźbionych Pomorza Zachodniego. Zeszyty Problemowe Postępów Nauk Rolniczych 272, 7—18 (in Polish).

Chudecki, Z., Koćmit, A., Niedźwiecki, E., 1993. Przejawy i skutki erozji wodnej w strefie czołowo-morenowej Wyżyny Ińskiej w świetle wieloletnich badań. In: Kostrzewski, A. (Ed.), Geoekosystem obszarów nizinnych, Komitet Naukowy Przy Prezydium PAN "Człowiek i Środowisko", Zeszyty Naukowe, vol. 6. Wrocław—Warszawa—Kraków, pp. 31—33 (in Polish).

Churska, Z., 1973. Zagrożenie erozją gleb doliny Drwęcy i obszarów sąsiednich w granicach województwa bydgoskiego. Acta Universitatis Nicolai Copernici Geografia 9 (31), 187—231 (in Polish).

Chytrý, M., Jarošík, V., Pyšek, P., Hájek, O., Knollová, I., Tichý, L., Danihelka, J., 2008. Separating habitat invasibility by alien plants from the actual level of invasion. Ecology 89 (6), 1541—1553.

Cieślak, M., Dombrowski, A., 1993. The effect of forest size on breeding bird communities. Acta Ornithologica 27 (2), 97—111.

Costanza, R., D'Arge, R., de Groot, R., Farber, S., Grasso, M., Hannon, B., Limburg, K., Naeem, S., O'Neill, R.V., Paruelo, J., Raskin, G.R., Sutton, P., van den Belt, M., 1997. The value of the world's ecosystem services and natural capital. Nature 387, 253—260.

Costanza, R., 2008. Ecosystem Services: multiple classification system are needed. Biological Conservation 141 (2), 350—352.

Cox, D.R., Lewis, P.A.W., 1966. The Statistical Analysis of Series of Events. Ser. Methuen's Monographs on Applied Probability and Statistics. Methuen, London.

Crane, E., Walker, P., 1985. Important honeydew sources and their honeys. Bee World 66 (3), 105—112.

Crane, E., 1990. Bees and Beekeeping: Science, Practice and World Resources. Heinemann Newnes, Oxford.

Crossman, N.D., Burkhard, B., Nedkov, S., Willemen, L., Petz, K., Palomo, I., Drakou, E.G., Martín-Lopez, B., McPhearson, T., Boyanova, K., Alkemade, R., Egoh, B., Dunbar, M.B., Maes, J., 2013. A blueprint for mapping and modelling ecosystem services. Ecosystem Services 4, 4—14.

Czapiewski, K.Ł., Niewęgłowska, G., Stolbova, M., 2008. Obszary o niekorzystnym gospodarowaniu w rolnictwie. Stan obecny i wnioski na przyszłość. Raport nr 95. Instytut Ekonomiki Rolnictwa i Gospodarki Żywnościowej — Państwowy Instytut Badawczy, Warszawa (in Polish).

Czyżyk, P., Żurkowski, M., Ciepluch, Z., Struziński, T., Czajka, W., 2007. Parametry populacyjne jelenia szlachetnego (Cervus elaphus L.) w Leśnym Kompleksie Promocyjnym "Lasy Mazurskie". Część I. Ocena masy poroża i masy tuszy byków pozyskanych w wyniku odstrzałów selekcyjnych. Sylwan 9, 41—50 (in Polish).

Daily, G.C., Ehrlich, P.R., 1995. Preservation of biodiversity in small rainforest patches: rapid evaluations using butterfly trapping. Biodiversity & Conservation 4 (1), 35—55.

Daily, G.C. (Ed.), 1997. Nature's Services: Societal Dependence on Natural Ecosystems. Island Press, Washington.

Dajdok, Z., Wuczyński, A., 2005. Zróżnicowanie biocenotyczne, funkcje i problemy ochrony drobnych cieków śródpolnych. In: Tomiałojć, L., Drabiński, A. (Eds.), Środowiskowe aspekty gospodarki wodnej. Komitet Ochrony Przyrody PAN, Wydział Inżynierii Kształtowania Środowiska i Geodezji AR we Wrocławiu, Wrocław, pp. 227—252 (in Polish).

Dale, V.H., Beyeler, S.C., 2001. Challenges in the development and use of ecological indicators. Ecological Indicators 1 (1), 3—10.

Daniel, T.C., Muhar, A., Arnberger, A., Aznar, O., Boyd, J.W., Chan, K.M.A., Costanza, R., Elmqvist, T., Flint, C.G., Gobster, P.H., Grêt-Regamey, A., Lave, R., Muhar, S., Penker, M., Ribe, R.G., Schauppenlehner, T., Sikor, T., Soloviy, I., Spierenburg, M., Taczanowska, K., Tam, J., von der Dunk, A., 2012. Contributions of cultural services to the ecosystem services agenda. Proceedings of the National Academy of Sciences of the United States of America 109 (23), 8812—8819.

de Groot, R.S., Wilson, M.A., Boumans, R.M.J., 2002. A typology for the classification, description and valuation of ecosystem functions, goods and services. Ecological Economics 41 (3), 393—408.

de Groot, R.S., Alkemade, R., Braat, L., Hein, L.G., Willemen, L., 2010a. Challenges in integrating the concept of ecosystem services and values in landscape planning, management and decision-making. Ecological Complexity 7 (3), 260—272.

de Groot, R.S., Fisher, B., Christie, M., Aronson, J., Braat, L., Gowdy, J., Haines-Young, R., Maltby, E., Neuville, A., Polasky, S., Portela, R., Ring, I., 2010b. Integrating the ecological and economic dimensions in biodiversity and ecosystem service valuation. In: Kumar, P. (Ed.), The Economics of Ecosystems and Biodiversity: Ecological and Economic Foundations. Earthscan, London and Washington, pp. 1—40.

de Groot, R.S., 1992. Functions of Nature: Evaluation of Nature in Environmental Planning, Management and Decision Making. Wolters-Noordhoff, Amsterdam.

de Groot, R., 2006. Function-analysis and valuation as a tool to assess land use conflicts in planning for sustainable, multi-functional landscapes. Landscape and Urban Planning 75 (3—4), 175—186.

Decocq, G., Andrieu, E., Brunet, J., Chabrerie, O., De Frenne, P., De Smedt, P., Deconchat, M., Diekmann, M., Ehrmann, S., Giffard, B., Gorriz Mifsud, E., Hansen, K., Hermy, M., Kolb, A., Lenoir, J., Liira, J., Moldan, F., Prokofieva, I., Rosenqvist, L., Varela, E., Valdés, A., Verheyen, K., Wulf, M., 2016. Ecosystem services from small forest patches in agricultural landscapes. Current Forestry Reports 2 (1), 30—44.

Degórski, M., Solon, J., 2014. Ecosystem services as a factor strengthening regional development trajectory. Ekonomia i Środowisko 4 (51), 48—57.

Degórski, M., 2002. Przestrzenna zmienność właściwości gleb bielicoziemnych Środkowej i Północnej Europy a geograficzne zróżnicowanie czynników pedogenicznych. Prace Geograficzne, vol. 182. IGiPZ PAN, Warszawa (in Polish).

Degórski, M., 2007. Spatial Variability in Podzolic Soils of Central and Northern Europe. EPA/600/R-07/059. U.S. Environmental Protection Agency, Washington, D.C.

Degórski, M., 2010. Wykorzystanie świadczeń ekosystemów w rozwoju regionów. Ekonomia i Środowisko 1 (37), 85—97 (in Polish).

Dembner, S.A., Perlis, A. (Eds.), 1999. Non-wood Forest Products and Income Generation. FAO Forestry Department, Rome. Unasylva 198 (50).

Demianowicz, Z., Hłyń, M., Jabłoński, B., Maksymiuk, I., Podgórska, J., Ruszkowska, B., Szklanowska, K., Zimna, J., 1960. Wydajność miodowa ważniejszych roślin miododajnych w warunkach Polski. Część I. Pszczelnicze Zeszyty Naukowe 4 (2), 87—104 (in Polish).

Denisow, B., Wrzesień, M., 2007. The anthropogenic refuge areas for bee flora in agricultural landscape. Acta Agrobotanica 60 (1), 147—157.

Denisow, B., Wrzesień, M., 2015. The importance of field-margin location for maintenance of food niches for pollinators. Journal of Apicultural Science 59 (1), 27—37.

Denisow, B., 2011. Pollen production of selected ruderal plant species in the Lublin area. Rozprawy Naukowe, vol. 35. Wydawnictwo Uniwersytetu Przyrodniczego, Lublin.

DeWalt, S.J., Denslow, J.S., Ickes, K., 2004. Natural-enemy release facilitates habitat expansion of the invasive tropical shrub *Clidemia hirta*. Ecology 85 (2), 471—483.

Di Castri, F., 1990. On invading species and invaded ecosystems: the interplay of historical chance and biological necessity. In: Di Castri, F., Hansen, A.J., Debussche, M. (Eds.), Biological Invasions in Europe and Mediterranean Basin, vol. 65. Kluwer, Dordrecht, pp. 3—16.

Díaz, S., Purvis, A., Cornelissen, J.H.C., Mace, G.M., Donoghue, M.J., Ewers, R.M., Jordano, P., Pearse, W.D., 2013. Functional traits, the phylogeny of function, and ecosystem service vulnerability. Ecology and Evolution 3 (9), 2958–2975.

Díaz, S., Pascual, U., Stenseke, M., Martín-López, B., Watson, R.T., Molnár, Z., Hill, R., Chan, K.M.A., Baste, I.A., Brauman, K.A., Polasky, S., Church, A., Lonsdale, M., Larigauderie, A., Leadley, P.W., van Oudenhoven, A.P.E., van der Plaat, F., Schröter, M., Lavorel, S., Aumeeruddy-Thomas, Y., Bukvareva, E., Davies, K., Demissew, S., Erpul, G., Failler, P., Guerra, C.A., Hewitt, C.L., Keune, H., Lindley, S., Shirayama, Y., 2018. Assessing nature's contributions to people. Science 359 (6373), 270–272.

Dick, J., Maes, J., Smith, R.I., Paracchini, M.L., Zulian, G., 2014. Cross-scale analysis of ecosystem services identified and assessed at local and European level. Ecological Indicators 38, 20–30.

Diekötter, T., Billeter, R., Crist, T.O., 2008. Effects of landscape connectivity on the spatial distribution of insect diversity in agricultural mosaic landscapes. Basic and Applied Ecology 9 (3), 298–307.

Diercke International Atlas, 2010. Last Ice Age (Map). Westermann Schulbuch, Braunschweig, p. 26.

Dolek, M., Geyer, A., 2002. Conserving biodiversity on calcareous grasslands in the Franconian Jura by grazing: a comprehensive approach. Biological Conservation 104 (3), 351–360.

Drzewiecki, W., Mularz, S., 2005. Model USPED jako narzędzie prognozowania efektów erozji i depozycji materiału glebowego. Roczniki Geomatyki 3 (2), 45–54 (in Polish).

Duncker, P.S., Raulund-Rasmussen, K., Gundersen, P., Katzensteiner, K., Jong, J.D., Ravn, H.P., Smith, M., Eckmüllner, O., Spiecker, H., 2012. How forest management affects ecosystem services, including timber production and economic return: synergies and trade-offs. Ecology and Society 17 (4).

Dylikowa, A., 1973. Geografia Polski. Krainy geograficzne. Państwowe Zakłady Wydawnictw Szkolnych, Warszawa (in Polish).

Dziadowiec, H., Gonet, S. (Eds.), 1999. Przewodnik metodyczny do badań materii organicznej gleb. Prace Komisji Naukowych Polskiego Towarzystwa Gleboznawczego, vol. 120. Warszawa (in Polish).

Dzięciołowski, R., 1970. Variation in red deer (Cervus elaphus L.) food selection in relation to environment. Ekologia Polska 18, 635–645.

Dzwonko, Z., Loster, S., 2001. Wskaźnikowe gatunki roślin starych lasów i ich znaczenie dla ochrony przyrody i kartografii roślinności. In: Roo-Zielińska, E., Solon, J. (Eds.), Typologia zbiorowisk i kartografia roślinności w Polsce — rozważania nad stanem współczesnym, Prace Geograficzne, vol. 178. IGiPZ PAN, Warszawa, pp. 119–132 (in Polish).

Edwards, D., Jay, M., Jensen, F.S., Lucas, B., Marzano, M., Montagné, C., Peace, A., Weiss, G., 2012. Public preferences for structural attributes of forests: towards a pan-European perspective. Forest Policy and Economics 19, 12–19.

EEA, 1999. Environmental Indicators: Typology and Overview. Technical report No 25/1999. European Environment Agency, Copenhagen.

EEA, 2002. Europe's Biodiversity — Biogeographical Regions and Seas. EEA Report No 1/2002. https://www.eea.europa.eu/publications/report_2002_0524_154909.

EEA, 2012. Corine Land Cover 2012 (Vector Data). European Environment Agency. https://www.eea.europa.eu/data-and-maps/data/clc-2012-vector.

EEA, 2013. Air Quality in Europe — 2013. report. No 9/2013. European Environment Agency, Copenhagen.

Egoh, B., Reyers, B., Rouget, M., Bode, M., Richardson, D.M., 2009. Spatial congruence between biodiversity and ecosystem services in South Africa. Biological Conservation 142 (3), 553–562.

Egoh, B., Drakou, E.G., Dunbar, M.B., Maes, J., Willemen, L., 2012. Indicators for Mapping Ecosystem Services: A Review. JRC Scientific and Policy Reports. European Commission Joint Research Centre, Institute for Environment and Sustainability, Publications Office of the European Union, Luxembourg.

Ehrlich, P.R., Ehrlich, A., 1970. Population, Resources, Environment: Issues in Human Ecology. W.H. Freeman, San Francisco.

Ehrlich, P.R., Ehrlich, A.H., 1981. Extinction: The Causes and Consequences of the Disappearance of Species. Random House, New York.

Ehrlich, P.R., Mooney, H.A., 1983. Extinction, substitution, and ecosystem services. BioScience 33 (4), 248—254.

Elliot, C.C.H., 1988. The assessment of on-farm losses due to birds and rodents in eastern Africa. Insect Science and its Application 9 (6), 717—720.

Elmqvist, T.E., Maltby, E., Barker, T., Mortimer, M., Perrings, C., Aronson, J., de Groot, R., Fitter, A., Mace, G., Norberg, J., Sousa Pinto, I., Ring, I., Jax, K., Grimm, V., Leemans, R., Salles, J.M., 2010. Biodiversity, ecosystems and ecosystem services. In: Kumar, P. (Ed.), The Economics of Ecosystems and Biodiversity: Ecological and Economic Foundations. Taylor and Francis, Earthscan, pp. 41—112.

Elton, C.S., 1958. The Ecology of Invasions by Animals and Plants. Methuen & Co., London.

Ernst & Ernst, 1972—1978. Social Responsibility Disclosure: Survey of Fortune 500 Annual Reports. Ernst & Ernst, Cleveland.

Faith, D.S., Walker, P.A., 1996. How do indicator groups provide information about the relative biodiversity of different sets of areas? On hotspots, complementarity, and pattern-based approaches. Biodiversity Letters 3 (1), 18—25.

Falencka-Jabłońska, M., 2012. Walory przyrodnicze polskich lasów i ich uzdrowiskowo-turystyczne wykorzystanie. Inżynieria Ekologiczna 30, 60—69 (in Polish).

Falińska, K., 2004. Ekologia roślin. Wydawnictwo Naukowe PWN, Warszawa (in Polish).

Faliński, J.B., 2004. Inwazje w świecie roślin — mechanizmy, zagrożenia, projekt badawczy. Phytocoenosis, 16 (N.S.). Seminarium Geobotanicum 10, 5—31 (in Polish).

Feld, C.K., da Silva, P.M., Sousa, J.P., De Bello, F., Bugter, R., Grandin, U., Hering, D., Lavorel, S., Mountford, O., Pardo, I., Pärtel, M., Römbke, J., Sandin, L., Jones, B., Harrison, P., 2009. Indicators of biodiversity and ecosystem services: a synthesis across ecosystems and spatial scales. Oikos 118 (12), 1862—1871.

Felipe-Lucia, M.R., Comín, F.A., Escalera-Reyes, J., 2015. A framework for the social valuation of ecosystem services. Ambio 44 (4), 308—318.

Filipiak, K., 2003. Ocena wykorzystania rolniczej przestrzeni produkcyjnej w Polsce w ujęciu regionalnym. Pamiętnik Puławski 132, 73—79 (in Polish).

Fisher, B., Costanza, R., Turner, R.K., Morling, P., 2008. Defining and Classifying Ecosystem Services for Decision Making. CSERGE Working Paper EDM 07-04. Centre for Social and Economic Research on the Global Environment, Norwich.

Fisher, B., Turner, R.K., Morling, P., 2009. Defining and classifying ecosystem services for decision making. Ecological Economics 68 (3), 643—653.

Fleishman, E., Thomson, J.R., Mac Nally, R., Murphy, D.D., Fay, J.P., 2005. Using indicator species to predict species richness of multiple taxonomic groups. Conservation Biology 19 (4), 1125—1137.

Foley, J.A., DeFries, R., Asner, G.P., Barford, C., Bonan, G., Carpenter, S.R., Chapin, F.S., Coe, M.T., Daily, G.C., Gibbs, H.K., Helkowski, J.H., Holloway, T., Howard, E.A., Kucharik, C.J., Monfreda, C., Patz, J.A., Prentice, I.C., Ramankutty, N., Snyder, P.K., 2005. Global consequences of land use. Science 309, 570—574.

Food and Agriculture Organization of the United Nations, 2016. FAOSTAT. Livestock Primary. http://faostat3.fao.org/download/Q/QL/E.

Fox, M., Fox, B., 1986. The susceptibility of natural communities to invasion. In: Groves, R.H., Burdon, J.J. (Eds.), Ecology of Biological Invasions: An Australian Perspective. Australian Academy of Science, Canberra, pp. 57—66.

Freeman, R.E., 2010. Strategic Management: A Stakeholder Approach. Cambridge University Press, New York.

Frélichová, J., Vačkář, D., Pártl, A., Loučková, B., Harmáčková, Z.V., Lorencová, E., 2014. Integrated assessment of ecosystem services in the Czech Republic. Ecosystem Services 8, 110—117.

Friedrich, M., Kromer, B., Spurk, M., Hofmann, J., Kaiser, K.F., 1999. Paleo-environment and radiocarbon calibration as derived from Lateglacial/Early Holocene tree-ring chronologies. Quaternary International 61 (1), 27—39.

García-Nieto, A.P., García-Llorente, M., Iniesta-Arandia, I., Martín-López, B., 2013. Mapping forest ecosystem services: from providing units to beneficiaries. Ecosystem Services 4, 126—138.

García-Nieto, A.P., Quintas-Soriano, C., García-Llorente, M., Palomo, I., Montes, C., Martín-López, B., 2015. Collaborative mapping of ecosystem services: the role of stakeholders' profiles. Ecosystem Services 13, 141—152.

Gawrysiak, L., Łopatka, A., Stuczyński, T., 2004. Numeryczna mapa retencji wodnej gleb w skali 1:25000. Zintegrowany System Informacji o Rolniczej Przestrzeni Produkcyjnej Województwa Podlaskiego dla Potrzeb Ochrony Gruntów. IUNiG, Puławy.

Gerlach, T., 1976. Bombardująca działalność kropel deszczu i jej znaczenie w przemieszczaniu gleby na stokach. Studia Geomorphologica Carpatho-Balcanica 10, 125—137 (in Polish).

Gerula, D., Węgrzynowicz, P., Semkiw, P., 2007. Analiza sektora pszczelarskiego w Polsce dla opracowania 3-letniego Programu Wsparcia Pszczelarstwa w Polsce w latach 2007—2010. Puławy (in Polish).

Giese, L.A.B., Aust, W.M., Kolka, R.K., Trettin, C.C., 2003. Biomass and carbon pools of disturbed riparian forests. Forest Ecology and Management 180, 493—508.

Głowaciński, Z. (Ed.), 2001. Polska czerwona księga zwierząt. Kręgowce. Państwowe Wydawnictwo Rolnicze i Leśne, Warszawa (in Polish).

Goma, H.C., Rahim, K., Nangendo, G., Riley, J., Stein, A., 2001. Participatory studies for agro-ecosystem evaluation. Agricultural, Ecosystems & Environment 87 (2), 179—190.

Gómez-Baggethun, E., de Groot, R., Lomas, P.L., Montes, C., 2010. The history of ecosystem services in economic theory and practice: from early notions to markets and payment schemes. Ecological Economics 69 (6), 1209—1218.

Górniak, A., 2000. Klimat województwa podlaskiego. IMGW Oddział w Białymstoku, Białystok (in Polish).

Gould, W.A., Walker, M.D., 1999. Plant communities and landscape diversity along a Canadian Arctic river. Journal of Vegetation Science 10 (4), 537—548.

Gould, R.K., Ardoin, N.M., Woodside, U., Satterfield, T., Hannahs, N., Daily, G.C., 2014. The forest has a story: cultural ecosystem services in Kona, Hawai'i. Ecology and Society 19 (3), 55.

Grabińska, B., 2011. Uwarunkowania naturalne i antropogeniczne rozmieszczenia ssaków łownych w Polsce. Prace Geograficzne, vol. 228. IGiPZ PAN, Warszawa (in Polish).

Grau, J., Jung, R., Münker, B., 1983. Beeren, Wildgemüse, Heilkräuter. Mosaik Verlag, München (in German).

Green, I.-M., Folke, C., Turner, R.K., Bateman, I., 1994. Primary and secondary values of wetland ecosystems. Environmental and Resource Economics 4 (1), 55—74.

Greenleaf, S.S., Williams, N.M., Winfree, R., Kremen, C., 2007. Bee foraging ranges and their relationship to body size. Oecologia 153 (3), 589—596.

Grêt-Regamey, A., Weibel, B., Bagstad, K.J., Ferrari, M., Geneletti, D., Klug, H., Schirpke, U., Tappeiner, U., 2014. On the effects of scale for ecosystem services mapping. PLoS One 9 (12), e112601.

Grilli, G., Jonkisz, J., Ciolli, M., Lesinski, J., 2016. Mixed forests and ecosystem services: investigating stakeholders' perceptions in a case study in the Polish Carpathians. Forest Policy and Economics 66, 11–17.

Grime, J.P., 1979. Plant Strategies and Vegetation Processes. John Wiley & Sons, Chichester–New York–Brisbane–Toronto.

Grzyb, S., Prończuk, J., 1995. Podział i waloryzacja siedlisk łąkowych oraz ocena ich potencjału produkcyjnego. In: Ogólnopolska Konferencja Łąkarstwa nt. "Kierunki rozwoju łąkarstwa na tle aktualnego poziomu wiedzy w najważniejszych jego działach", Warszawa, 27–28 września 1994. SGGW, Warszawa, pp. 51–63 (in Polish).

Haase, G., 1976. Zur Bestimmung und Erkundung von Naturraumpotentialen. Geographische Gesellschaft der DDR, Mitteilungsblatt 13, 5–8 (in German).

Haase, G., 1978. Zur Ableitung und Kennzeichnung von Naturraumpotentialen. Petermanns Geographische Mitteilungen, Jg. 122 (2), 113–125 (in German).

Haines-Young, R., Potschin, M., 2010a. Proposal for a Common International Classification of Ecosystem Goods and Services (CICES) for Integrated Environmental and Economic Accounting (V1) 21st March 2010. Report to the European Environment Agency, Contract No: EEA/BSS/07/007. University of Nottingham, Nottingham.

Haines-Young, R., Potschin, M., 2010b. The links between biodiversity, ecosystem services and human well-being. In: Raffaelli, D., Frid, C. (Eds.), Ecosystem Ecology: A New Synthesis, BES Ecological Reviews Series. CUP, Cambridge, pp. 110–139.

Haines-Young, R., Potschin, M., 2013. CICES V4.3 – Revised Report Prepared Following Consultation on CICES Version 4, August–December 2012. EEA Framework Contract No EEA/IEA/09/003. University of Nottingham, Nottingham.

Haines-Young, R.H., Potschin, M.B., 2018. Common International Classification of Ecosystem Services (CICES) V5.1. And Guidance on the Application of the Revised Structure (Nottingham).

Haines-Young, R., Potschin, M., Kienast, F., 2012. Indicators of ecosystem service potential at European scales: mapping marginal changes and trade-offs. Ecological Indicators 21, 39–53.

Hammer, Ø., Harper, D.A.T., Ryan, P.D., 2001. PAST: paleontological statistics software package for education and data analysis. Palaeontologia Electronica 4 (1), 9 pp.

Haragsim, O., 1966. Medovice a vcely. Statni zemedelske nakl, Praha (in Czech).

Harrison, P.A., Berry, P.M., Simpson, G., Haslett, J.R., Blicharska, M., Bucur, M., Dunford, R., Egoh, B., Garcia-Llorente, M., Geamănă, N., Geertsema, W., Lommelen, E., Meiresonne, L., Turkelboom, F., 2014. Linkages between biodiversity attributes and ecosystem services: a systematic review. Ecosystem Services 9, 191–203.

Hassan, R., Scholes, R., Ash, N. (Eds.), 2005. Ecosystems and Human Well-Being: Current State and Trends. Findings of the Condition and Trends Working Group, vol. 1. Island Press, Washington, Covelo, London.

Hauck, J., Görg, C., Varjopuro, R., Ratamaki, O., Maes, J., Wittmer, H., Jax, K., 2013. Maps have an air of authority: potential benefits and challenges of ecosystem service maps at different levels of decision making. Ecosystem Services 4, 25–32.

Hefting, M.M., Clement, J.-C., Bienkowski, P., Dowrick, D., Guenat, C., Butturini, A., Topa, S., Pinay, G., Verhoeven, J.T.A., 2005. The role of vegetation and litter in the nitrogen dynamics of riparian buffer zones in Europe. Ecological Engineering 24 (5), 465–482.

Hein, L., van Koppen, K., de Groot, R.S., van Ierland, E.C., 2006. Spatial scales, stake-holders and the valuation of ecosystem services. Ecological Economics 57 (2), 209—228.

Heink, U., Hauck, J., Jax, K., Sukopp, U., 2016. Requirements for the selection of ecosystem service indicators — the case of MAES indicators. Ecological Indicators 61 (Part 1), 18—26.

Helfenstein, J., Kienast, F., 2014. Ecosystem service state and trends at the regional to national level: a rapid assessment. Ecological Indicators 36, 11—18.

Helliwell, D.R., 1969. Valuation of wildlife resources. Regional Studies 3 (1), 41—47.

Hermy, M., Honnay, O., Firbank, L., Grashof-Bokdam, C., Lawesson, J.E., 1999. An ecological comparison between ancient and other forest plant species of Europe, and the implications for forest conservation. Biological Conservation 91 (1), 9—22.

Heroldová, M., Tkadlec, E., 2011. Harvesting behaviour of three central European rodents: identifying the rodent pest in cereals. Crop Protection 30 (1), 82—84.

Holt, A.R., Mears, M., Maltby, L., Warren, P., 2015. Understanding spatial patterns in the production of multiple urban ecosystem services. Ecosystem Services 16, 33—46.

Hood, W.G., Naiman, R.J., 2000. Vulnerability of riparian zones to invasion by exotic vascular plants. Plant Ecology 148 (1), 105—114.

Hopf, H.S., Morley, G.E.J., Humphries, J.R.O., 1976. Rodent Damage to Growing Crops and to Farm and Village Storage in Tropical and Subtropical Regions — Results of a Postal Survey 1972—73. Centre for Overseas Pest Research and Tropical Production Institute, London.

Hueting, R., 1970. Moet de Natuur worden gekwantificeerd? Economisch Statistische Berichten 2730, 80—84 (in Dutch).

Isbell, F., Calcagno, V., Hector, A., Connolly, J., Harpole, W.S., Reich, P.B., Scherer-Lorenzen, M., Schmid, B., Tilman, D., van Ruijven, J., Weigelt, A., Wilsey, B.J., Zavaleta, E.S., Loreau, M., 2011. High plant diversity is needed to maintain ecosystem services. Nature 477, 199—202.

ISCU-UNESCO-UNO, 2008. Ecosystem Change and Human Well-Being: Research and Monitoring Priorities Based on the Millennium Ecosystem Assessment. International Council for Science, Paris.

IUSS Working Group WRB, 2015. World reference base for soil resources 2014, update 2015. International Soil Classification System for Naming Soils and Creating Legends for Soil Maps. World Soil Resources Report, vol. 106. FAO, Rome.

Jackson, W.B., 1977. Evaluation of rodent depredations to crops and stored products. EPPO Bulletin 7 (2), 439—458.

Jacobs, S., Spanhove, T., de Smet, L., van Daele, T., van Reeth, W., van Gossum, P., Stevens, M., Schneiders, A., Panis, J., Demolder, H., Michels, H., Thoonen, M., Simoens, I., Peymen, J., 2016. The ecosystem service assessment challenge: reflections from Flanders-REA. Ecological Indicators 61 (Part 2), 715—727.

Jagodziński, A.M., 2011. Wyniki — Retencja węgla w biomasie drzew i drzewostanów. In: Raport końcowy z realizacji tematu badawczego "Bilans węgla w biomasie drzew głównych gatunków lasotwórczych Polski". Część III B. Kórnik (in Polish).

Jamrozy, G., 2008. Ocena występowania i tendencji zmian liczebności dużych i średnich ssaków w polskich parkach narodowych. Sylwan 152 (2), 36—44 (in Polish).

Jańczak, J. (Ed.), 1999. Atlas jezior Polski. Tom III. Jeziora Pojezierza Mazurskiego i Polski południowej. Bogucki Wydawnictwo Naukowe, Poznań (in Polish).

Jarić, S., Mačukanović-Jocić, M., Mitrović, M., Pavlović, P., 2013. The melliferous potential of forest and meadow plant communities on Mount Tara (Serbia). Environmental Entomology 42 (4), 724—732.

Jędrzejewska, B., Jędrzejewski, W., 2001. Ekologia Zwierząt Drapieżnych Puszczy Białowieskiej. Wydawnictwo Naukowe PWN, Warszawa (in Polish).

Jędrzejewski, W., Jędrzejewska, B., 1993. Predation on rodents in Białowieża primeval forest, Poland. Ecography 16 (1), 47—64.

Jędrzejewski, W., Jędrzejewska, B., Brzeziński, M., 1993a. Winter habitat selection and feeding habits of polecats (*Mustela putorius*) in the Białowieża national park, Poland. Zeitschrift für Säugetierkunde 58 (2), 75—83.

Jędrzejewski, W., Zalewski, A., Jędrzejewska, B., 1993b. Foraging by pine marten Martes martes in relation to food resources in Białowieża national park. Acta Theriologica 38 (4), 405—426.

Jelonek, T., Tomczak, A., 2011. Gęstość drewna. In: Raport końcowy z realizacji tematu badawczego "Bilans węgla w biomasie drzew głównych gatunków lasotwórczych Polski". Część II. Przegląd literatury, Poznań, pp. 31—88 (in Polish).

Kalinowska, K., 1961. Zanikanie jezior polodowcowych w Polsce. Przegląd Geograficzny 33 (3), 511—518 (in Polish).

Kamieniarz, R., Panek, M., 2008. Zwierzęta łowne w Polsce na przełomie XX i XXI wieku. Stacja Badawcza — OHZ PZŁ w Czempiniu (in Polish).

Kandziora, M., Burkhard, B., Müller, F., 2013. Mapping provisioning ecosystem services at the local scale using data of varying spatial and temporal resolution. Ecosystem Services 4, 47—59.

Karg, J., 1989. Zróżnicowanie liczebności i biomasy owadów latających krajobrazu rolniczego zachodniej Wielkopolski. Roczniki Akademii Rolniczej w Poznaniu, Rozprawy Naukowe 188, 1—78 (in Polish).

Karg, J., 2004. Importance of midfield shelterbelts for over-wintering entomofauna (Turew area, West Poland). Polish Journal of Ecology 52 (4), 421—431.

Keane, R.M., Crawley, M.J., 2002. Exotic plant invasions and the enemy release hypothesis. Trends in Ecology & Evolution 17 (4), 164—170.

Kienast, F., Bolliger, J., Potschin, M., de Groot, R., Verburg, P.H., Heller, I., Wascher, D., Haines-Young, R., 2009. Assessing landscape functions with broad-scale environmental data: insights gained from a prototype development for Europe. Environmental Management 44 (6), 1099—1120.

Kikulski, J., 2009. Turystyczno-rekreacyjne funkcje lasów w Polsce — obraz społecznych potrzeb w zakresie przepływu informacji. Sylwan 153 (1), 62—72 (in Polish).

Kikulski, J., 2011. Prowadzenie gospodarki leśnej a rekreacyjne użytkowanie lasu. Sylwan 155 (4), 269—278 (in Polish).

King, R.T., 1966. Wildlife and man. New York Conservationist 20 (6), 8—11.

Kistowski, M., 1996. Metoda oceny potencjału krajobrazu obszarów młodoglacjalnych. Przegląd Geograficzny 68 (3—4), 367—386 (in Polish).

Kleiber, H.P., Knies, J., Niessen, F., 2000. The Late Weichselian glaciation of the Franz Victoria Trough, northern Barents Sea: ice sheet extent and timing. Marine Geology 168 (1—4), 25—44.

Koćmit, A., Podlasiński, M., Roy, M., Tomaszewicz, T., Chudecka, J., 2006. Water erosion in the catchment basin of the Jeleni Brook. Journal of Water and Land Development 10, 121—131.

Kolejka, J., 2001. Krajinné plánování a využití GIS. In: Létal, A., Szczyrba, Z., Vysoudil, M. (Eds.), Česká geografie v období rozvoje informačních technologií. Univerzita Palackého, Olomouc, pp. 79—90 (in Czech).

Kołtowski, Z., 2006. Wielki Atlas Roślin Miododajnych. Przedsiębiorstwo Wydawnicze Rzeczpospolita SA, Warszawa (in Polish).

Komosińska, H., Podsiadło, E., 2002. Ssaki kopytne. Wydawnictwo Naukowe PWN, Warszawa (in Polish).

Kondracki, J., 1998. Uwagi o ewolucji morfologicznej Pojezierza Mazurskiego. Prace i Studia Geograficzne 24, 57—95 (in Polish).

Konecka-Betley, K., Czępińska-Kamińska, D., Janowska, E., 1999. Systematyka i karto-grafia gleb. Wydawnictwo SGGW, Warszawa (in Polish).

Konrad, R., Wäckers, F.L., Romeis, J., Babendreier, D., 2009. Honeydew feeding in the solitary bee *Osmia bicornis* as affected by aphid species and nectar availability. Journal of Insect Physiology 55 (12), 1158—1166.

Konvicka, M., Fric, Z., Benes, J., 2006. Butterfly extinctions in European states: do so-cioeconomic conditions matter more than physical geography? Global Ecology and Biogeography 15 (1), 82—92.

Kornaś, J., 1990. Plant invasions in Central Europe: historical and ecological aspects. In: di Castri, F., Hansen, A.J., Debussche, M. (Eds.), Biological Invasions in Europe and the Mediterranean Basin, vol. 65. Kluwer, Dordrecht, pp. 19—36.

Koschke, L., Fürst, C., Lorenz, M., Witt, A., Frank, S., Makeschin, F., 2013. The inte-gration of crop rotation and tillage practices in the assessment of ecosystem services provision at the regional scale. Ecological Indicators 32, 157—171.

Kostrowicki, A.S. (Ed.), 1988. Studium geoekologiczne rejonu jezior wigierskich. Prace Geograficzne, vol. 147. IGiPZ PAN, Warszawa (in Polish).

Kostrowicki, A.S., 1999. Świat zwierzęcy. In: Starkel, L. (Ed.), Geografia Polski. Środowisko przyrodnicze. Wydawnictwo Naukowe PWN, Warszawa, pp. 475—493 (in Polish).

Kostrzewski, A., Zwoliński, Z., Andrzejewski, L., Florek, W., Mazurek, M., Niewiarowski, W., Podgórski, Z., Rachlewicz, G., Smolska, E., Stach, A., Szmańda, J., Szpikowski, J., 2008. Współczesny morfosystem strefy młodoglacjalnej. Landform Analysis 7, 7—11 (in Polish).

Kowalkowski, A., 1986. Evolution of holocene soils in Poland. Quaestiones Geographicae 11/12, 93—120.

Kowalska, A., Affek, A., Solon, J., Degórski, M., Grabińska, B., Kołaczkowska, E., Kruczkowska, B., Regulska, E., Roo-Zielińska, E., Wolski, J., Zawiska, I., 2017. Po-tential of cultural ecosystem services in postglacial landscape from the beneficiaries' perspective. Ekonomia i Środowisko 60 (1), 236—245.

Kozarski, S., 1981. Stratygrafia i chronologia Vistulianu Niziny Wielkopolskiej. Ser. Geo-grafia, 6. PAN Oddział Poznański, Państwowe Wydawnictwo Naukowe, Warszawa—Poznań (in Polish).

Kozarski, S., 1986. Skale czasowe a rytm zdarzeń geomorfologicznych Vistulianu na Niżu Polskim. Czasopismo Geograficzne 57 (2), 247—270 (in Polish).

Kożuchowski, K., Wibig, J., 1988. Kontynentalizm pluwialny w Polsce: zróżnicowanie geograficzne i zmiany wieloletnie. Acta Geographica Lodziensia 55. Łódzkie Towa-rzystwo Naukowe, Łódź (in Polish).

Kremen, C., Williams, N.M., Aizen, M.A., Gemmill-Herren, B., LeBuhn, G., Minckley, R., Packer, L., Potts, S.G., Roulston, T., Steffan-Dewenter, I., Vázquez, D.P., Winfree, R., Adams, L., Crone, E.E., Greenleaf, S.S., Keitt, T.H., Klein, A.-M., Regetz, J., Ricketts, T.H., 2007. Pollination and other ecosystem services produced by mobile organisms: a conceptual framework for the effects of land-use change. Ecology Letters 10 (4), 299—314.

Kremen, C., 2005. Managing ecosystem services: what do we need to know about their ecology? Ecology Letters 8 (5), 468—479.

Kruczkowska, B., Solon, J., Wolski, J., 2017. Mapping ecosystem services — a new regional-scale approach. Geographia Polonica 90 (4), 503—520.

Krzymowska-Kostrowicka, A., 1997. Geoekologia turystyki i wypoczynku. Wydawnictwo Naukowe PWN, Warszawa (in Polish).

Krzysztofiak, A., 2001. Struktura zgrupowań pszczół (Apoidea, Hymenoptera) w różnowiekowych drzewostanach świerkowo-sosnowych Wigierskiego Parku

Narodowego. Zeszyty Naukowe Akademii Bydgoskiej im. Kazimierza Wielkiego w Bydgoszczy. Studia Przyrodnicze 15, 113—215 (in Polish).

Krzysztofiak, L., 2012. Ochrona przyrody a użytkowanie jezior na obszarze LSROR "Pojezierze Suwalsko-Augustowskie". LGR Pojezierze Suwalsko-Augustowskie, Suwałki (in Polish).

Krzywicki, T., Pochocka-Szwarc, K., 2014. Geologiczno-środowiskowe warunki utworzenia Geoparku Kanał Augustowski — Augustowskie Sandry. In: Pochocka-Szwarc, K. (Ed.), Dynamika lądolodów plejstoceńskich na obszarze Sokólszczyzny i Równiny Augustowskiej. XXI Konferencja "Stratygrafia Plejstocenu Polski". Państwowy Instytut Geologiczny — PIB, Warszawa, pp. 60—67 (in Polish).

Krzywosz, T., Kamiński, M., 2012. Wpływ kormorana na populacje ryb w zbiornikach wodnych na obszarze LSROR "Pojezierze Suwalsko-Augustowskie". LGR Pojezierze Suwalsko-Augustowskie, Suwałki (in Polish).

Krzywosz, T., Traczuk, P., 2011. Impact of the cormorant and angling catches on ichthyological biodiversity of lakes. In: Jankun, M., Furgała-Selezniow, G., Woźniak, M., Wiśniewska, A.M. (Eds.), Water Biodiversity Assessment and Protection. Faculty of Environmental Protection and Fisheries, University of Warmia and Mazury, Olsztyn, pp. 155—164.

Kuc, M., Piszczek, M., Janusz, A., 2014. Wielkość i wartość skupu oraz eksportu grzybów i owoców leśnych w latach 2007—2011 oraz ich znaczenie dla społeczeństwa i gospodarki. Studia i Materiały CEPL w Rogowie 16 (38/1), 143—152 (in Polish).

Kulczyk, S., Derek, M., Woźniak, E., 2016. How much is the "wonder of nature" worth? The valuation of tourism in the Great Masurian Lakes using travel cost method. Ekonomia i Środowisko 4 (59), 235—249.

Kunáková, L., 2016. Assessing of landscape potential for water management regarding its surface water (using the example of the micro-region Minčol). Ekológia 35 (2), 148—159.

Laflen, J.M., Moldehauer, W.C., 2003. Pioneering Soil Erosion Prediction. The USLE Story, vol. 1. World Association of Soil and Water Conservation, Special Publication, Beijing, China.

Lake, J.C., Leishman, M.R., 2004. Invasion success of exotic plants in natural ecosystems: the role of disturbance, plant attributes and freedom from herbivores. Biological Conservation 117 (2), 215—226.

Lal, R., 2004. Soil carbon sequestration impacts on global climate change and food security. Science 304 (5677), 1623—1627.

Lal, R., 2005. Forest soils and carbon sequestration. Forest Ecology and Management 220 (1—3), 242—258.

Lal, R., 2008. Carbon sequestration. Philosophical Transactions of the Royal Society of London B 363 (1492), 815—830.

Lamarque, P., Tappeiner, U., Turner, C., Steinbacher, M., Bardgett, R.D., Szukics, U., Schermer, M., Lavorel, S., 2011. Stakeholder perceptions of grassland ecosystem services in relation to knowledge on soil fertility and biodiversity. Regional Environmental Change 11 (4), 791—804.

Landers, D.H., Nahlik, A.M., 2013. Final Ecosystem Goods and Services Classification System (FEGS-CS). EPA/600/R-13/ORD-004914. U.S. Environmental Protection Agency, Office of Research and Development, Washington, D.C.

Landvik, J.Y., Bondevik, S., Elverhøi, A., Fjeldskaar, W., Mangerud, J., Salvigsen, O., Siegert, M.J., Svendsen, J.I., Vorren, T.O., 1998. The last glacial maximum of Svalbard and the Barents Sea area: ice sheet extent and configuration. Quaternary Science Reviews 17 (1—3), 43—75.

Larsen, L.I., Jensen, J.N., 2000. An Overview of Selected International and National Plans for the Conservation and Management of Marine Biological Diversity. A Report for the

Danish National Forest and Nature Agency. International Council for the Exploration of the Sea, Copenhagen.

Lavorel, S., Grigulis, K., Lamarque, P., Colace, M.-P., Garden, D., Girel, J., Pellet, G., Douzet, R., 2011. Using plant functional traits to understand the landscape distribution of multiple ecosystem services. Journal of Ecology 99 (1), 135—147.

Lawton, J.H., Gaston, K.J., 2001. Indicator species. In: Encyclopedia of Biodiversity, vol. 3. Academic Press, pp. 437—450.

Lee, H., Lautenbach, S., 2016. A quantitative review of relationships between ecosystem services. Ecological Indicators 66, 340—351.

Legendre, P., Legendre, L., 1998. Numerical Ecology. Developments in Environmental Modelling, vol. 24. Elsevier Science, Amsterdam.

Leitão, A.B., Ahern, J., 2002. Applying landscape ecological concepts and metrics in sustainable landscape planning. Landscape and Urban Planning 59 (2), 65—93.

Leopold, A., 1949. A Sand County Almanac and Sketches from Here and There. Oxford University Press, New York.

Levene, H., 1960. Robust tests for equality of variances. In: Olkin, I., Ghurye, S.G., Hoeffding, W., Madow, W.G., Mann, H.B. (Eds.), In Contributions to Probability and Statistics: Essays in Honor of Harold Hotelling. Stanford University Press, Palo Alto, pp. 278—292.

Lewan, L., Söderqvist, T., 2002. Knowledge and recognition of ecosystem services among the general public in a drainage basin in Scania, Southern Sweden. Ecological Economics 42 (3), 459—467.

Limburg, K.E., O'Neill, R.V., Costanza, R., Farber, S., 2002. Complex system and valuation. Ecological Economics 41 (3), 409—420.

Lindner, L., Marks, L., Nita, M., 2013. Climatostratigraphy of interglacials in Poland: middle and upper Pleistocene lower boundaries from a Polish perspective. Quaternary International 292, 113—123.

Lipski, C., Kostuch, R., 2005. Charakterystyka procesów erozyjnych gleb na przykładzie zlewni wybranych rzek w Karpatach. Infrastruktura i Ekologia Terenów Wiejskich 3, 95—105 (in Polish).

Liquete, C., Kleeschulte, S., Dige, G., Maes, J., Grizzetti, B., Olah, B., Zulian, G., 2015. Mapping green infrastructure based on ecosystem services and ecological networks: a Pan-European case study. Environmental Science & Policy 54, 268—280.

Liquete, C., Cid, N., Lanzanova, D., Grizzetti, B., Reynaud, A., 2016. Perspectives on the link between ecosystem services and biodiversity: the assessment of the nursery function. Ecological Indicators 63, 249—257.

Lisicki, S., 1994. Objaśnienia do Szczegółowej Mapy Geologicznej Polski 1:50 000. Arkusze: Sejny (110) i Veisiejai (111). Państwowy Instytut Geologiczny, Warszawa (in Polish).

Lõhmus, A., Lõhmus, P., Vellak, K., 2007. Substratum diversity explains landscape-scale covariation in the species richness of bryophytes and lichens. Biological Conservation 135 (3), 405—414.

Lonsdorf, E., Kremen, C., Ricketts, T., Winfree, R., Williams, N., Greenleaf, S., 2009. Modelling pollination services across agricultural landscapes. Annals of Botany 103 (9), 1589—1600.

Loreau, M., Hector, A., 2001. Partitioning selection and complementarity in biodiversity experiments. Nature 412, 72—76.

Lorenc, H. (Ed.), 2005. Atlas Klimatu Polski. Instytut Meteorologii i Gospodarki Wodnej, Warszawa (in Polish).

Lubinski, D.J., Korsun, S., Polyak, L., Forman, S.L., Lehman, S.J., Herlihy, F.A., Miller, G.H., 1996. The last deglaciation of the Franz Victoria trough, northern Barents Sea. Boreas 25 (2), 89—100.

Luck, G.W., Daily, G.C., Ehrlich, P.R., 2003. Population diversity and ecosystem services. Trends in Ecology & Evolution 18 (7), 331—336.

Luck, G.W., Chan, K.M.A., Fay, J.P., 2009. Protecting ecosystem services and biodiversity in the world's watersheds. Conservation Letters 2 (4), 179—188.

Lugnot, M., Martin, G., 2013. Biodiversity provides ecosystem services: scientific results versus stakeholders' knowledge. Regional Environmental Change 13 (6), 1145—1155.

Maass, J.M., Balvanera, P., Castillo, A., Daily, G.C., Mooney, H.A., Ehrlich, P., Quesada, M., Miranda, A., Jaramillo, V.J., García-Oliva, F., Martínez-Yrizar, A., Cotler, H., López-Blanco, J., Pérez-Jiménez, A., Búrquez, A., Tinoco, C., Ceballos, G., Barraza, L., Ayala, R., Sarukhán, J., 2005. Ecosystem services of tropical dry forests: insights from long-term ecological and social research on the Pacific Coast of Mexico. Ecology and Society 10 (1), 17.

Mace, G.M., Norris, K., Fitter, A.H., 2012. Biodiversity and ecosystem services: a multi-layered relationship. Trends in Ecology & Evolution 27 (1), 19—26.

Mačukanović-Jocić, M.P., Jarić, S.V., 2016. The melliferous potential of Apiflora of southwestern Vojvodina (Serbia). Archives of Biological Sciences 68 (1), 81—91.

Maczka, K., Matczak, P., Pietrzyk-Kaszyńska, A., Rechciński, M., Olszańska, A., Cent, J., Grodzińska-Jurczak, M., 2016. Application of the ecosystem services concept in environmental policy — a systematic empirical analysis of national level policy documents in Poland. Ecological Economics 128, 169—176.

Maes, J., Egoh, B., Willemen, L., Liquete, C., Vihervaara, P., Schägner, J.P., Grizzetti, B., Drakou, E.G., La Notte, A., Zulian, G., Bouraoui, F., Paracchini, M.L., Braat, L., Bidoglio, G., 2012a. Mapping ecosystem services for policy support and decision making in the European Union. Ecosystem Services 1 (1), 31—39.

Maes, J., Paracchini, M.L., Zulian, G., Dunbar, M.B., Alkemade, R., 2012b. Synergies and trade-offs between ecosystem service supply, biodiversity, and habitat conservation status in Europe. Biological Conservation 155, 1—12.

Maes, J., Teller, A., Erhard, M., Liquete, C., Braat, L., Berry, P., Egoh, B., Puydarrieux, P., Fiorina, C., Santos, F., Paracchini, M.L., Keune, H., Wittmer, H., Hauck, J., Fiala, I., Verburg, P.H., Condé, S., Schägner, J.P., San Miguel, J., Estreguil, C., Ostermann, O., Barredo, J.I., Pereira, H.M., Stott, A., Laporte, V., Meiner, A., Olah, B., Royo Gelabert, E., Spyropoulou, R., Petersen, J.E., Maguire, C., Zal, N., Achilleos, E., Rubin, A., Ledoux, L., Brown, C., Raes, C., Jacobs, S., Vandewalle, M., Connor, D., Bidoglio, G., 2013. Mapping and Assessment of Ecosystems and Their Services. An Analytical Framework for Ecosystem Assessments under Action 5 of the EU Biodiversity Strategy to 2020. European Union, Luxembourg (Discussion paper — Final, April 2013). Technical Report — 2013 — 067.

Maes, J., Teller, A., Erhard, M., Murphy, P., Paracchini, M.L., Barredo, J.I., Grizzetti, B., Cardoso, A., Somma, F., Petersen, J.-E., Meiner, A., Gelabert, E.R., Zal, N., Kristensen, P., Bastrup-Birk, A., Biala, K., Romao, C., Piroddi, C., Egoh, B., Fiorina, C., Santos, F., Naruševičius, V., Verboven, J., Pereira, H., Bengtsson, J., Gocheva, K., Marta-Pedroso, C., Snäll, T., Estreguil, C., Miguel, J.S., Braat, L., Grêt-Regamey, A., Perez-Soba, M., Degeorges, P., Beaufaron, G., Lillebø, A., Malak, D.A., Liquete, C., Condé, S., Moen, J., Östergård, H., Czúcz, B., Drakou, E.G., Zulian, G., Lavalle, C., 2014. MAES — Mapping and Assessment of Ecosystems and Their Services. Indicators for ecosystem assessments under Action 5 of the EU Biodiversity Strategy to 2020 (2nd Report — Final, February 2014). Technical Report — 2014 — 08. European Union, Luxembourg.

Maes, J., Fabrega, N., Zulian, G., Barbosa, A., Vizcaino, P., Ivits, E., Polce, C., Vandecasteele, I., Rivero, I.M., Guerra, C., Castillo, C.P., Vallecillo, S., Baranzelli, C., Barranco, R., Batista e Silva, F., Jacobs-Crisoni, C., Trombetti, M., Lavalle, C., 2015.

Mapping and Assessment of Ecosystems and Their Services. JRC Science and Policy Report. Trends in Ecosystems and Ecosystem Services in the European Union between 2000 and 2010.

Maes, J., Zulian, G., Thijssen, M., Castell, C., Baró, F., Ferreira, A.M., Melo, J., Garrett, C.P., David, N., Alzetta, C., Geneletti, D., Cortinovis, C., Zwierzchowska, I., Louro Alves, F., Souto Cruz, C., Blasi, C., Alós Ortí, M.M., Attorre, F., Azzella, M.M., Capotorti, G., Copiz, R., Fusaro, L., Manes, F., Marando, F., Marchetti, M., Mollo, B., Salvatori, E., Zavattero, L., Zingari, P.C., Giarratano, M.C., Bianchi, E., Duprè, E., Barton, D., Stange, E., Perez-Soba, M., van Eupen, M., Verweij, P., de Vries, A., Kruse, H., Polce, C., Cugny-Seguin, M., Erhard, M., Nicolau, R., Fonseca, A., Fritz, M., Teller, A., 2016. Mapping and Assessment of Ecosystems and Their Services. Urban Ecosystems (4th Report — Final, May 2016). Technical Report — 2016 — 102. European Union, Luxembourg.

Magurran, A.E., 2004. Measuring Biological Diversity. Blackwell Publishing, Oxford.

Mäkipää, R., 1999. Response patterns of *Vaccinium myrtillus* and *V. vitis-idaea* along nutrient gradients in boreal forest. Journal of Vegetation Science 10 (1), 17—26.

Maksymiuk, I., 1960. Nektarowanie lipy drobnolistnej *Tilia Cordata* Mill. w Rezerwacie Obrożyska koło Muszyny. Pszczelnicze Zeszyty Naukowe 4 (2), 105—125 (in Polish).

Manikowska, B., 1999. Gleby kopalne i okresy pedogenetyczne w ewolucji środowiska Polski Środkowej po zlodowaceniu warciańskim. In: Dzieduszyńska, D., Turkowska, K. (Eds.), Rola plejstoceńskich procesów peryglacjalnych w modelowaniu rzeźby Polski, Acta Geographica Lodziensia, vol. 76, pp. 41—100 (in Polish).

Mann, K.H., 1969. The dynamic of aquatic ecosystems. Advances in Ecological Research 6, 1—81.

Marchese, A., Arciola, C.R., Barbieri, R., Silva, A.S., Nabavi, S.F., Tsetegho Sokeng, A.J., Izadi, M., Jafari, N.J., Suntar, I., Daglia, M., Nabavi, S.M., 2017. Update on monoterpenes as antimicrobial agents: a particular focus on p-cymene. Materials 10 (8), E947.

Marsh, G.P., 1864. Man and Nature; or, Physical Geography as Modified by Human Action. Charles Scribner, New York.

Matuszkiewicz, J.M., 1993. Krajobrazy roślinne i regiony geobotaniczne Polski. Prace Geograficzne, vol. 158. IGiPZ PAN, Warszawa (in Polish).

Matuszkiewicz, W., 2001. Przewodnik do oznaczania zbiorowisk roślinnych Polski. Vademecum Geobotanicum, vol. 3. Państwowe Wydawnictwo Naukowe, Warszawa (in Polish).

Matuszkiewicz, J.M., 2008. Potential natural vegetation of Poland. IGiPZ PAN, Warszawa. http://www.igipz.pan.pl/potential-vegetation-zgik.html.

McGarigal, K., Marks, B.J., 1995. FRAGSTATS: Spatial Pattern Analysis Program for Quantifying Landscape Structure. General Technical Reports PNW-GTR-351. USDA Forest Service, Pacific Northwest Research Station, Portland.

McGarigal, K., 2015. FRAGSTATS Help. https://www.umass.edu/landeco/research/fragstats/documents/fragstats.help.4.2.pdf.

MEA, 2003. Ecosystems and Human Well-Being: A Framework for Assessment. Millennium Ecosystem Assessment. Island Press, Washington.

MEA, 2005. Ecosystems and human well-being: current state and trends. Vol. 1. Findings of the Condition and Trends. Working Group of the Millennium Ecosystem Assessment. Island Press, Washington, Covelo, London.

Mehtälä, J., Vuorisalo, T., 2010. High aesthetic valuation of urban thrush nightingales in 19th century Helsinki. Landscape and Urban Planning 98 (2), 117—123.

Meli, P., Benayas, J.M.R., Balvanera, P., Ramos, M.M., 2014. Restoration enhances wetland biodiversity and ecosystem service supply, but results are context-dependent: a meta-analysis. PLoS One 9 (4), e93507.

Metzger, M.J., Rounsevell, M.D.A., Acosta-Michlik, L., Leemans, R., Schröter, D., 2006. The vulnerability of ecosystem services to land change. Agriculture, Ecosystems & Environment 114 (1), 69−85.

Meusel, H., Jäger, E. (Eds.), 1992. Vergleichende Chorologie der Zentraleuropäischen Flora, vol. 3. Gustav Fischer, Jena−Stuttgart−New York (in German).

Michener, C.D., 2007. The Bees of the World. The Johns Hopkins University Press, Baltimore.

Migoń, P., 2006. Geomorfologia. Wydawnictwo Naukowe PWN, Warszawa (in Polish).

Miguntanna, N.S., Egodawatta, P., Kokot, S., Goonetilleke, A., 2010. Determination of a set of surrogate parameters to assess urban stormwater quality. The Science of the Total Environment 408 (24), 6251−6259.

Miina, J., Hotanen, J.-P., Salo, K., 2009. Modelling the abundance and temporal variation in the production of bilberry (*Vaccinium myrtillus* L.) in Finnish mineral soil forests. Silva Fennica 43 (4), 577−593.

Misiukiewicz, W., 2014. Wigierski Park Narodowy. In: Jamrozy, G. (Ed.), Ssaki polskich parków narodowych. Instytut Bioróżnorodności Leśnej, Magurski Park Narodowy, Kraków−Krempna, pp. 178−191 (in Polish).

Mitchell, M., Vanberg, J., Sipponen, M., 2010. Commercial Inland Fishing in Member Countries of the European Inland Fisheries Advisory Commission (EIFAC): Operational Environments, Property Rights Regimes and Socio-Economic Indicators. Country Profiles. EIFAC Ad Hoc Working Party on Socio-Economic Aspects of Inland Fisheries. http://www.fao.org/docrep/015/an222e/an222e.pdf.

Mitchell, M.G.E., Suarez-Castro, A.F., Martinez-Harms, M., Maron, M., McAlpine, C., Gaston, K.J., Johansen, K., Rhodes, J.R., 2015. Reframing landscape fragmentation's effects on ecosystem services. Trends in Ecology & Evolution 30 (4), 190−198.

Mitchell-Jones, A.J., Amori, G., Bogdanowicz, W., Kryštufek, B., Reijnders, P.J.H., Spitzenberger, F., Stubbe, M., Thissen, J.B.M., Vohralik, V., Zima, J., 1999. The Atlas of European Mammals. T & AD Poyser Natural History, London.

Mojski, J.E., 2005. Ziemie polskie w czwartorzędzie. Zarys morfogenezy. Państwowy Instytut Geologiczny, Warszawa (in Polish).

Molan, P.C., 2001. Potential of honey in the treatment of wounds and burns. American Journal of Clinical Dermatology 2 (1), 13−19.

Molnár, D.K., Julien, P.Y., 1998. Estimation of upland erosion using GIS. Computer and Geosciences 24 (2), 183−192.

Mononen, L., Auvinen, A.-P., Ahokumpu, A.-L., Rönkä, M., Aarras, N., Tolvanen, H., Kamppinen, M., Viirret, E., Kumpula, T., Vihervaara, P., 2016. National ecosystem service indicators: measures of social-ecological sustainability. Ecological Indicators 61 (Part 1), 27−37.

Morri, E., Pruscini, F., Scolozzi, R., Santolini, R., 2014. A forest ecosystem services evaluation at the river basin scale: supply and demand between coastal areas and upstream lands (Italy). Ecological Indicators 37 (Part A), 210−219.

Moser, D., Sauberer, N., Willner, W., 2011. Generalisation of Drought Effects on Ecosystem Goods and Services over the Alps. Vienna Institute for Nature Conservation & Analyses, Vienna.

Mouchet, M.A., Lamarque, P., Martín-López, B., Crouzat, E., Gos, P., Byczek, C., Lavorel, S., 2014. An interdisciplinary methodological guide for quantifying associations between ecosystem services. Global Environmental Change 28, 298−308.

MRiRW, 2007. Załącznik D. Uzasadnienie dla delimitacji i poziomu wsparcia finansowego dla działania pt. "Wspieranie działalności rolniczej na obszarach o niekorzystnych warunkach gospodarowania (ONW)". Warszawa (in Polish).

Mücher, C.A., Bunce, R.G.H., Jongman, R.H.G., Klijn, J.A., Koomen, A.J.M., Metzger, M.J., Wascher, D.M., 2003. Identification and Characterisation of Environments and Landscapes in Europe. Alterra-rapport 832. Alterra, Wageningen.

Mücher, C.A., Wascher, D.M., Klijn, J.A., Koomen, A.J.M., Jongman, R.H.G., 2006. A new European landscape map as an integrative framework for landscape character assessment. In: Bunce, R.G.H., Jongman, R.H.G. (Eds.), Landscape Ecology in the Mediterranean: Inside and Outside Approaches. Proceedings of the European IALE Conference. IALE Publication Series 3, pp. 233—243.

Müller, D., Schröder, B., Müller, J., 2009. Modelling habitat selection of the cryptic Hazel Grouse Bonasa bonasia in a montane forest. Journal of Ornithology 150 (4), 717—732.

Naidoo, R., Balmford, A., Costanza, R., Fisher, B., Green, R.E., Lehner, B., Malcolm, T.R., Ricketts, T.H., 2008. Global mapping of ecosystem services and conservation priorities. Proceedings of the National Academy of Sciences of the United States 105 (28), 9495—9500.

Naveh, Z., 2007. From biodiversity to ecodiversity: a landscape-ecology approach to conservation and restoration. Restoration Ecology 2: 180—189. In: Naveh, Z. (Ed.), Transdisciplinary Challenges in Landscape Ecology and Restoration Ecology — an Anthology, Landscape Series, vol. 7. Springer, Dordrecht, pp. 117—134.

Neef, E., 1966. Zur Frage des gebietswirtschaftlichen Potentials. Forschungen und Fortschritte 40 (3), 65—70 (in German).

Nelson, E., Mendoza, G., Regetz, J., Polasky, S., Tallis, H., Cameron, D.R., Chan, K.M.A., Daily, G.C., Goldstein, J., Kareiva, P.M., Lonsdorf, E., Naidoo, R., Risketts, T.H., Shaw, M.R., 2009. Modeling multiple ecosystem services, biodiversity conservation, commodity production, and tradeoffs at landscape scales. Frontiers in Ecology and the Environment 7 (1), 4—11.

Nemec, K.T., Raudsepp-Hearne, C., 2013. The use of geographic information systems to map and assess ecosystem services. Biodiversity & Conservation 22 (1), 1—15.

Nestby, R., Percival, D., Martinussen, I., Opstad, N., Rohloff, J., 2011. The European blueberry (Vaccinium myrtillus L.) and the potential for cultivation. A review. The European Journal of Plant Science and Biotechnology 5 (1), 5—16.

Niewiadomski, W., Skrodzki, M., 1964. Nasilenie spływów i zrywów a system rolniczego zagospodarowania stoków. Zeszyty Naukowe WSR w Olsztynie 17 (2), 269—291 (in Polish).

Nogué, S., Long, P.R., Eycott, A.E., de Nascimento, L., Fernández-Palacios, J.M., Petrokofsky, G., Vandvik, V., Willis, K.J., 2016. Pollination service delivery for European crops: challenges and opportunities. Ecological Economics 128, 1—7.

Norton, L.R., Inwood, H., Crowe, A., Baker, A., 2012. Trialling a method to quantify the 'cultural services' of the English landscape using countryside survey data. Land Use Policy 29 (2), 449—455.

Nowocień, E., 2007. Zagadnienia erozji gleb. In: Woch, F. (Ed.), Wademekum klasyfikatora gleb, Instytut Uprawy Nawożenia i Gleboznawstwa - Państwowy. Instytut Badawczy, Puławy, pp. 302—326 (in Polish).

Nüblein, F., 1988. Das praktische Handbuch der Jagdkunde. BLV Verlagsgesellschaft, München-Wien-Zürich (in German).

Odum, E.P., 1953. Fundamentals of Ecology. Saunders, Philadelphia.

Okarma, H., Tomek, A., 2008. Łowiectwo. Wydawnictwo Edukacyjno-Naukowe H₂0, Kraków (in Polish).

Okruszko, T., Duel, H., Acreman, M., Grygoruk, M., Flörke, M., Schneider, C., 2011. Broad-scale ecosystem services of European wetlands — overview of the current situation and future perspectives under different climate and water management scenarios. Hydrological Sciences Journal 56 (8), 1501—1517.

Operat ochrony fauny (fauna lądowa), 2013. Plan Ochrony dla Wigierskiego Parku Narodowego i obszaru Natura 2000 Ostoja Wigierska PLH200004 w granicach Parku wraz z
aneksem dotyczącym fragmentów obszaru Natura 2000 Ostoja Wigierska położonych
poza granicami Parku. Wigierski Park Narodowy, Warszawa, Gdańsk, Suwałki (unpublished, in Polish).

Operat ochrony fauny, 1999. Plan Ochrony Wigierskiego Parku Narodowego. Wigierski
Park Narodowy, Warszawa (unpublished, in Polish).

Operat ochrony zasobów i ekosystemów wodnych. Plan Ochrony dla Wigierskiego
Parku Narodowego i obszaru Natura 2000 Ostoja Wigierska PLH200004, 2014.
Wigierski Park Narodowy, Warszawa, Białystok, Olsztyn, Suwałki (unpublished, in
Polish).

Osborn, F., 1948. Our Plundered Planet. Little, Brown and Company, Boston.

Ostaszewska, K., 2005. Ewolucja krajobrazu naturalnego. In: Richling, A., Ostaszewska, K.
(Eds.), Geografia fizyczna Polski. Wydawnictwo Naukowe PWN, Warszawa,
pp. 309—315 (in Polish).

Ostler, J., Roper, T.J., 1998. Changes in size, status, and distribution of badger *Meles meles* L.
setts during a 20-year period. Zeitschrift für Säugetierkunde 63 (4), 200—209.

Oswald, M., Jonsson, A., Wibeck, V., Asplund, T., 2013. Mapping energy crop cultivation
and identifying motivational factors among Swedish farmers. Biomass and Bioenergy 50,
25—34.

Paduch, R., Kandefer-Szerszeń, M., Trytek, M., Fiedurek, J., 2007. Terpenes: substances
useful in human healthcare. Archivum Immunologiae et Therapiae Experimentalis 55
(5), 315—327.

Panek, M., Bresiński, W., 2002. Red fox *Vulpes vulpes* density and habitat use in a rural area
of western Poland in the end of 1990s, compared with the turn of 1970s. Acta Theriologica 47 (4), 433—442.

Paracchini, M.L., Zulian, G., Kopperoinen, L., Maes, J., Schägner, J.P., Termansen, M.,
Zandersen, M., Perez-Soba, M., Scholefield, P.A., Bidoglio, G., 2014. Mapping cultural
ecosystem services: a framework to assess the potential for outdoor recreation across the
EU. Ecological Indicators 45, 371—385.

Pawlikowski, T., 2010. Dynamics of bee communities (Hymenoptera: Apoidea: Apiformes)
in heath and grassland patches during secondary succession of the Peucedano-Pinetum
series in the Toruń Basin. Ecological Questions 13, 29—33.

Pereira, E., Queiroz, C., Pereira, H.M., Vicente, L., 2005. Ecosystem services and human
well-being: a participatory study in a mountain community in Portugal. Ecology and
Society 10 (2), 14.

Pieczyński, P., 2012. Ukształtowanie powierzchni parku. Wigry. Kwartalnik Wigierskiego
Parku Narodowego 38 (2) (in Polish).

Pielowski, Z., 1999. Sarna. Oficyna Edytorska "Wydawnictwo Świat", Warszawa (in
Polish).

Pietrzak, M., 1998. Syntezy krajobrazowe — założenia, problemy, zastosowania. Bogucki
Wydawnictwo Naukowe, Poznań (in Polish).

Pimentel, D., Harvey, C., Resosudarmo, P., Sinclair, K., Kurz, D., McNair, M., Crist, S.,
Shpritz, L., Fitton, L., Saffouri, R., Blair, R., 1995. Environmental and economic costs
of soil erosion and conservation benefits. Science 267 (5201), 1117—1123.

Pistocchi, A., Cassani, G., Zani, O., 2002. Use of the USPED Model for Mapping Soil
Erosion and Managing Best Land Conservation Practices. International Congress on
Environmental Modelling and Software, Lugano, Switzerland, pp. 163—168.

Piwowarczyk, J., Kronenberg, J., Dereniowska, M.A., 2013. Marine ecosystem services in
urban areas: do the strategic documents of Polish coastal municipalities reflect their
importance? Landscape and Urban Planning 109, 85—93.

Planty-Tabacchi, A.M., Tabacchi, E., Naiman, R.J., Deferrari, C., Décamps, H., 1996. Invasibility of species-rich communities in riparian zones. Conservation Biology 10 (2), 598—607.

Plieninger, T., Dijks, S., Oteros-Rozas, E., Bieling, C., 2013. Assessing, mapping, and quantifying cultural ecosystem services at community level. Land Use Policy 33, 118—129.

Polyak, L., Forman, S.L., Herlihy, F.A., Ivanov, G., Krinitsky, P., 1997. Late Weichselian deglacial history of the Svyataya (Saint) Anna Trough, northern Kara Sea, Arctic Russia. Marine Geology 143 (1—4), 169—188.

Pramova, E., Locatelli, B., Brockhaus, M., Fohlmeister, S., 2012. Ecosystem services in the national adaptation programmes of action. Climate Policy 12 (4), 393—409.

Preś, J., Rogalski, M., 1997. Wartość pokarmowa pasz z użytków zielonych w różnych uwarunkowaniach ekologicznych. Zeszyty Problemowe Postępów Nauk Rolniczych 453, 39—48 (in Polish).

Prusinkiewicz, Z., Bednarek, R., 1999. Gleby. In: Starkel, L. (Ed.), Geografia Polski. Środowisko przyrodnicze. Wydawnictwo Naukowe PWN, Warszawa, pp. 373—396 (in Polish).

Przewoźniak, M., 1991. Krajobrazowy system interakcyjny strefy nadmorskiej w Polsce. Rozprawy i Monografie, vol. 172. Uniwersytet Gdański, Gdańsk (in Polish).

Puchalski, T., Prusinkiewicz, Z., 1975. Ekologiczne podstawy siedliskoznawstwa. Państwowe Wydawnictwo Rolnicze i Leśne, Warszawa (in Polish).

Qing-fan, M., Song, W., 2000. Effects of community succession dynamics on forest biodiversity in eastern mountainous area of Heilongjiang province. Journal of Forestry Research 11 (3), 210—212.

Quijas, S., Schmid, B., Balvanera, P., 2010. Plant diversity enhances provision of ecosystem services: a new synthesis. Basic and Applied Ecology 11 (7), 582—593.

Rabe, S.-E., Koellner, T., Marzelli, S., Schumacher, P., Grêt-Regamey, A., 2016. National ecosystem services mapping at multiple scales — the German exemplar. Ecological Indicators 70, 357—372.

Rakotomalala, R., 2005. TANAGRA: un logiciel gratuity pour l'enseignement et la recherche. Actes de EGC'2005, vol. RNTI-E-3, 697—702 (in French).

Rąkowski, G. (Ed.), 2005. Rezerwaty przyrody w Polsce Północnej. Instytut Ochrony Środowiska, Warszawa (in Polish).

Randall, R.G., Minns, C.K., Kelso, J.R.M., 1995. Fish production in freshwaters: are rivers more productive than lakes? Canadian Journal of Fisheries and Aquatic Sciences 52 (3), 631—643.

Raudsepp-Hearne, C., Peterson, G.D., Bennett, E.M., 2010. Ecosystem service bundles for analyzing tradeoffs in diverse landscapes. Proceedings of the National Academy of Sciences 107 (11), 5242—5247.

Raymond, C.M., Bryan, B.A., MacDonald, D.H., Cast, A., Strathearn, S., Grandgirard, A., Kalivas, T., 2009. Mapping community values for natural capital and ecosystem services. Ecological Economics 68 (5), 1301—1315.

Rempel, R.S., Kaukinen, D., Carr, A.P., 2012. Patch Analyst and Patch Grid. Ontario Ministry of Natural Resources. Centre for Northern Forest Ecosystem Research, Thunder Bay, Ontario.

Renard, K.G., Foster, G.R., Weesies, G.A., McCool, D.K., Yoder, D.C., 1997. Predicting Soil Erosion by Water: A Guide to Conservation Planning with the Revised Universal Soil Loss Equation (RUSLE). USDA Agricultural Research Service. Agriculture Handbook 703.

Reynaud, A., Lanzanova, D., 2017. A global meta-analysis of the value of ecosystem services provided by lakes. Ecological Economics 137, 184—194.

Richardson, D.M., Allsopp, N., D'Antonio, C.M., Milton, S.J., Rejmánek, M., 2000a. Plant invasions — the role of mutualisms. Biological Reviews of the Cambridge Philosophical Society 75 (1), 65—93.

Richardson, D.M., Pyšek, P., Rejmánek, M., Barbour, M.G., Panetta, F.D., West, C.J., 2000b. Naturalization and invasion of alien plants: concepts and definitions. Diversity and Distributions 2 (6), 93—107.

Richling, A., Solon, J. (Eds.), 2001. Z badań nad strukturą i funkcjonowaniem Wigierskiego Parku Narodowego. Wydawnictwo Akademickie "Dialog", Warszawa (in Polish).

Richling, A., Solon, J., 2011. Ekologia krajobrazu. Wydawnictwo Naukowe PWN, Warszawa (in Polish).

Richling, A., Solon, J., Malinowska, E., 2001. Zasoby i walory krajobrazowe Wigierskiego Parku Narodowego. In: Richling, A., Solon, J. (Eds.), Z badań nad strukturą i funkcjonowaniem Wigierskiego Parku Narodowego. Wydawnictwo Akademickie "Dialog", Warszawa, pp. 209—222.

Ricketts, T.H., Regetz, J., Steffan-Dewenter, I., Cunningham, S.A., Kremen, C., Bogdanski, A., Gemmill-Herren, B., Greenleaf, S., Klein, A.M., Mayfield, M.M., Morandin, L.A., Ochieng', A., Potts, S.G., Viana, B.F., 2008. Landscape effects on crop pollination services: are there general patterns? Ecology Letters 11 (5), 499—515.

Ridder, B., 2008. Questioning the ecosystem services argument for biodiversity conservation. Biodiversity & Conservation 17 (4), 781—790.

Riera, J.L., Magnuson, J.J., Vande Castle, J.R., MacKenzie, M.D., 1998. Analysis of large-scale spatial heterogeneity in vegetation indices among North American landscapes. Ecosystems 1 (3), 268—282.

Rodríguez-Loinaz, G., Alday, J.G., Onaindia, M., 2015. Multiple ecosystem services landscape index: a tool for multifunctional landscapes conservation. Journal of Environmental Management 147, 152—163.

Roo-Zielińska, E., Grabińska, B., 2012. Ecosystem services — classification and different approaches at various levels of biosphere organisation — a literature review. Geographia Polonica 85 (2), 65—81.

Roo-Zielińska, E., Matuszkiewicz, J.M., 2016. Ancient and recent (post-agricultural) forest communities as indicators of environmental conditions in north-eastern Poland (Masuria and Kurpie region). Geographia Polonica 89 (3), 287—309.

Roo-Zielińska, E., Solon, J., Degórski, M., 2007. Ocena stanu i przekształceń środowiska przyrodniczego na podstawie wskaźników geobotanicznych, krajobrazowych i glebowych. Monografie, vol. 9. IGiPZ PAN, Warszawa (in Polish).

Roo-Zielińska, E., Affek, A., Kowalska, A., Grabińska, B., Kruczkowska, B., Wolski, J., Solon, J., Degórski, M., Kołaczkowska, E., Regulska, E., Zawiska, I., 2016. Potential of provisioning and regulating ecosystem services in postglacial landscape. Ekonomia i Środowisko 4 (59), 274—291.

Roo-Zielińska, E., 2014. Wskaźniki ekologiczne zespołów roślinnych Polski. Wydawnictwo Akademickie Sedno, Warszawa (in Polish).

Rotnicki, K., Starkel, L., 1999. Przekształcenie rzeźby w holocenie. In: Starkel, L. (Ed.), Geografia Polski. Środowisko przyrodnicze. Wydawnictwo Naukowe PWN, Warszawa, pp. 137—159 (in Polish).

Rudnicki, A., Waluga, J., Waluś, T., 1971. Rybactwo jeziorowe. Państwowe Wydawnictwo Rolnicze i Leśne, Warszawa (in Polish).

Rybak-Chmielewska, H., Szczęsna, T., Waś, E., Jaśkiewicz, K., Teper, D., 2013. Characteristics of Polish unifloral honeys IV. Honeydew honey, mainly *Abies alba* L. Journal of Apicultural Science 57 (1), 51—59.

Ryszkowski, L., Bałazy, S. (Eds.), 1994. Functional Appraisal of Agricultural Landscape in Europe (Seminar 1992). Research Center for Agricultural and Forest Environment PAS, Poznań.

Saastamoinen, O., Matero, J., Horne, P., Kniivilä, M., Haltia, E., Vaara, M., Mannerkoski, H., 2014. Classification of Boreal Forest Ecosystem Goods and Services in Finland. Reports and Studies in Forestry and Natural Sciences, vol. 11. University of Eastern Finland, Joensuu.

Sagie, H., Morris, A., Rofè, Y., Orenstein, D.E., Groner, E., 2013. Cross-cultural perceptions of ecosystem services: a social inquiry on both sides of the Israeli-Jordanian border of the Southern Arava Valley Desert. Journal of Arid Environments 97, 1—11.

Sandalj, M., Treydte, A.C., Ziegler, S., 2016. Is wild meat luxury? Quantifying wild meat demand and availability in Hue, Vietnam. Biological Conservation 194, 105—112.

Schallenberg, M., de Winton, M.D., Verburg, P., Kelly, D.J., Hamill, K.D., Hamilton, D.P., 2013. Ecosystem services of lakes. In: Dymond, J.R. (Ed.), Ecosystem Services in New Zealand. Conditions and Trends. Manaaki Whenua Press, Lincoln, New Zealand, pp. 203—225.

Schindler, S., von Wehrden, H., Poirazidis, K., Wrbka, T., Kati, V., 2013. Multiscale performance of landscape metrics as indicators of species richness of plants, insects and vertebrates. Ecological Indicators 31, 41—48.

Scholte, S.S.K., van Teeffelen, A.J.A., Verburg, P.H., 2015. Integrating socio-cultural perspectives into ecosystem service valuation: a review of concepts and methods. Ecological Economics 114, 67—78.

Schulp, C.J.E., Thuiller, W., Verburg, S.H., 2014. Wild food in Europe: a synthesis of knowledge and data of terrestrial wild food as an ecosystem service. Ecological Economics 105, 292—305.

Schumacher, E.F., 1973. Small is Beautiful: A Study of Economics as if People Mattered. Blond and Briggs, London.

Science for Environment Policy, 2015. Ecosystem Services and the Environment. In-Depth Report 11 Produced for the European Commission, DG Environment by the Science Communication Unit. UWE, Bristol.

Scott, R.W., Huff, F.A., 1996. Impacts of the Great Lakes on regional climate conditions. Journal of Great Lakes Research 22 (4), 845—863.

Seamans, G.S., 2013. Mainstreaming the environmental benefits of street trees. Urban Forestry and Urban Greening 12 (1), 2—11.

Sears, P.B., 1956. The processes of environmental change by man. In: Thomas, W.L. (Ed.), Man's Role in Changing the Face of the Earth, vol. 2. University of Chicago Press, Chicago.

Semkiw, P., Ochal, J., 2009. Analiza sektora pszczelarskiego w Polsce dla opracowania Krajowego Programu Wsparcia Pszczelarstwa w latach 2010—2013. Instytut Sadownictwa i Kwiaciarstwa, Oddział Pszczelnictwa, Puławy (in Polish).

Semkiw, P., 2015. Sektor pszczelarski w Polsce w 2015 roku. Instytut Ogrodnictwa, Zakład Pszczelnictwa, Puławy (in Polish).

Shannon, C.E., Weaver, W., 1949. The Mathematical Theory of Communication. University of Illinois Press, Chicago.

Sidorovich, V.E., Jędrzejewska, B., Jędrzejewski, W., 1996. Winter distribution and abundance of mustelids and beavers in the river valleys of Białowieża Primeval Forest. Acta Theriologica 41 (2), 155—170.

Sikorski, P., Pawlikowski, P., Solon, J., Skrajna, T., Wołkowycki, D., Wierzba, M., 2013a. Plan ochrony dla Wigierskiego Parku Narodowego i obszaru Natura 2000 Ostoja Wigierska PLH200004 w granicach Parku. Operat ochrony lądowych ekosystemów nieleśnych, torfowiskowych i bagiennych, Warszawa, Białystok, Olsztyn, Suwałki (unpublished, in Polish).

Sikorski, P., Solon, J., Żołnierczuk, M., Sikorska, D., 2013b. Plan ochrony dla Wigierskiego Parku Narodowego i obszaru Natura 2000 Ostoja Wigierska. Aktualizacja "Operatu ochrony zasobów i walorów krajobrazowych", Warszawa, Białystok, Olsztyn, Suwałki (unpublished) (in Polish).

Simberloff, D., Von Holle, B., 1999. Positive interactions of nonindigenous species: invasional meltdown? Biological Invasions 1 (1), 21−32.

Skorupski, M., Małek, S., Patalan-Misiaszek, I., Zwydak, M., Nowiński, M., 2011. Badanie zawartości węgla w ściółce i glebie. In: Raport końcowy z realizacji tematu badawczego "Bilans węgla w biomasie drzew głównych gatunków lasotwórczych Polski". Część III A. Wyniki. Poznań, pp. 6−76 (in Polish).

Skrzypczak, A., Mamcarz, A., Gierej, A., 2011. Fishery types of lakes in the context of lake eutrophication and changes in the structure of ichthyofauna. In: Jankun, M., Furgała-Selezniow, G., Woźniak, M., Wiśniewska, A.M. (Eds.), Fish Management in a Variable Water Environment. University of Warmia and Mazury in Olsztyn, Olsztyn, pp. 105−120.

Smirnova, O.V., Bobrovsky, M.V., Khanina, L.G., 2018. European Russian Forests. Their Current State and Features of Their History. Ser. Plant and Vegetation, vol. 15. Springer, Dordrecht.

Smolska, E., 2002. The intensity of soil erosion in agricultural areas in North-Eastern Poland. Landform Analysis 3, 25−33.

Snäll, T., Berglund, H., Bengtsson, J., Moen, J., 2015. Mapping multiple ecosystem services (Sweden). In: Mapping and Assessment of Forest Ecosystems and Their Services − Applications and Guidance for Decision Making in the Framework of MAES. JRC Science for Policy report, pp. 44−47.

Söderström, B., Svensson, B., Vessby, K., Glimskär, A., 2001. Plants, insects and birds in semi-natural pastures in relation to local habitat and landscape factors. Biodiversity & Conservation 10 (11), 1839−1863.

Sodhi, N.S., Lee, T.M., Sekercioglu, C.H., Webb, E.L., Prawiradilaga, D.M., Lohman, D.J., Pierce, N.E., Diesmos, A.C., Rao, M., Ehrlich, P.R., 2010. Local people value environmental services provided by forested parks. Biodiversity & Conservation 19 (4), 1175−1188.

Soini, K., Vaarala, H., Pouta, E., 2012. Residents' sense of place and landscape perceptions at the rural-urban interface. Landscape and Urban Planning 104 (1), 124−134.

Sokołowski, A.W., 1966. Fitosocjologiczna charakterystyka borów iglastych w Puszczy Augustowskiej. Prace IBL 306, 107−125 (in Polish).

Sokołowski, A.W., 1968. Zespoły roślinne Nadleśnictwa Suwałki w Puszczy Augustowskiej. Prace IBL 349, 171−213 (in Polish).

Sokołowski, A.W., 1980. Zbiorowiska leśne północno-wschodniej Polski. Monographiae Botanicae, vol. 60. Polskie Towarzystwo Botaniczne, Warszawa (in Polish).

Sokołowski, A.W., 1988. Fitosocjologiczna charakterystyka zbiorowisk roślinnych Wigierskiego Parku Narodowego. Prace IBL 673, 3−80 (in Polish).

Solon, J., Roo-Zielińska, E., Affek, A., Kowalska, A., Kruczkowska, B., Wolski, J., Degórski, M., Grabińska, B., Kołaczkowska, E., Regulska, E., Zawiska, I., 2017. Świadczenia ekosystemowe w krajobrazie młodoglacjalnym. Ocena potencjału i wykorzystania. Instytut Geografii i Przestrzennego Zagospodarowania PAN, Wydawnictwo Akademickie SEDNO, Warszawa (in Polish).

Solon, J., 1983. The local complex of phytocenoses and the vegetation landscape − fundamental units of the spatial organization of the vegetation above the phytocenose level. Acta Botanica Hungarica 29 (1−4), 377−384.

Solon, J., 1993. Changes in the vegetation landscape in the Pińczów environs (S Poland). Phytocoenologia 21 (4), 387−409.

Solon, J., 2003. Scots pine forests of the Vaccinio-Piceetea class in Europe: forest sites studied. Polish Journal of Ecology 51 (4), 421−439.

Solon, J., 2004a. The comparison of landscape metrics in different scales — the raster and vector approaches. Ekológia-Bratislava 23 (Suppl. 1), 320—332.

Solon, J., 2004b. Zastosowanie koncepcji potencjałów krajobrazowych dla oceny stopnia spójności krajobrazu. Problemy Ekologii Krajobrazu 14, 29—43 (in Polish).

Solon, J., 2008. Koncepcja "Ecosystem Services" i jej zastosowania w badaniach ekologiczno-krajobrazowych. Problemy Ekologii Krajobrazu 21, 25—44 (in Polish).

Spracklen, D.V., Bonn, B., Carslaw, K.S., 2008. Boreal forests, aerosols and the impacts on clouds and climate. Philosophical Transactions of the Royal Society of London A 366 (1885), 4613—4626.

Stallman, H.R., 2011. Ecosystem services in agriculture: determining suitability for provision by collective management. Ecological Economics 71, 131—139.

Staniszewski, P., Janeczko, E., 2012. Problemy udostępniania lasów w kontekście użytkowania zasobów runa. Studia i Materiały Centrum Edukacji Przyrodniczo-Leśnej 14 (32.3), 161—170 (in Polish).

State Forest Policy, 1997. Ministerstwo Ochrony Środowiska, Zasobów Naturalnych i Leśnictwa, Warszawa (in Polish).

Steiniger, S., Lange, T., Burghardt, D., Weibel, R., 2008. An approach for the classification of urban building structures based on discriminant analysis techniques. Transactions in GIS 12 (1), 31—59.

Stopa-Boryczka, M., Boryczka, J., Wawer, J., Grabowska, K., Dobrowolska, M., Osowiec, M., Błażek, E., Skrzypczuk, J., Grzęda, M., 2013. Atlas współzależności parametrów meteorologicznych i geograficznych w Polsce. XXX. Klimat północno-wschodniej Polski według podziału fizycznogeograficznego J. Kondrackiego i J. Ostrowskiego. WGiSR UW, Warszawa (in Polish).

Su, S., Xiao, R., Jiang, Z., Zhang, Y., 2012. Characterizing landscape pattern and ecosystem service value changes for urbanization impacts at an eco-regional scale. Applied Geography 34, 295—305.

Svendsen, J.I., Alexanderson, H., Astakhov, V.I., Demidov, I., Dowdeswell, J.A., Funder, S., Gataullin, V., Henriksen, M., Hjort, C., Houmark-Nielsen, M., Hubberten, H.W., Ingólfsson, Ó., Jakobsson, M., Kjær, K.H., Larsen, E., Lokrantz, H., Lunkka, J.P., Lyså, A., Mangerud, J., Matiouchkov, A., Murray, A., Möller, P., Niessen, F., Nikolskaya, O., Polyak, L., Saarnisto, M., Siegert, C., Siegert, M.J., Spielhagen, R.F., Stein, R., 2004. Late Quaternary ice sheet history of northern Eurasia. Quaternary Science Reviews 23, 1229—1271.

Svendsen, J.I., Briner, J.P., Mangerud, J., Young, N.E., 2015. Early break-up of the Norwegian channel ice stream during the last glacial maximum. Quaternary Science Reviews 107, 231—242.

Swenson, J.E., Angelstam, P., 1993. Habitat separation by sympatric forest grouse in Fennoscandia in relation to boreal forest succession. Canadian Journal of Zoology 71 (7), 1303—1310.

Swift, M.J., Izac, A.-M.N., van Noordwijk, M., 2004. Biodiversity and ecosystem services in agricultural landscapes — are we asking the right questions? Agriculture, Ecosystems & Environment 104 (1), 113—134.

Swinton, S.M., Lupi, F., Robertson, G.P., Hamilton, S.K., 2007. Ecosystem services and agriculture: cultivating agricultural ecosystems for diverse benefits. Ecological Economics 64 (2), 245—252.

Syrbe, R.-U., Walz, U., 2012. Spatial indicators for the assessment of ecosystem services: providing, benefiting and connecting areas and landscape metrics. Ecological Indicators 21, 80—88.

Szklanowska, K., 1973. Bory jako baza pożytkowa pszczół. Pszczelnicze Zeszyty Naukowe 17, 51—85 (in Polish).

Szklanowska, K., 1979. Nektarowanie i wydajność miodowa ważniejszych roślin runa lasu liściastego. Pszczelnicze Zeszyty Naukowe 23, 123—130 (in Polish).

Szneidrowski, M. (Ed.), 2014. Plan ochrony dla Wigierskiego Parku Narodowego i obszaru Natura 2000 Ostoja Wigierska PLH200004. Operat ochrony ekosystemów leśnych, Warszawa, Białystok, Olsztyn, Suwałki.

Szwarczewski, P., Kupryjanowicz, M., 2008. Etapy rozwoju zagłębień bezodpływowych w okolicach Sejn. In: Wacnik, A., Madeyska, E. (Eds.), Polska północno-wschodnia w holocenie. Człowiek i jego środowisko. Botanical Guidebooks 30, W. Szafer Institute of Botany PAS, Kraków, pp. 195—205 (in Polish).

Śniegocki, R., 2018. Co nam dają drzewa? Regionalna Dyrekcja Lasów Państwowych, Poznań. http:///www.poznan.lasy.gov.pl/swieto-drzewa (in Polish).

Święchowicz, J., 2013. Wskaźnik średniej rocznej erozyjności deszczu i spływu powierzchniowego (R) na przykładzie stacji meteorologicznej w Łazach k. Bochni (region klimatu Pogórza Karpackiego). Landform Analysis 24, 85—95 (in Polish).

Świtoniak, M., 2014. Use of soil profile truncation to estimate influence of accelerated erosion on soil cover transformation in young morainic landscapes, North-Eastern Poland. Catena 116, 173—184.

Taki, H., Okochi, I., Okabe, K., Inoue, T., Goto, H., Matsumura, T., Makino, S., 2013. Succession influences wild bees in a temperate forest landscape: the value of early successional stages in naturally regenerated and planted forests. PLoS One 8 (2), e56678.

Tallis, H., Mooney, H., Andelman, S., Balvanera, P., Cramer, W., Karp, D., Polasky, S., Reyers, B., Ricketts, T., Running, S., Thonicke, K., Tietjen, B., Walz, A., 2012. A global system for monitoring ecosystem service change. BioScience 62 (11), 977—986.

Tamhane, A.C., 1977. Multiple comparisons in model I one-way ANOVA with unequal variances. Communications in Statistics — Theory and Methods 6 (1), 15—32.

TEEB, 2010. The Economics of Ecosystems and Biodiversity. Ecological and Economic Foundation. Earthscan, London and Washington.

Terkenli, T.S., 2001. Towards a theory of the landscape: the Aegean landscape as a cultural image. Landscape and Urban Planning 57 (3—4), 197—208.

Thompson, L.M., Troeh, F.R., 1973. Soils and Soil Fertility. McGraw-Hill, New York.

Tokarska-Guzik, B., 2005. The Establishment and Spread of Alien Plant Species (Kenophytes) in the Flora of Poland. Wydawnictwo Uniwersytetu Śląskiego, Katowice.

Turkelboom, F., Raquez, P., Dufrene, M., Raes, L., Simoens, I., Jacobs, S., Stevens, M., De Vreese, R., Panis, J., Hermy, M., Thoonen, M., Liekens, I., Fontaine, C.M., Dendoncker, N., van der Biest, K., Casaer, J., Heyrman, H., Meiresonne, L., Keune, H., 2014. CICES going local: ecosystem services classification adapted for a highly populated country. In: Jacobs, S., Dendoncker, N., Keune, H. (Eds.), Ecosystem Services — Global Issues, Local Practices. Elsevier, Boston, pp. 223—247.

Turkelboom, F., Thoonen, M., Jacobs, S., García-Llorente, M., Martín-López, B., Berry, P., 2016. Ecosystem service trade-offs and synergies. In: Potschin, M.B., Jax, K. (Eds.), OpenNESS Ecosystem Services Reference Book. EC FP7 Grant Agreement No. 308428.

Turkowski, K., 2011. Utility and market value of lakes. In: Jankun, M., Furgała-Selezniow, G., Woźniak, M., Wiśniewska, A.M. (Eds.), Fish Management in a Variable Water Environment. University of Warmia and Mazury in Olsztyn, Olsztyn, pp. 87—104.

Turner, R.K., Adger, W.N., Brouwer, R., 1998. Ecosystem services value, research needs, and policy relevance: a commentary. Ecological Economics 25, 61—65.

Turner, R.K., Paavola, J., Cooper, P., Farber, S., Jessamy, V., Georgiou, S., 2003. Valuing nature: lessons learned and future research directions. Ecological Economics 46, 493—510.

Turner, K.G., Vestergaard Odgaard, M., Bøcher, P.K., Dalgaard, T., Svenning, J.-C., 2014. Bundling ecosystem services in Denmark: trade-offs and synergies in a cultural landscape. Landscape and Urban Planning 125, 89—104.

Tzoulas, K., James, P., 2010. People's use of, and concerns about, green space networks: a case study of Birchwood, Warrington New Town, UK. Urban Forest and Urban Greening 9 (2), 121—128.

Uggla, H., Mirowski, Z., Grabarczyk, S., Nożyński, A., Rytelewski, J., Solarski, H., 1968. Proces erozji wodnej w terenach pagórkowatych północno-wschodniej części Polski. Roczniki Gleboznawcze 18 (2), 415—447 (in Polish).

UK, N.E.A., 2011. The UK National Ecosystem Assessment: Synthesis of the Key Findings. UNEP-WCMC, Cambridge.

UK, N.E.A., 2014. The UK National Ecosystem Assessment Follow-On: Synthesis of the Key Findings. UNEP-WCMC/LWEC, UK.

Ulrich, B., Meiwes, K., König, N., Khanna, K., 1984. Untersuchungsverfahren und Kriterien zur Bewertung der Versauerung und ihrer Folgen in Waldböden. Forst und Holz 39 (1), 278—286 (in German).

US EPA, 2015. National Ecosystem Services Classification System (NESCS): Framework Design and Policy Application. Final Report. EPA-800-R-15-002. United States Environmental Protection Agency, Washington, DC.

van Berkel, D.B., Verburg, P.H., 2014. Spatial quantification and valuation of cultural ecosystem services in an agricultural landscape. Ecological Indicators 37 (Part A), 163—174.

van Reeuwijk, L.P., 2002. Procedures for Soil Analysis. Technical Paper no. 9. International Soil and Reference Information Centre, Wageningen.

Verhagen, W., Van Teeffelen, A.J.A., Compagnucci, A.B., Poggio, L., Gimona, A., Verburg, P.H., 2016. Effects of landscape configuration on mapping ecosystem service capacity: a review of evidence and a case study in Scotland. Landscape Ecology 31 (7), 1457—1479.

Vihervaara, P., Kumpula, T., Tanskanen, A., Burkhard, B., 2010. Ecosystem services — a tool for sustainable management of human — environment systems. Case study Finnish Forest Lapland. Ecological Complexity 7 (3), 410—420.

Villardón, V.J.L., 2015. Multbiplot: A Package for Multivariate Analysis Using Biplots. Departamento de Estadística, Universidad de Salamanca. http://biplot.usal.es/multbiplot/introduction.html.

Vogt, W., 1948. Road to Survival. William Sloan Associates, New York.

Vos, C.C., Grashof-Bokdam, C.J., Opdam, P.F.M., 2014. Biodiversity and Ecosystem Services: Does Species Diversity Enhance Effectiveness and Reliability? A Systematic Literature Review. Statutory Research Tasks Unit for Nature & the Environment (WOT Natuur & Milieu). WOt-Technical Report, vol. 25. Wageningen.

Wall, D.H., Bardgett, R.D., Behan-Pelletier, V., Herrick, J.E., Jones, T.H., Ritz, K., Six, J., Strong, D.R., van der Putten, W.H. (Eds.), 2012. Soil Ecology and Ecosystem Services. Oxford University Press.

Wallace, K., 2008. Ecosystem services: multiple classifications or confusion? Biological Conservation 141 (2), 353—354.

Wanic, T., Brożek, S., Lasota, J., Zwydak, M., 2011. Różnorodność gleb olsów i łęgów. Roczniki Gleboznawcze 62 (4), 109—123 (in Polish).

Warakomska, Z., 1972. Badania nad wydajnością pyłkową roślin. Pszczelnicze Zeszyty Naukowe 16, 63—96 (in Polish).

Ward, O.G., Wurster-Hill, D.H., 1990. Nyctereutes procyonoides. The American Society of Mammologists, Mammalian species 358, 1—5.

Weisner, J., Bergmann, H.H., Klaus, S., Müller, F., 1977. Siedlungsdichte und *Ha-bitatstruktur* des Haselhuhns (*Bonasa bonasia*) im Waldgebiet von Bialowieza (Polen). Journal für Ornithologie 118 (1), 1—20 (in German).

Westrich, P., 1996. Habitat requirements of central European bees and the problems of partial habitats. In: Matheson, A., Buchmann, S.L., O'Toole, C., Westrich, P., Williams, I.H. (Eds.), The Conservation of Bees. Linnean Society Symposium Series, vol. 18. Academic Press, London, pp. 1—16.

Wicik, B., 2005. Gleby. In: Richling, A., Ostaszewska, K. (Eds.), Geografia fizyczna Polski. Wydawnictwo Naukowe PWN, Warszawa, pp. 201—244 (in Polish).

Wiggering, H., Müller, F. (Eds.), 2004. Umweltziele und Indikatoren. Springer, Berlin, Heidelberg, New York (in German).

Wilk, T., Jujka, M., Krogulec, J., Chylarecki, P. (Eds.), 2010. Ostoje ptaków o znaczeniu międzynarodowym w Polsce. OTOP, Marki (in Polish).

Willner, W., Roleček, J., Korolyuk, A., Dengler, J., Chytrý, M., Janišová, M., Lengyel, A., Aćić, S., Becker, T., Ćuk, M., Demina, O., Jandt, U., Kącki, Z., Kuzemko, A., Kropf, M., Lebedeva, M., Semenishchenkov, Y., Šilc, U., Stančić, Z., Staudinger, M., Vassilev, K., Yamalov, S., 2019. Formalized classification of semi-dry grasslands in central and eastern Europe. Preslia 91, 25—49.

Wilson, C.L., Matthews, W.H., 1970. Man's Impact on the Global Environment: Assessment and Recommendations for Action. Study of Critical Environmental Problems (SCEP). MIT Press, Cambridge, Massachusetts.

Winfree, R., Fox, J.W., Williams, N.M., Reilly, J.R., Cariveau, D.P., 2015. Abundance of common species, not species richness, drives delivery of a real-world ecosystem service. Ecology Letters 18 (7), 626—635.

Winfree, R., 2010. The conservation and restoration of wild bees. Annals of the New York Academy of Sciences 1195, 169—197.

Wischmeier, W.H., Smith, D.D., 1978. Predicting Rainfall Erosion Losses. A Guide to Conservation Planning. USDA Agricultural Handbook 537, Washington D.C.

Witek, T., Górski, T., 1977. Przyrodnicza bonitacja rolniczej przestrzeni produkcyjnej w Polsce. Wydawnictwa Geologiczne, Warszawa (in Polish).

Witek, T. (Ed.), 1981. Waloryzacja rolniczej przestrzeni produkcyjnej Polski według gmin. IUNG, Puławy (in Polish).

Witek, T. (Ed.), 1994. Waloryzacja rolniczej przestrzeni produkcyjnej Polski według gmin. Suplement. IUNG, Puławy (in Polish).

Witkowska-Żuk, L., 2008. Atlas roślinności lasów. Multico Oficyna wydawnicza, Warszawa (in Polish).

Wolfe, L.M., 2002. Why alien invaders succeed: support for the escape-from-enemy hypothesis. The American Naturalist 160 (6), 705—711.

Wołos, A., Draszkiewicz-Mioduszewska, H., Mickiewicz, M., 2015a. Wielkość i charakterystyka jeziorowej produkcji rybackiej w 2014 roku. In: Mickiewicz, M., Wołos, A. (Eds.), Zrównoważone korzystanie z zasobów rybackich na tle ich stanu w 2014 roku. Wydawnictwo Instytutu Rybactwa Śródlądowego, Olsztyn, pp. 9—20 (in Polish).

Wołos, A., Draszkiewicz-Mioduszewska, H., Trella, M., 2015b. Charakterystyka presji i połowów wędkarskich w jeziorach użytkowanych przez gospodarstwa rybackie w 2013 roku. In: Mickiewicz, M., Wołos, A. (Eds.), Zrównoważone korzystanie z zasobów rybackich na tle ich stanu w 2014 roku. Wydawnictwo Instytutu Rybactwa Śródlądowego, Olsztyn, pp. 159—171 (in Polish).

Woś, A., 2010. Klimat Polski w drugiej połowie XX wieku. Wydawnictwo Naukowe UAM, Poznań (in Polish).

Wójcik, J., 2004. Stan uszkodzenia lasów w Polsce w 2003 roku na podstawie badań monitoringowych. Biblioteka Monitoringu Środowiska, Warszawa, pp. 55—64 (in Polish).

Wuana, R.A., Okieimen, F.E., 2011. Heavy metals in contaminated soils: a review of sources, chemistry, risks and best available strategies for remediation. International Scholarly Research Notices, ISRN Ecology, 2011, 402647. https://doi.org/10.5402/2011/402647.

Yang, G., Ge, Y., Xue, H., Yang, W., Shi, Y., Peng, C., Du, Y., Fan, X., Ren, Y., Chang, J., 2015. Using ecosystem service bundles to detect trade-offs and synergies across urban—rural complexes. Landscape and Urban Planning 136, 110—121.

Záhlavová, L., Konvička, M., Fric, Z., Hula, V., Filipová, L., 2009. Landscape heterogeneity and species richness and composition: a middle scale study. Ekológia (Bratislava) 28 (4), 346—362.

Zawadzka, D., Zawadzki, J., 2006. Ptaki jako gatunki wskaźnikowe różnorodności biologicznej i stopnia naturalności lasów. Studia i Materiały Centrum Edukacji Przyrodniczo-Leśnej 8 (4.14), 249—262 (in Polish).

Zawisza, J., 1973. Prognoza rozwoju gospodarki rybnej na obszarze kraju, zasoby rybne i zasady eksploatacji. Sympozjum ZG PZW. Warszawa (in Polish).

Zhang, Z., Gao, J., 2016. Linking landscape structures and ecosystem service value using multivariate regression analysis: a case study of the Chaohu Lake Basin, China. Environmental Earth Science 75 (3), 1—16.

Zhang, W., Ricketts, T.H., Kremen, C., Carney, K., Swinton, S.M., 2007. Ecosystem services and dis-services in agriculture. Ecological Economics 64 (2), 253—260.

Zhang, Y., Wang, R., Kaplan, D., Liu, J., 2015. Which components of plant diversity are most correlated with ecosystem properties? A case study in a restored wetland in northern China. Ecological Indicators 49, 228—236.

Ziemnicki, S., 1978. Erozja wodna. Państwowe Wydawnictwo Rolnicze i Leśne, Warszawa (in Polish).

Zimmermann, K., Fric, Z., Filipová, L., Konvička, M., 2005. Adult demography, dispersal and behaviour of Brenthis ino (Lepidoptera: Nymphalidae): how to be a successful wetland butterfly. European Journal of Entomology 102 (4), 699—706.

Legal Acts (Chronologically)

Council Directive 92/43/EEC of 21 May 1992 on the conservation of natural habitats and of wild fauna and flora (Official Journal of the European Union L 206, 22.07.1992, p. 7). https://eur-lex.europa.eu/legal-content/EN/TXT/?uri=CELEX:31992L0043.

Hunting Law of 13 October 1995 (Journal of Laws from 2015 item 2168) (in Polish). http://prawo.sejm.gov.pl/isap.nsf/download.xsp/WDU20150002168/U/D20152168Lj.pdf.

European Landscape Convention of 20 October 2000 in Florence (Council of Europe, European Treaty Series 176).

European Commission Directorate General for Maritime Affairs and Fisheries, 2011. EU intervention in Inland fisheries. EU wide report — final version. https://ec.europa.eu/fisheries/publications/eu-intervention-inland-fisheries_en.

Regulation of the Minister of the Environment of 9 September 2011 on the list of non-native species of plants and animals, which in the case of release into the environment can threaten native species or natural habitats (Journal of Laws of 2011 No. 210 item 1260) (in Polish). http://prawo.sejm.gov.pl/isap.nsf/download.xsp/WDU20112101260/O/D20111260.pdf.

Regulation of the Minister of the Interior and Administration of 17 November 2011 on the database of topographic object and database of general geographic objects as well as standard cartographic documentation. Description of databases of topographic and general geographic objects as well as technical standards for mapping. Volume 1 (Journal of Laws of 2011 No. 279 item 1642) (in Polish). http://prawo.sejm.gov.pl/isap.nsf/download.xsp/WDU20112791642/O/D20111642-01.pdf; http://prawo.sejm.gov.pl/isap.nsf/download.xsp/WDU20112791642/O/D20111642-02.pdf; http://prawo.sejm.gov.pl/isap.nsf/download.xsp/WDU20112791642/O/D20111642-03.pdf.

Regulation of the Minister of the Environment of 13 April 2010 on Natural Habitats and Species of Community Interest, as Well as the Criteria for Selection of Areas Eligible for Recognition or Designation as a Natura 2000 Sites (Journal of Laws of 2014, item 1713) (in Polish). http://prawo.sejm.gov.pl/isap.nsf/download.xsp/WDU20140001713/O/D20141713.pdf.

Regulation (EU) No 1143/2014 of the European Parliament and of the Council of 22 October 2014 on the prevention and management of the introduction and spread of invasive alien species (Official Journal of the European Union L 317, 4.11.2014, p. 35). https://eur-lex.europa.eu/legal-content/EN/TXT/?uri=uriserv:OJ.L_.2014.317.01.0035.01.ENG.

Index